T0201560

"There are very few people who truly understand both telco and Web and can weave strategic narratives between the two. Paul is one of those guys. His explanation of modern and emergent Web trends and technologies is compulsory reading for all those who take mobile innovation seriously."

Gregory Gorman, Co-Founder Mobile 2.0 Silicon Valley, Wireless Entrepreneur

"Connected Services is a must-read for telco strategists who need to get up to speed on how the world of software and the web 2.0 works. Paul is a rare breed of strategist, one that can go from writing Android apps to debating future strategies with telco boards – he is the best storyteller when it comes to understanding both telecoms and software worlds.

Andreas Constantinou, Research Director, VisionMobile

"This book is a must read for those charged with leading innovation in a world of connected services where telco and Internet collide. Paul does a great job of articulating the interplay of diverse technologies and systems, creating a cogent narrative on how best to understand modern Web trends and interpret them into great products and service strategies in this challenging and rapidly evolving world."

Jason Goecke, VP of Innovation, Voxeo Labs

"Paul is one of the rare thought leaders who is trusted by all of his peers. His previous book explained the technology of the mobile apps space. This book is the guide on how to understand and interpret the trends of modern Web technologies and methods, plus insights into how to capitalize on the opportunities in the telco universe. An ironclad must-read for all in mobile."

Tomi T. Ahonen, Author & Consultant, Hong Kong, latest book: Insider's Guide to Mobile

"Paul is a true architect of innovation being as able to communicate a vision as he is to describing the detail of how to deliver it in practice. This book provides valuable insight to anyone who wishes to first appreciate the potential of pervasive communications and data services and then implement for commercial benefit. Rare insight is provided from hands-on experience with multiple stakeholders in the ecosystem as it has evolved from fixed and mobile to today's system where boundaries between man and machine communications are blurred and the role of historically dominant stakeholders is in a state of flux. Compelling reading for anyone serious about realising the potential of next-generation communications."

Geoff McGrath, Managing Director, McLaren Applied Technology

CONNECTED SERVICES

CONNECTED SERVICES

A GUIDE TO THE INTERNET TECHNOLOGIES SHAPING THE FUTURE OF MOBILE SERVICES AND OPERATORS

Paul Golding
CEO, Wireless Wanders

A John Wiley & Sons, Ltd., Publication

This edition first published 2011
© 2011 John Wiley & Sons, Ltd.

Registered office
John Wiley & Sons Ltd, The Atrium, Southern Gate, Chichester, West Sussex, PO19 8SQ, United Kingdom

For details of our global editorial offices, for customer services and for information about how to apply for permission to reuse the copyright material in this book please see our website at www.wiley.com.

The right of the author to be identified as the author of this work has been asserted in accordance with the Copyright, Designs and Patents Act 1988.

All rights reserved. No part of this publication may be reproduced, stored in a retrieval system, or transmitted, in any form or by any means, electronic, mechanical, photocopying, recording or otherwise, except as permitted by the UK Copyright, Designs and Patents Act 1988, without the prior permission of the publisher.

Wiley also publishes its books in a variety of electronic formats. Some content that appears in print may not be available in electronic books.

Designations used by companies to distinguish their products are often claimed as trademarks. All brand names and product names used in this book are trade names, service marks, trademarks or registered trademarks of their respective owners. The publisher is not associated with any product or vendor mentioned in this book. This publication is designed to provide accurate and authoritative information in regard to the subject matter covered. It is sold on the understanding that the publisher is not engaged in rendering professional services. If professional advice or other expert assistance is required, the services of a competent professional should be sought.

Library of Congress Cataloging-in-Publication Data

Golding, Paul, 1968-
Connected services : a guide to the Internet technologies shaping the future of mobile services and operators / Paul Golding.
p. cm.
Includes bibliographical references and index.
ISBN 978-0-470-97455-1 (hardback)
1. Internet–Technological innovations. 2. World Wide Web. 3. Ubiquitous computing. 4. Mobile computing. 5. Online social networks. I. Title.
TK5105.875.I57G647 2011
004.67′8–dc23

2011014521

A catalogue record for this book is available from the British Library.

ISBN hbk: 9780470974551
ISBN eBook: 9781119976455
ISBN oBook: 9781119976448
ISBN epub: 9781119977476
ISBN eMobi: 9781119977483

Typeset in 10/12pt Times by Aptara Inc., New Delhi, India
Printed and bound in Singapore by Markono Print Media Pte Ltd

Biography

Paul is a widely respected and unassuming technologist in mobile and Web with 16 patents and a leading book in mobile applications, now available in Chinese. He has been Chief Architect, CTO and various senior tech roles for companies across the world, from start-ups to multi-nationals. He started one of the first mobile apps companies in Europe. He leads platform initiatives for O2's new "intrapreneurial" business unit and set up the O2 Incubator. His tech activities include big data and cloud telephony. He specializes in far-horizon business and product strategies whilst still being a hands-on architect and developer. http://wirelesswanders.com

Contents

Foreword

In 2005, when the world crossed-over to Web 2.0, business professionals and managers in every industry – from traditional retail to high tech media and telecom – felt the first powerful tremors of the strategic shifts taking place, sweeping away traditional business models and altering competitive landscapes.

Many companies – often leaders in their industry – were faced with more questions than answers, more risks and challenges than opportunities. At that time, Tim O'Reilly, the acknowledged Silicon Valley founder of Web 2.0 asked me for a strategy guidebook published in 2008, to explain in plain language the business implications of Web 2.0. My book's goal was to make sure that – even if you were unfamiliar with Silicon Valley buzzwords and your company wasn't started up with Google, Flickr, Facebook, Twitter or Apple DNA – you could use Web 2.0 business thinking to multiply your customers, partners and revenue streams and turn your networks into valuable digital assets.

Mobile 2.0 – smartphone apps combined with viral distribution over vast online social and professional networks – has thoroughly shaken and stirred the business world once again. The year 2011 is a time of great opportunity for those who are agile in orchestrating the right combination of capabilities and putting Web and Mobile 2.0 best practices into place.

Read this book and you'll benefit from the author's global, strategic and hands-on experience in the mobile world. He is as comfortable cutting code at a hackday event with start-ups in San Francisco as he is presenting strategic advice to the board of the world's largest telcos in Europe and across the globe. His first CTO position was in a Hong-Kong start-up designing location-based services for China. His Chief Architect role at Motorola covered the Middle East, Africa and Latin America.

You'll find out the best practices you need to achieve success in:

1. Creating open platforms and ecosystems that exploit the network effect
2. Embracing "Big Data" – creating value from unthinkably large amounts of data
3. Creating "augmented web" experiences by exploiting the Internet of Things

In exploring the theme of open Web platforms and ecosystems, the author gives a detailed guide of the Twitter ecosystem where over 75 per cent of the traffic (Tweets) flow through open APIs into a rich ecosystem of over 50,000 apps built by a massive community of developers attracted to Twitter's low-friction programming interfaces (APIs).

In exploring the theme of "Big Data", the author gives us another example of Web thinking where the "data geeks" of Amazon create value by finding patterns in data sets that are unthinkably large and can only be processed by inventing a new type of data store, called Dynamo. He argues how Amazon, ostensibly a "retailer", is actually a software powerhouse whose real business is converting website visitors into customers, a task that is all about data-crunching on an unthinkable scale.

Preparing us for the Internet of Things, the author first lays the groundwork with a discussion of "linked data", which is how the Web is evolving from a system of links pointing to Web pages readable by humans to a system of links pointing to data about objects, digital and real, readable by machines. He then charts the explosion of intelligent sensors that will provide us with streams of data from real objects, arguing that the nexus between linked data and sensory data will be smartphones.

This is why I'm excited to see the publication of Paul Golding's book. The Mobile 2.0 revolution has just begun and with this book in hand, you and your company can be part of the wave of innovative business models and success stories to come. This book is a great starting point for anyone who wants the inside scoop on what's going on, emerging best practices and how to position and think exponentially for success.

Amy Shuen
Best-selling author of *Web 2.0: A Strategy Guide* and a Yale,
Harvard, Berkeley-trained strategy consultant, multinational board adviser
and award-winning business school professor at Wharton, Berkeley, CEIBS,
HKU, Chalmers, Ecole des Ponts and Ecole Polytechnique.

Preface

I wrote this book primarily for a telco industry audience, at all levels, technical and managerial. I wrote it to help stimulate innovation through "Web thinking" inside telcos, knowing that:

1. Many folk inside telcos still don't get the underpinning principles, patterns and technologies of Web 2.0 and its evolutionary offshoots, like the Internet of Things, which I will explain.
2. Some telcos, though not all, must fight to become platform providers, following the same open innovation patterns as Web platforms, like Facebook, Twitter and Google.
3. Regardless of vision, all telcos can benefit hugely by adopting many of the established and emergent Web patterns, methods and technologies. This does not mean that they should become "Web companies", but more "Web-enhanced Telcos", for want of a better description.

I've been involved with the digital mobile industry more or less from its start (1990) and I've been working with the Web since its beginnings, actively trying to converge the two in all manner of ways, some credible and some incredible. I've worked with mobile suppliers, mobile innovators, mobile operators, web companies, start-ups and entrepreneurs across the entire globe. I have worked with technology, sales, marketing and biz dev folk, wearing all of those hats too. I know mobile and the Web pretty well. I've programmed for them, built products for them, worked on the science, tried to make money, started companies, advised companies, proposed standards, got my hands dirty in all kinds of ways, had all kinds of successes and failures. Phew! I survived all that and now consult for various clients, big and small, who want to do something that matters with mobile and Web software. I'm saying this so that you will feel confident that I'm the right kind of person to write this book for you.

In working with telcos, I find an industry that has mostly failed to innovate in any meaningful way when it comes to the Web and the digital economy. Operator folk usually think of the issue in the wrong way, as though the Web is just another distribution channel rather than a major infrastructural force in the modern industrial landscape. Many senior folk have always known that they can't really ignore the Web, but have followed the pattern of "our business has been successful without it, what's there to learn?"

The answer is: a lot!

What this book attempts to do is to explain, mostly to telco folk and their cohorts, how the modern Web (2.0+) really works and where it's headed. This is an attempt to explain how the Internet is *the* future of mobile, with "http://" being the dial tone for modern connected services. There really is no such thing as "mobile Internet". There's just the Internet. How it evolves will affect and determine how mobile evolves. The Internet will be at the heart of every way that we communicate, share, interact and conduct our lives. Connected services, in the main, will be Internet ones.

For those of you who have read my previous book *Next Generation Wireless Applications*, the book you find in your hands is kind of what that book would be if I re-wrote it for the current post Web 2.0 age. However, much of the previous book explained the technology of the Web at a detailed level that isn't necessary or useful here. You can still refer back to it, if you want a detailed explanation of Internet and Web protocols relevant to mobile. In the current book, I want to focus much more heavily on the technological principles, rather than the mechanics, taking you on a tour of the big ideas that are shaping our connected lives in dramatic ways.

I wrote this book because my experiences of consulting for mobile operators informed me how far adrift many of the telco folk are in understanding what's really happening with emergent memes on the Web (themes[1]), such as "Big Data", "platforms" and "real-time Web".

This book is written for those adventurous spirits who want to be in the innovation race and who don't need to be convinced of why the modern Web is important, but just want to know how it works and what to do about it, or with it. A big obstacle for many in the telco innovation race is that they simply don't understand what's happening in the post Web 2.0 landscape. This book should fix that, or at least get you headed in the right direction.

That said, no one really knows where this innovation race is eventually headed, including me. However, there is one interesting aspect of the connected services evolution that can't be overlooked. It's called the Internet. Ubiquity today has become about access to the Web and its services wherever we go – call it "everyware".

Here's a summary of what I'm going to look at:

Chapter 1: Connected Services: The Collision of Internet with Telco – Here I explore what I mean by a "Connected Service" and how the real backbone of connected services is software, not networks. The dial tone of connected services is "http://" I explore the common architectural pattern for connected services on the Web, which is "open platforms". Successful platforms enable digital ecosystems to flourish.

Chapter 2: The Web 2.0 Services Ecosystem, How It Works and Why – Low-cost and easily programmed software is at the heart of Web 2.0, especially the LAMP stack and its derivatives, which we explore in this chapter. I explore common software patterns using this stack and then outline the ongoing importance of the Web browser. We explore how the Web has evolved from being informational to being social and how the Web has become a highly programmable platform by virtue of open APIs. I conclude by looking at the role of the smartphone as the nexus of modern Web and Mobile software trends. It is the ultimate "connected service" device.

Chapter 3: The Web Operating System, The Future (Mobile) Services Platform – A "Web Operating System" allows the developer of Web and connected services to "hand-off" much of the underlying plumbing to a set of existing services that provide key common functions. In this chapter, I explore the meaning and shape of the Web OS and its strategic implications for telcos.

[1] A "Meme" is a word coined by Richard Dawkins when talking about "the survival of the fittest" in terms of ideas passed on rather than genes. It is used fashionably in the online world.

Chapter 4: Big-Data and Real-Time Web – Big Data is a collection of ideas, trends and technologies that enable Web ventures to exploit the value in massive data sets that exceed the confines of the conventional storage and processing limits of single computers. Big Data is about making value out of unthinkably large amounts of data. In this chapter, we take a tour of the Big Data landscape, decoding some of its components and buzzwords, and also debunking some of its myths. I look at real examples of Big Data technology and think about its application to telcos.

Chapter 5: Real-Time and Right-Time Web – The movement of data on the Web has migrated from an on-demand pull mode to a "just in time" push mode. Data increasingly flows across the Web as it becomes available—in real-time. In this chapter, I explore the real-time nature of the modern Web and how, when combined with real- "Big Data" and smartphone platforms, it enables the right data to be delivered at the right time, leading to the "right-time Web". I conclude by looking at why telcos need to move quickly to ensure that telco platforms are fully integrated into the right-time Web.

Chapter 6: Modern Device Platforms – In thinking about device platforms, it is sometimes more useful to think of connected devices rather than mobile devices. "Being mobile", is increasingly about being able to stay connected to a number of key data streams at all times. To stay "connected" at the informational level, smartphones are the key platforms. I explore the dominant platforms of iOS, Android and Mobile Web, each important in their own way. I give a detailed summary of HTML5 and its associated technologies and standards, all of which will deliver a substantial step increase in the power of the mobile Web.

Chapter 7: The Augmented Web – There is no doubt that Augmented Reality (AR) services are going to occupy an important place in our digital lives. We interact regularly with both the digital and the physical world. Using one to augment the other is a natural progression and mobile platforms are the natural intersection points. In this chapter I explore the key components of the emergent "Augmented Web", including a discussion of how HTML5 and standards will accelerate its adoption. I also explore how sensors are going to become the next frontier of the Web, enabled by convergence with mobile platforms and cloud-computing.

Chapter 8: Cloud Computing, Saas and PaaS – Cloud computing is one of the key enablers of connected services, underpinning the software paradigms of Software- and Platform-as-a-Service. I explore all of these paradigms in some detail and conclude with a discussion of how telcos must develop meaningful strategies for cloud-computing, PaaS and SaaS, both as providers and consumers of these technologies.

Chapter 9: Operator Platform: Network as a Service – Network as a Service (NaaS) is where a telco exposes existing network enablers via APIs, usually associated with the core capabilities of the network. I describe NaaS patterns and strategies in this chapter, including an important and in-depth discussion of developers, the "customers" of NaaS services. I also include some promising and insightful examples of NaaS.

Chapter 10: Harnessing Web 2.0 Start-Up Methods for Telcos – Although highly successful in their own right, telcos can still benefit from understanding how modern Web ventures work, which I explore in detail in this chapter. I look at what we mean by "Scalable Web start-ups" and how they tend to exhibit a common set of approaches towards exploiting Web 2.0 as a platform for doing business. These approaches span technological, cultural, organizational and commercial concerns, all of which offer important lessons and opportunities for telcos.

Thank you for reading this book. I hope that you find it useful. Please contact me with any questions and comments.

Paul Golding
http://wirelesswanders.com
Twitter: @pgolding
Email: paul@wirelesswanders.com

1

Connected Services: The Collision of Internet with Telco

```
For coolnames.each do |c|
display c.coolword
call ubiquity
End
```

- Any digital service that brings people together in a meaningful way – to engage, transact, share, and so on – is a "connected service". This includes digital communications in telcos and on the Web.
- The real backbone of connected services is software, not networks. The dial tone of connected services is "http://".
- A common architectural pattern for connected services on the Web is open platforms. Successful platforms enable digital ecosystems to flourish.
- Using standard telco business models to explore the value of a service isn't congruent with the value of Web platform services.
- When building a platform, user experience remains important.

1.1 Connected What?

An uninformed observer, or visitor from a distant galaxy, would be forgiven for thinking that telcos ought to have been at the heart of the Internet revolution that has swept through much of the developed world these past 10 years, following the tipping point of the Web. After all, telco is all about networks, as is the Web. Telco is all about connecting people, as is...

You've guessed it – The Web!

Instead of Google, Yahoo, Facebook, Flickr, our alien visitor might expect to see the icons of the Web to be O2, Verizon, Orange...

But they won't. Figure 1.1 shows what they will see (you will probably recognize most of them):

Connected Services: A Guide to the Internet Technologies Shaping the Future of Mobile Services and Operators, First Edition. Paul Golding.
© 2011 John Wiley & Sons, Ltd. Published 2011 by John Wiley & Sons, Ltd.

Figure 1.1 Web logos – spot the telco!

Users surfing the "mobile web" often arrive at their digital destination via the on-ramp of Google search. Users finding their way across town often arrive at their physical location via Google Maps. Developers are hacking with Google's Android. Users finding old friends, and making new ones, are doing so via Facebook. Business folk are connecting and networking via LinkedIn. And on the list goes, dominated by companies that all appear to have one thing in common – they were born on the Web.

But don't they have something else in common? That's right. They all appear to be obsessed with *connecting* people, to other people, to data, to places, to whatever – to things? Furthermore, many of these ventures were born in universities, although often not in the labs. They were born in the dorms and sometimes in the coffee houses. This is a key symbol of the Web 2.0's innovation *culture*. There you go – I did it – I used a dreaded "C" word.[1] And I mean *culture*, not connecting.

Reluctant to use that particular "C" word as I am, as it generally sends corporate minds in a spin ("What is culture?" "How do we change culture?"), I am not going to shy away from talking about "non-tech" stuff in this book wherever it serves to make a valuable point. You see, in my experience, technological enterprise – the art and science of really getting something done, something worth doing with tech – is not done in isolation of people, attitudes and verve. This point, perhaps more than any other, might explain why the Internet is not dominated

[1] Doesn't every book related to telco have to have a list of words beginning with C? The five Cs? The three Cs? Not sure what the magic number is these days.

by telcos. They are different types of enterprise creature, if you will, or should I say ecosystem (more on that later).

It's not as if telcos didn't have the money to build substantial Web ventures. Well, some of them tried, and failed. It's not as if they didn't have lots of "technical people" either, or, more importantly, lots of paying people who make up those incredibly large customer bases that would be the envy of any Web start-up and most Web ventures. Maybe they didn't have the right cultural conditions. They mostly still don't. And the only reason I mention this now is that Web 2.0 is as much about culture – the way people think and behave *by habit* – as it is about technology and business patterns. If you work in a telco and you still don't get this point, then I recommend reading this book on an airplane, one destined to Silicon Valley where you can hang out with Web ventures and see how they really work.[2]

Sure, 99 per cent of this book is going to be about tech stuff, but that's almost irrelevant if you don't set up the conditions to make the tech work for you. I know what I'm talking about. My first book – *Next Generation Wireless Applications* – explained much of the Web 1.0 tech, the emergent Web 2.0 stuff, and its mobile offshoots in great detail. That was back in 2004 (and I started writing in late 2002). I wrote the follow up in 2007/8 sprinkled liberally with 2.0-isms.

Both these books were bought mostly by folk in the telco ecosystem – and then mostly ignored. I know, because I held numerous workshops based on the books' themes. I got the feedback firsthand, which was almost always a room full of "Why would we do that?" and other "Why?" questions that added up to a unanimous "We don't get this . . ." message, which is cultural, not technical. Culture is embedded in language and, if you don't speak the lingo, you really won't get the culture, not in any depth.

I set up one of the first mobile ISVs in Europe, back in 1997. I built the first Mobile Portal ever (Zingo) back in 1998, which we (i.e. with my client Lucent Technologies) took to Netscape as their mobile play – imagine that, a telco supplier (Lucent) pushing product to a Web darling. Whilst acting as Motorola's Chief Applications Architect 2005–7, I set up their "Mashing Room" lab to build hacks that would demonstrate the intersection of mobile and Web 2.0 – "Mobile 2.0," if you will. We built a telephony mash-up not too dissimilar to Google Voice (previously Grand Central). I spent much of those two years evangelizing various Mobile 2.0 themes to operators globally. Again, my enthusiasm and ideas were mostly met with blank stares.

Which brings me to the next "C" word – COLLISION!

That's pretty much what's happened. Web 2.0 has hit the telco world, almost taking them by surprise, even though it's been a gradual creeping up, like the vine that slowly grapples a wall (and pulls it down). The overwhelming sentiment is that "these guys" – that is, the Web companies – are slowly eating our lunch, and they're doing it using our networks (bit pipes). What's more, they appear to be doing what we do, don't they? Connecting people!

No need to debate this point. Let's get straight to the killer question:

"What can be done about it?"

This brings me to the final "C" word of the series (noting my dear readers that every seasoned evangelist has to tell a story using 3 or 5 Cs at least once in their career):

"CONNECTED services!"

[2] This is not a flippant point. I took one senior telco guy on such a trip and he came back "converted."

This phrase happens to be one I've heard used by O2, one of the companies I consulted for when I was writing this book. But they're not unique in their ambition, which is to become something other than just a "mobile company" in order to avoid the inevitable descent to dumb bit-pipe, should they, or any other telco, not want to end up there, which is debatable (see Section 1.3 Six Models for Potential Operator Futures).

The phrase "Connected Services," is supposed to cast a wide net, and one that frees us from the constraints of a telephony network. Any digital service that brings people together in any meaningful way – to engage, transact, share, and so on – is a *connected service*. In that way, Twitter is a connected service. Facebook is a connected service. Even search is a connected service. I don't want to get too prissy about definitions, as experience has taught me that such distractions are exactly that – distractions. A quick skim of the contents page will tell you the sorts of stuff I mean by connected services. The issue for operators is that telephony is a very old technology that hasn't changed much. And, while people will always want to talk, at least for the foreseeable future, we can see that more and more people are finding ways to connect without voice, like the examples just given, which all take place on a giant platform called Web 2.0, quite separate from telco networks, which just carry the traffic to and from these various Web platforms.

So, what can be done about it?

This isn't one of those "get rich quick" books. There's no easy answer. . . .

Actually, there is, which is to do something different from what you've been doing. That's the easy answer, incomplete as it is. Nonetheless, many operators remain in limbo, trying to gain the freedom to innovate that evades them and blesses the innovators at the extreme ends of the "freedom to innovate" spectrum – the cash-rich Googles and VC-funded companies at one end and the cash-starved boot-strapping bedroom start-ups at the other.

As I keep telling my colleagues in the industry: "Think, try, fail, tune, deliver . . ."

You've got to stop pondering about all this stuff, stop thinking about a "them (Web) and us (telco)," and start building stuff, putting it out there and tuning as you go. This is the agile way, the Web way (see How Chapter 10). The battle-hardened roadmap process for deploying and running vast arrays of network infrastructure, supporting millions of customers and running giant marketing campaigns serves very little purpose on the frontiers of the Web. It doesn't matter if you're a 100-year-old company that dug up roads to wire the nation, when it comes to the Web, you're a start-up – it's still a frontier world where more is still unknown than known and where we continue to be surprised by the rampant success of "new" ideas (like Twitter) and emergent categories (like Social Networking). In this regard, most Web ventures are still start-ups, whether launched in a dorm or from the labs of the 100-year-old giant. And in the world of start-ups operating in the unknown, agility is king, as are other memes, as the Web-geeks call them, like platforms, real-time and "Big Data," all of which we shall explore in enough tantalizing detail to get you motivated to try something different.

This book is about the ingredients, patterns and technologies that will enable connected services to work in the Web that's emerging post Web 2.0. Is that Web 3.0? Well, I don't want to mess our heads with yet more conceptual claptrap, but if you think it's time you really got to grips with Web 2.0, then you're a bit late. But don't worry. Whether it's Web 3.0, the Semantic Web, the Internet of Things, the Real-time Web, or all of these things, you'll know which is which by the end of this book. You'll also have enough feel for these ideas to go do something new and interesting, maybe start the next billion Euro industry.

Most of what I have to write about in this book is the underlying technological patterns emerging right now, such as "Big Data," which is the ability to make value from unthinkably large amounts of data that would have previously languished on arrays of disks and vaults of tape sitting somewhere, potentially gathering dust.

So let's crack on. Let's set the scene for moving beyond the collision of Web with Telco to a place of congruency – the world of connected services.

1.2 Ubiquity: IP Everywhere or Software Everyware?

I spent much of the ink in my previous two books explaining mobile and IP networks from soup to nuts. I'm going to follow the Web hacker's motto:

Don't Repeat Yourself... or DRY...

Most of the networks stuff I wrote about in my last book remains current, so go take a look. While interesting, it's really not that relevant to this story, so don't worry if you don't know it. However, networking and related protocols still underpin the Web and, for the unfamiliar, I still maintain that a solid understanding of certain principles, like the way HTTP works, will carry you a long way in understanding and accessing new ideas on the Web.

For years the mobile industry got us all in a frenzy about ubiquity. I think slogans were coined about it: "Anytime, anyplace, any... something," I struggle to remember. Yeah, we get it. We really do – an IP connection that is! Almost everywhere we go, we can grab an IP connection thanks to the huge and ongoing investment in wireless broadband networks. In most advanced markets, it's difficult to go anywhere without the ability to connect to an IP endpoint: WiFi, 3G, 4G.

What this really boils down to is the ability to make "http://" work everywhere, which is like the "dial tone" for connected services. This is the kind of ubiquity that's become important – software services everywhere – call it everyware![3] Sure, this low-level bit-shifting network stuff still dominates the telco world. And for good reason. That's what they do, and they do it well. But we're not here to consider the "Telco versus Web business model" debate ("we make money and they don't... blah, blah.") We're here to explore connected services – making the most of the telco and Web worlds *combined*, or collided. This is not about telcos becoming Web companies, which they will often be the first to say is unlikely. This is about making better, more relevant and future-proof telcos by harnessing and exploiting the technologies, patterns and capabilities of the Web. The opportunities have never been more promising and exciting than today, thanks to a significant evolution of Web technologies and patterns.

What the folk in the telcos often don't get is how the Web world *really* works, especially post Web 2.0. Sure, they get HTTP and all those "Webby" protocols at a distance, but in my experience they have failed to keep up. Where we are today is a world awash with Web-centric software and applications, many of them free of charge, that enable coders to work in even bigger chunks, like writing a novel paragraphs at a time. That might sound crazy, but that's

[3] Thanks to Mike Ellis for that buzzword!

how it works. Moreover, a hacker can take someone else's paragraphs, even whole chapters, and craft them into a new story – that is, an application or service.

And that's where DRY comes into play. If it's already been said, or written, then why say it again? Just run with it. This reflects the idea of re-using stuff as a principle. Much of the progress in Web software has been possible by this re-use principle, which we can widen to include: the open source ethos, things like "social coding", open APIs, mash-ups and so on. Don't worry if these concepts are still alien to you. They won't be for long. Read on.

The essence of ubiquity in the post Web 2.0 era is in the ability to connect via software. It's incredibly low friction. If you can think of an idea for a service, then it isn't long before you can articulate the idea in software and start a conversation with users via their Web browser or mobile app. Talking and thinking in Web software is the new ubiquity. All that low-level stuff that makes it work – the IP stuff – is just taken for granted.

1.3 Six Models for Potential Operator Futures

Having worked on dozens of projects and having had dealings with all kinds of folk, I've concluded that no one really knows how the telco world is going to pan out in the face of current pressures. One thing for sure is that we will always need physical-layer networks. That future is guaranteed for someone. It's not unlike the utilities worlds or the ISP world. Most survive in the realms of efficient operations, value-engineering, rigid cost controls, customer care efficiencies, effective marketing and so on. Some differentiate with niche services (e.g. Heroku), hyper-effective support (e.g. Rackspace), and so on.

But what of these "Connected Services" futures? What might they look like? During a consulting gig for the world's biggest texting infrastructure provider, I postulated some possible future operating models, extending out to the year 2015. This was just to guide their thinking in terms of how their infrastructure products might need to evolve and adapt. Again, I like to follow the lean model of just getting something shipped and then working with real feedback rather than working on grand theories in a vacuum. In that spirit, I threw up these six models as a starting point for the discussion. Once you have something to discuss, ideas can begin to form and crystallize. As part of the backdrop for proposing the models, I identified a number of key trends, challenges and opportunities facing telcos, as shown in Figure 1.2.

These trends led me to propose six models for future telcos (or OpCos as I call them in these diagrams). I don't want to bore you with models for the sake of it. These models will serve to provide us with context as to how and where the Internet and telco worlds might intersect.

These models, shown in Figure 1.3, are not mutually exclusive. Some operators might well end up following all six. There might be more than six. They might also overlap in terms of definition, but each one has a sufficiently distinct set of attributes to make it useful as a potential pattern in the future of telco operating models. I'm not going to play out the models in detail, as this isn't supposed to be a business models book. I'll briefly summarize the essence of each model as a basis for thinking about the nature of connected services for each scenario that a telco might face.

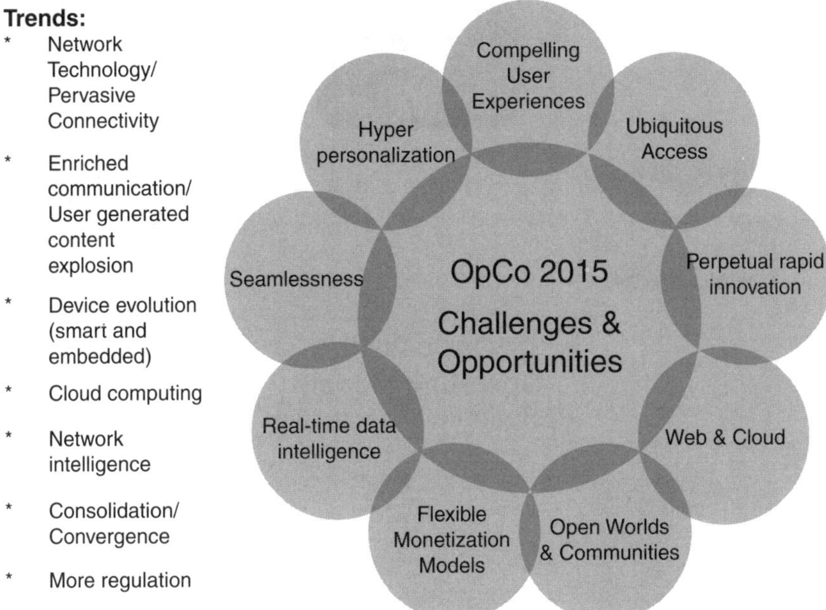

Figure 1.2 OpCo 2015 challenges and opportunities.

1.3.1 Access Provider

This is the pure network play – a bit-pipe. Not to be underestimated in terms of profitable futures, but not the most interesting of models within the connected services context, as it is blind to the services layer running atop. There are some interesting opportunities here for the use of modern Internet software techniques to run a bit-pipe a lot more cost effectively in the OSS/BSS domain, but that's a different story.

1.3.2 Connected Services Platform

The keyword here is "Platform," which is the first time I've introduced this term, but we will be exploring its meaning and implications in depth throughout this book. As you will soon discover, the idea of platforms is absolutely central to modern Web ventures. The idea of this OpCo model is to take the current telco network and enterprise infrastructure and turn it into a platform on which it is easy for other service providers to build new services that can mix and match internal components and capabilities (e.g. Billing) with external components, which means more or less any Web service.

This isn't a new idea, but I mean here something a lot more visionary and aggressive than opening up a few APIs to developers. I mean a radical extension of the business and technology architecture to support Software-as-a-Service and Platform-as-a-Service patterns, which might also include extension into new infrastructure opportunities, like cloud computing. I also

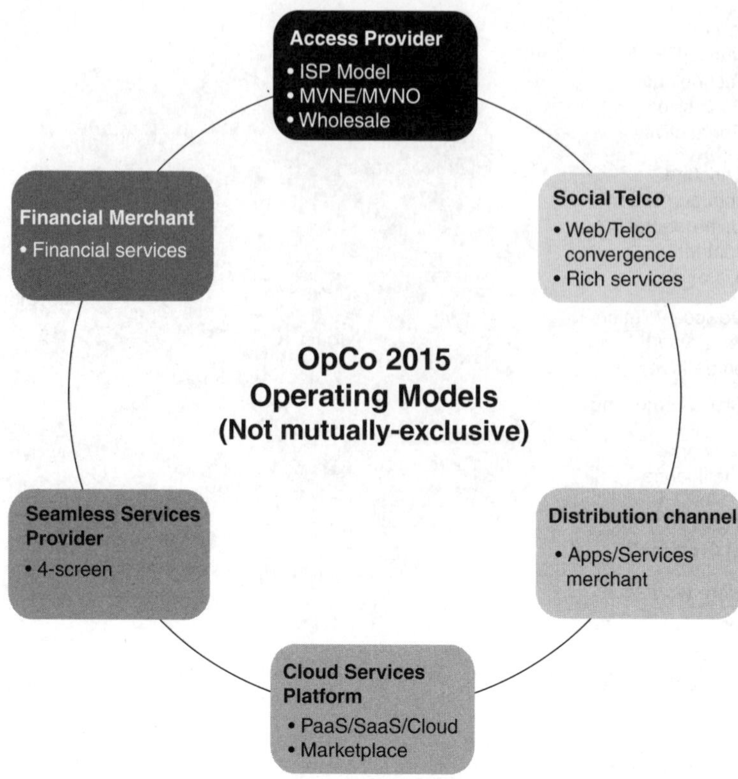

Figure 1.3 OpCo 2015 operating models.

include potential ideas such as Support-as-a-Service. After all, if you're good at supporting customers, then why not at least consider how to turn it into a revenue-generating service as opposed to a cost centre.

1.3.3 Distribution Channel

A major OpCo asset is the extensive user base in consumer and business markets. In theory, a carrier knows a lot about its users, which ought to be a highly exploitable asset in many ways. A carrier also has the means to extract payments and to maintain a relationship with the users. All combined, these are powerful ingredients for a compelling distribution channel, particularly for digital goods, which includes adverts, coupons and even virtual currencies. Many carriers already have retail stores and healthy e-commerce apparatus, all of which are extensible in theory to become low-friction digital distribution channels. The tuning and reconfiguration of the OpCo platform to become an efficient and targeted distribution channel has lots of potential.

A lot of potential for this model is in the adoption and exploitation of so-called two-sided (or N-sided) business models, which means getting money from upstream customers in the way that Google gets most of its money from the advertisers, not the users. Telcos mostly get their money from the downstream users, although this is slowly changing.

1.3.4 Seamless Services Provider

Which carrier hasn't considered moving into adjacencies, like financial services, home services and so on? Many have already tried it, with varying degrees of success. Operators are also moving towards multi-channel offerings: mobile, fixed, broadband, TV etc. Meanwhile, competitors in those adjacent businesses are also moving into their adjacencies, so the nature of competition is shifting all the time.

Inside this milieu of stretching-the-brand offerings, the winners will increasingly be those who can offer the best "joined up" experiences for customers. This doesn't necessarily mean fancy tricks (like music that flows like liquid from the TV to the street to the car). It is more about information and experiences being in the right place and mode as the user moves from one service to another. It's about giving a compelling user experience across a wide portfolio of services, which is not easy. As we shall explore later, the idea of the "Right-time Web," is all about the right information at the right time and place.

We shall get to this in due course, but "joined-up-ness" will increasingly only be possible via various syndicated connections on the Web. Telcos do have a lot of information about their users, but not as much as they think they do in comparison to the Web, where users increasingly leave substantial trails of digital footprints. In other words, a large part of the user context that a seamless provider will need access to is situated and evolving on the Web. To illustrate this point, I include a diagram (see Figure 1.4) from my previous book, just to emphasize the shift in "centre of gravity" from the telco to the Web.

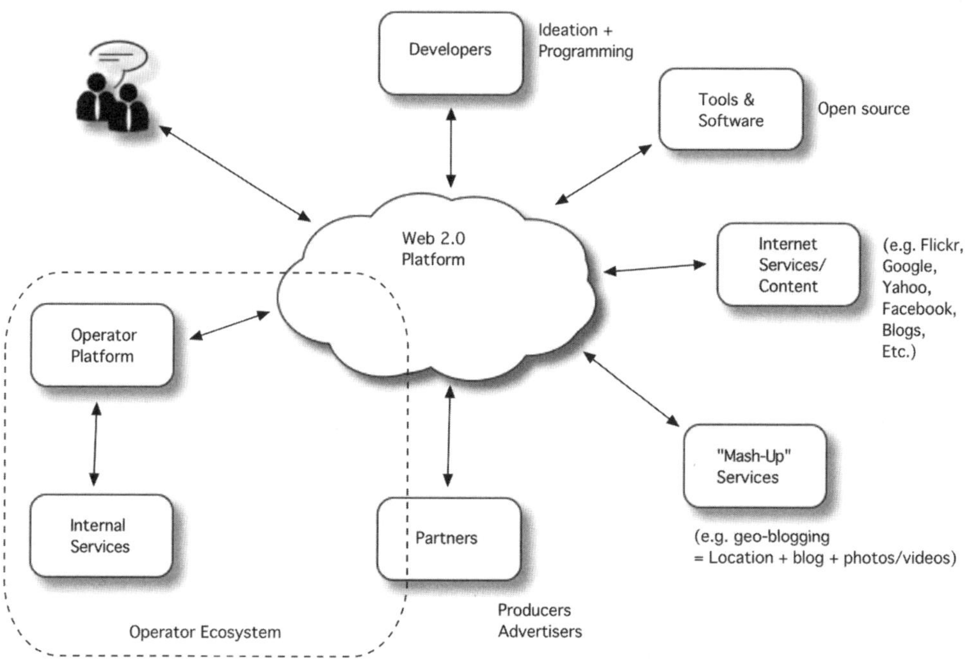

Figure 1.4 Shift of gravity towards Web platform.

Sadly, telcos spent years talking about "Context-aware Services," only to lose the march completely to the Web. Even now, many strategists inside operators still don't get that "Context," is what Web brands are all about, ever since they discovered the ad-funded model. Relevancy is king in that world. Relevancy is all about context. On the other hand, most telco services operate in exactly the same way no matter the user context. Making a call is making a call is making a call – it seldom adapts dynamically to the user's context. Going further still, the networks are oblivious to the content of those calls, which is like Google not bothering to exploit the content of search queries – can you imagine?

1.3.5 *Financial Merchant*

What does a telco do? One thing it does well is to add up lots of activities (rating) and then charge the user (charging), telling them what to pay (billing). For ages this capability has only been thought about as an adjunct to telephony, not a capability in its own right, capable of being extended to all kinds of ventures. Sure, various visionaries have seen the potential of extending the "money machine" to other applications, but they are generally hamstrung by the inflexible nature of the IT infrastructure used to build these systems.

This is in total contrast to the likes of Amazon and Paypal whose payment systems were born in the world of Web software ubiquity – just another application running on the LAMP stack, or similar.

However, with a similar approach, operators could substantially reduce current costs of running the money machine whilst freeing up considerable flexibility to pursue new models in an agile fashion. A lot of operators are keen to pursue this model, notwithstanding various regulatory restrictions and value-chain complexities that might impede their ambitions.

1.3.6 *Social Telco*

This model says that an operator will not sit back and concede its previously central role in communications to Web players. I think that the old-school thinking of "build a better internet" is over, whether it's called IMS or Wholesale Application Community (WAC).

Mobile phones have always shipped with address books. Unspotted by most, this combination of phone with address book made a unique communications device at the time (see Figure 1.5).

That's right, our mobile phones were actually *social networking* devices because they contained a list of our friends and associates along with their contact details, allowing social connections (calls and texts) to form, which is the essence of a social network. Conceptually at least, there is no difference between this address book of friends and the list of friends on Facebook, or LinkedIn. Mobile networks plus devices were the original social networks, but woefully under-exploited as such. Sadly, the address book didn't become the nexus of connectivity that it might (and should) have done, even though some of us (like myself) were talking about and demonstrating this idea from very early on. I wrote extensively about this in my first book, describing how the address book could and should become a fulcrum for all kinds of useful services. I even suggested the appropriate Web technologies and protocols[4] for making this happen.

[4] Such as the emergent Friends of a Friend (FOAF) data format.

Figure 1.5 Phone as a social networking device.

Operators sat back and did nothing, mostly. Again, this is said non-judgementally. This book is about the future, not lamenting about past and gross oversights and failures (e.g. WAP, MIDP) and missed opportunities (e.g. Social Networks) of the telco world. Pity! What a missed opportunity that was. Or was it?

This operating model says that there's no way that the socially-relevant services bet is off for telcos. I think that tier 1 players will have to embrace a lot of the current Web trends in order to remain competitive and relevant. It's not like they're sitting back anyway. Millions gets spent on various R&D programs and acquisitions every year, although often poorly executed. Some telco players and MVNOs will push this envelope hard and might set the standard. Who knows, we might even see a "Facebook Mobile Network," or equivalent, which exists entirely within the Facebook ecosystem. It is certainly possible and I have proposed such ideas in my consulting gigs.

1.3.7 Start Thinking Platforms

Whether it's distributing digital goods, enabling financial transactions or exploiting social connections, hopefully you might have noticed that in all these cases the network becomes a little bit more intelligent than just a switch, connecting one device to another. There's all kinds of intelligence in the connection, leading some commentators to posit the meta-model of "Smart Pipes," which I find somewhat oxymoronic as a term. I prefer to think of platforms, which is an exciting theme on the Web that I'm going to explore from multiple angles throughout this book. However, before we do that, let's just think a little about what our "telco platform" might look like for any of these OpCo models. Figure 1.6 begins to tell the story of operating a platform rather than a network.

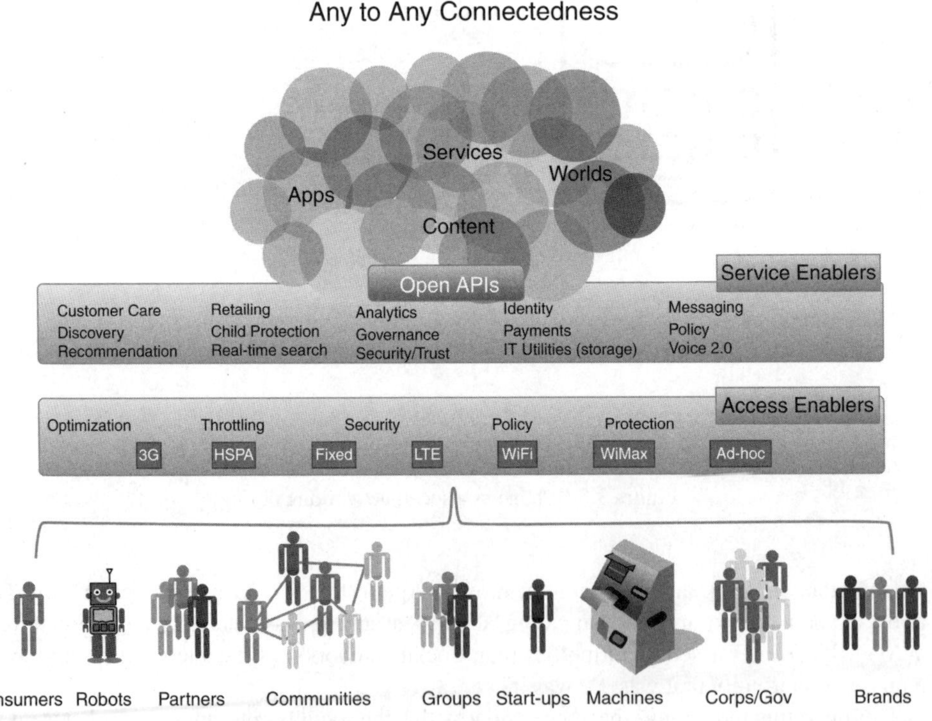

Figure 1.6 Any-to-any OpCo services platform.

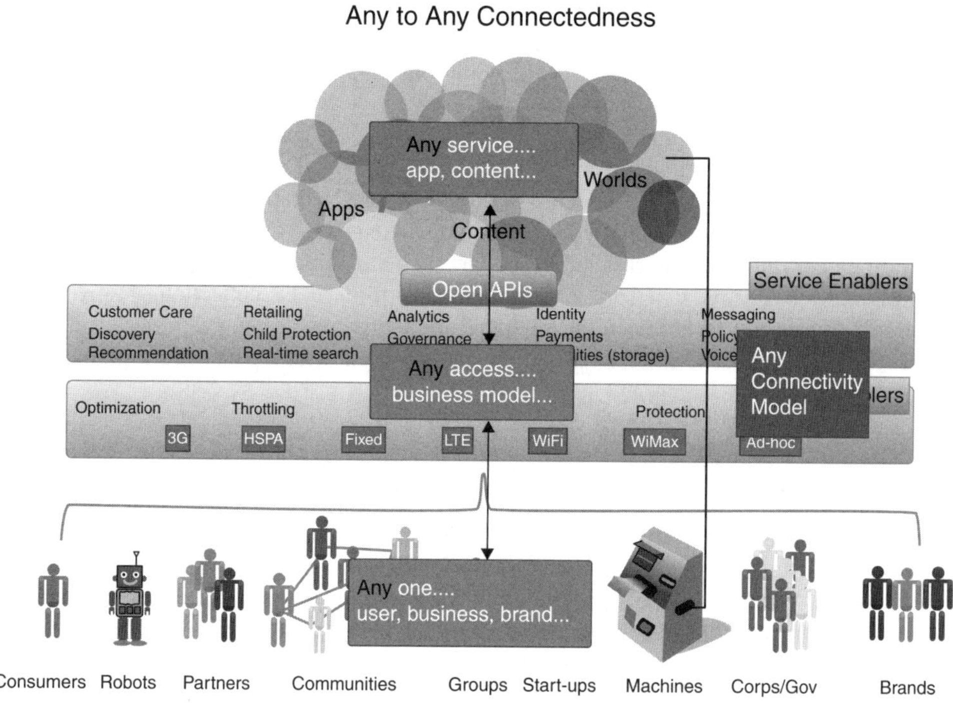

Figure 1.7 Any-to-any business models.

What the diagram shows is four layers:

1. **The users** of the services either running on or accessible via the platform, such as consumers, brands, communities and even machines.[5]
2. **The access enablers** that telcos traditionally think of as the network (e.g. HSPS, LTE) including support services like compression and security.
3. **The service enablers** that really add the intelligence to our services, enabling routing, personalization and all other manner of value-added functions in the service chain.
4. **The digital content** which includes applications and services that users want to access, either running on the platform or via the platform.

This platform model is how telcos should be thinking, no matter what the OpCo future they envisage. Platforms automatically bring to mind accessibility – allowing other parts of the value chain to seamlessly integrate with the telco's capabilities. This leads to a generic business model architecture, as shown in Figure 1.7.

Ultimately, the telco should have an ultra flexible and low-friction platform that enables agile construction of high reliability services that connect anyone (user, business or brand) with any digital service (app, content, etc.) via any access technology and within any business

[5] Note that "Robots" in the diagram means software acting on behalf of people, not physical robots.

model. What you might notice about the platform diagram above is how it still appears to be very telco centric, despite my earlier point about the shift of gravity towards the Web. Of course, as a telco, you are only concerned with your platform and your business, which is largely what the platform diagram depicts.

However, it is in the nature of the connectivity to the user layer and the services/content layer that the smart telco will build interfaces that work seamlessly to exploit the power of Web platforms. For example, if it makes sense to provide a service that enables brands to connect with mobile Facebook users in a way that only an operator could exploit, then this any-any-any platform model should be constructed to support such a possibility. The key point is that the telco should focus all of its efforts on building a platform, not a collection of stove-piped services that lack the flexibility to be deployed in other ways. Having a powerful platform allows the telco to engage in new types of business models with a wider set of partners in a wider number of value chains and networks.

1.3.8 Execution

Don't get too caught up in the opco models we've just looked at. They're not cast in stone. However, they are certainly plausible and a useful starting point for our exploration of the execution question. They do overlap, although each one has a fairly differentiated strategic emphasis. So, how does an operator evolve to any or all of these models? How do they execute?

Not by following business as usual, that's for sure. Here are some default behaviours that certainly won't do in the connected services universe that we now find ourselves in:

- Relying on current and traditional telco suppliers to follow the curve of innovation and adoption = fail.
- Relying on traditional customer insight models to guide service design and deployment = fail.
- Relying on the same key people who built the business for the past decade = fail.

To shoot for success, or even gain a chance of starting in the right direction with any of these models, with the possible exception of the first,[6] it is imperative to gain an insight into the new technological and socio-technological patterns emerging on the Web.

Make no mistake that apart from the first opco model (Access Provider), all of the other models are executable by non-traditional competitors, such as born-on-the-Web ventures, to varying degrees. Moreover, they are busy pursuing it, right now as you read this book. And they are mostly operating in the zones of innovation freedom that telcos find very hard to enjoy, not stuck in the limbo zone.

1.4 "Follow Me" Web – Social Networks and Social Software

Social networks are huge. They are getting bigger and better. For many Web surfers, social networks are the focal point of their digital lives. Facebook has over 500 million users, getting

---

[6] This model almost says to carry on business as usual. However, don't overlook that potential to use Web methodologies, technologies and patterns to enable operational innovation with this model.

bigger every day. Twitter has over 106 million users, increasing 300K per day at the time of writing this book. These figures might well change, but both of these "start-ups" are aiming for 1 billion users. That's huge, if they can get there. Some commentators have pointed out that the population of Facebook is bigger than the population of many countries combined. Facebook even has its own currency!

Facebook and the various social networks out there are more than just websites or "communities" (another popular "C" word). As Facebook founder Mark Zuckerberg said:

> "You don't make communities – you enable them."

The word "enable" is the key to Facebook's technological approach and to their success. Operators might do well to emulate it. Bit by bit, Facebook has provided the tools for users, partners and developers to enable various community behaviours that flourish on the Facebook platform. Facebook has exploited the network effect at every level, as has Google, Yahoo!, eBay, Skype, Wikipedia, Craigslist, Flickr and many others. MySpace didn't, failing to provide the access hooks that Facebook did, which is why it was rapidly over-shadowed by Facebook as Facebook's network effect kicked in. To quote my respected associate Amy Shuen in her insightful book: *Web 2.0: A Strategy Guide*:

> These enterprises have strategically combined different kinds of network effects - including direct, indirect, cross-network, and demand side, to multiply the overall positive impact of network value creation.

I wrote extensively about Network Effect in my first book and will return to it frequently in this book, but here I want to emphasize the underlying infrastructure for network effect in a social network, which is the Social Graph.

The social graph is a mathematical map of who's connected to who, see Figure 1.8.

The concept is easy to appreciate, but here's the really important idea about the social graph . . .

Increasingly, *the social graph is the Web.*

What do I mean by this? When the Web first materialized, and for much of its early growth, the gold rush was about connecting people with pages of text. Early Web protocols, like HTTP, were good for this purpose. They still are. These early pages were static, which means authored and created in advance by page creators, much the same way that every word in this book is committed to paper before you read it. It doesn't adapt or change according to the reader's needs and feedback.

Later on, programmers realized that HTML pages could be generated on-the-fly, or dynamically, by programs. That's what happens today with most websites, and various website software frameworks have been invented and gradually improved to make this easier. The page doesn't exist until your browser asks for it. A program on the server then generates the page and dishes it out to your browser. These pages can be personalized, which is how you can now use the web for personalized services like banking, shopping, travel bookings and so on.

In this transition from static to dynamic content and from publishing to interactive services, Web 2.0 came into being. A key theme (or meme) was the growing importance – and voice – of the users. No longer just a one-way publishing experience, the move to dynamic web allowed users to transition from consumers to producers of content. They could add their content, as

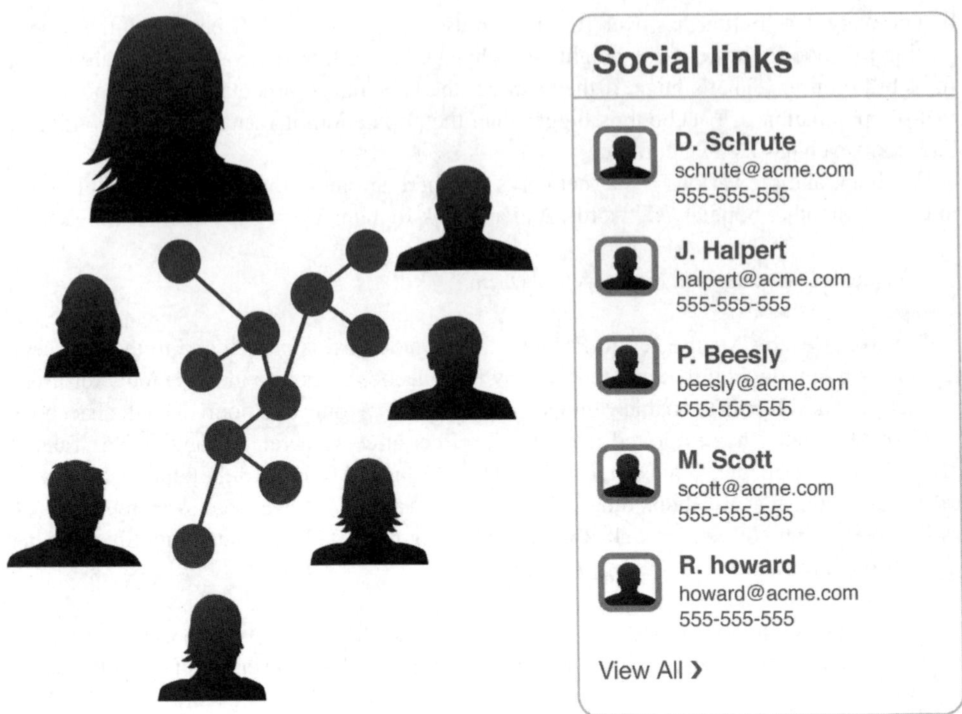

Figure 1.8 Social graph – who's connected to who.

sparked largely by the blogging revolution at the beginning of the 00s decade, followed by social networks, like MySpace.

Once users entered into the community of contributors, they rapidly took centre stage. And this is when the Web got social. It became rapidly clear – and is still rapidly evolving – that the connections between people really matter in Web services. People, and the attention they pay to each other in various modes, drive the network effect, which is a vital precursor to generating value on the Web.

Facebook and social networks have taken this idea to the next level. They have created the platform and tools to enable digital social connectivity, not only in ways that Facebook dictates, but in all kinds of new, unexpected and unintended ways. In other words, Web platforms, like Facebook, are more than just platforms for connectivity and services – they are also *platforms for innovation*.

One of the ways that this works is via the indirect network effect, which is why Facebook allows other services to exist on its platform, allowing these services to exploit the social graph. It is a brilliant strategy because it solves one of the key problems that all digital services suffer: discoverability.

The social graph is multi-dimensional. It doesn't just graph who knows who, it also graphs similar interests, habits, likes, dislikes, event attendance, fandom and so on. This graph of discoverability is essential to the success of a number of the operator models that I laid out in the previous section (see Section 1.3 Six Models for Potential Operator Futures): Distribution Channel, Connected Services Platform and Social Telco.

And it doesn't stop there. Facebook has recently moved its attention to exploiting the cross-network effect by opening up its Social Graph API, which is a protocol for external services to mesh their social graphs with Facebook's social graph (on a per user basis). Just as the cross-network connection for SMS sparked the texting boom in the UK (and eventually elsewhere), we can expect similar dynamics and value-creation from Facebook's move and similar initiatives like OpenSocial.org. This move to open and shared connections is accelerating. Facebook Connect is a service that allows a user to log-in to another service (which might have nothing to do with Facebook) using his or her Facebook log-in credentials. It turns out that this usually results in a much higher sign-up rate to the new service. In some cases, new sites are not even bothering with their own login system – they simply piggyback on Facebook, Twitter and other open authentication systems.

Similarly, Google had already released its "Friends Connect" service, which enables any web service to embed social graph information built on Google's social services. These are all examples of a kind of inverse-platform effect whereby one platform (such as Facebook) extends out to another (which could be an operator's portal) to become an on-ramp or enabler for that platform, which, once adopted and favoured by users, is hard to undo. This is a kind of "Trojan Horse" strategy that some Web players, like Google, have been keen to push hard for some years now.

The social graph is also dominant in Twitter, another service that simply has no value without the concept of social connections. It is important to grasp that the formation of social connections in Twitter and similar services is both explicit and central to the function of the service. A Twitter user has to explicitly – and publicly – "follow" another user. A user declares to the Twitter platform, and all of its users, that "I follow X," and "I follow Y." Once these "follow me" connections are formed, the Twitter platform keeps track of what each user does and then makes sure that this activity is passed along the appropriate parts of the social graph towards the user's followers.

This is unlike telephony where a user can connect to a number ephemerally and in private. There is no lasting connection and no concept of "following." The social graph in an operator network is entirely incidental to its operation. In social networks and increasingly the (social) Web generally, the social graph is pivotal to the service. Social telcos and Connected Services Platforms will have to find a way to put the social graph at the centre of their businesses. This might be a telco-centric social graph, a social network's social graph, or a hybrid of the two, which is the more likely scenario. That said, operators must push hard to innovate on top of the social graph that is inherent in mobile network operation.

As it turns out, building large scale software systems around social-graphs is not without substantial technical challenges. Each time a user carries out an action on the platform, the platform has to propagate this action along the social graph in its various dimensions, and in a timely manner. In the case of Twitter, this happens in near real-time. Moreover, this challenge rapidly mushrooms as users along the graph forward actions and outcomes to their followers, creating a cascade of updates. This requires constant evaluation of connections in the graph, which turns out to be a thorny computer processing problem capable of bringing Web servers and databases to their knees very easily, especially as the network grows with exponential impact on resource requirements.

Twitter was inspired – perhaps forced by performance concerns – to create its own database technology, called FlockDB, just to solve this problem at scale. I mention this because it reveals another interesting and dominant technological pattern in leading Web ventures, which is the drive to create novel types of technologies to maintain a competitive advantage. Operators are currently not equipped to handle such challenges. Despite operating very technically advanced platforms – that is, cellular networks – telcos are not technology companies. They lack the sorts of people who can build new technology platforms. For example, it is virtually unheard of for a telco to build a new database technology, whereas it has happened often in recent years on the Web, with Google, Amazon and their cohorts cooking up all kinds of new storage techniques and solutions (see Section 4.2 Some Key Examples of Big Data).

That said, operators will argue, quite correctly, that they haven't had to invent new types of technology because their suppliers have been doing this quite successfully for decades. Operators will also point out, quite correctly, that they have been addressing large-scale processing and storage problems for decades too, perhaps the fathers of "Big Data." All this is true. However, it misses an important point, which is that without the in-house expertise to recognize technological opportunities and without the pressure of finding new solutions to new business problems – that is, the pressure of product innovation – operators wouldn't know what opportunities they have missed as a result of not being technology companies. For example, it is clear to me that a different type of operator staffing and culture would have spotted the social networking opportunity some years ago and we would, contrary to my opening section (see The C Word, Again) be talking about telco giants and Web giants in the same breath. Alien visitors would indeed see telco logos dominating the Web.

To give another example of an opportunity that is difficult to exploit without at least some technological insights, one of the biggest recent advances in Web technologies has been the advent of "Big Data," which is the use of massively scalable data storage and processing engines to find patterns in data that would have previously been hard to find, economically and technically. It turns out that the ability to process unthinkably large amounts of data is perhaps the key to unlocking new value streams for Web ventures. The same probably applies to Telcos, which is a theme I shall explore when we consider Big Data patterns in depth in Chapter 4.

1.5 What are Platforms and Why are They Important?

I have already used the word "platform" quite a bit in this chapter, but without really exploring the idea in any depth. Let's do that now, because, as I hope you're beginning to appreciate, "platform thinking" has a place in telco futures. Platforms aren't a difficult idea to describe or understand, but can be hard to execute, often deceptively so. The software platform war probably began with the desktop operating system (with early winner Microsoft) and now continues with social networks like Facebook and with other platforms like the device platforms of Android and iOS.

A platform is different from a service or an application, in that a platform is something that other people build services and applications on top of. Platforms are not new. The obvious and pervasive example is an operating system, like Windows. However, there are other examples, perhaps a bit less obvious at first, like the Visa network and the Playstation 3. These are examples that are often quoted in the business literature, such as the intriguing writings of Andrei Hagiu, which I urge you to read.

In his fascinating book "*Invisible Engines*," which I highly recommend, he (and his co-authors) argue that in order to understand the successes of software platforms, we must first understand their role as a technological meeting ground where application developers and end users meet.

To me, this is the crux of the matter when we talk specifically about platforms. For now, I am not really concerned with examples like Visa network, which is an N-sided platform (users, retails, banks). Of course, such platforms might well prove to be of interest to "Financial Merchant" operators (see Section 1.3 Six Models for Potential Operator Futures), such as when exploring mobile wallet ideas and other financial-intermediary solutions. However, in this book, in order to narrow scope, I am going to focus mostly on software platforms that involve the Web, with software developers as one of the key platform users. Later on, we will get into the detail of who developers really are, in order to dispel the mythical notion of developers as folk who sit in messy bedrooms (or garages), more interested in breaking systems than creating value through enterprise where they exploit their works of creativity and utility. Developers are, and always have been, highly entrepreneurial and, increasingly, creators of massive "value nets" around their wares. They are rapidly becoming the industrial moguls of our age.

What initially deters operator stakeholders from exploring platforms seriously is when the ugly word "free" crops up, as it often does early in the conversation. Apple, Microsoft, and Google, for example, charge developers little or nothing for using their platforms and make most of their money from end users; Sony PlayStation and other game consoles, by contrast, subsidize users and make more money from developers, who pay royalties for access to the code they need to write games. More applications attract more users, and more users attract more applications. And more applications and more users lead to more profits.

Importantly, at least for this book, I'm only going to explore platforms that have the potential, at the very least, for an associated business model to emerge – a means to convert platform usage, however that gets metered, into revenue and profits. Some kind of commercial (and contractual) relationship is necessary and inherent in using the platform, both as an end user and a platform application developer. So, this rules out the open Web itself and similar infrastructure platforms that are, like most highways, completely open and essentially free of commercial restrictions. (This isn't entirely true, but we don't need to debate the point here.)

If we consider Facebook, for example, then a developer wishing to develop a Facebook application is required to sign up to Facebook's developer terms and conditions. These controls are obviously important for Facebook to manipulate the use of the platform to its commercial advantage. That said, successful platform ecosystems are always symbiotic. Both the platform provider and the platform developer stand to win if both are successful – a classic win-win dynamic that hopes to exploit the network effect of attracting more users to the mutually beneficial alliance of platform owner with platform service developer. In the case of Facebook, some of the applications, like the Zynga network of social games, have proven to be more profitable than Facebook itself.

Experience shows that an important realization about platforms is that you've got to know that you're in the platform business in the first place and then be really committed to making the platform work, which is usually synonymous with lowering friction of usage on both sides, especially the supply side. That sounds obvious, but not realizing that you're in the platform business, with all that it entails, is a common mistake. It's important to understand that for a platform to be successful its owners have to cater very well for the developers who want to build stuff atop of the platform. The history of software (and computers generally) is littered with the corpses of venturers who thought that they were in the product business when what they actually had was a platform. A recent example was the innovative product called Groove, which was an enterprise collaboration product that didn't require a central server. It used so-called peer-to-peer (P2P) technology, first brought to our attention by illegal music sharing sites like Napster.[7] Instead of music, Groove used P2P to keep information and content in sync directly between software instances running on separate desktop machines. For example, a team of project workers were able to share files between their PCs without a file server.

The team at Groove focused all their efforts into the central product experience. However, lots of other (external) developers were attracted to the product as a platform. This was because the problem of keeping data in sync between machines turns out to be a highly useful, yet thorny, problem to solve, and one that most developers would rather avoid tackling.[8] There were a queue of developers wanting and waiting to build their wares on top of the "Groove stack"[9] in order to achieve two things:

1. Offer their own killer product without the headache of having to write and maintain the awkward P2P synchronization infrastructure.
2. Benefit from the network effect of seeing (hopefully) more users on the Groove platform as a result of adding extra features and services that strengthened the core offering without involving the core team. Of course, in return, Groove could benefit from the indirect network effect of these other users becoming more interested in and committed to the Groove product as a result of these companion services.

When Groove began to understand how developers were viewing their product as a platform, they failed to take advantage of the opportunity. They remained wedded to the idea that they were really in the product business. They viewed the third-party stuff as a kind of "sideshow", rather than something that could have been core to their business strategy. In the end, the core product failed to gain enough traction (because it was missing the features that the third parties might have brought to the users) and the opportunity was missed. The third parties were denied the chance to ride on the platform because of prohibitive pricing and terms. The business model was a product model, not a platform model, even though it eventually became obvious that Groove should have been a platform. By then, it was too late.

There are numerous examples of half-hearted attempts to be a platform player, which almost always exhibit the same tendency, which is to alienate the very developers who could make the platform successful. Operators have excelled at alienating developers with unrealistic commercial terms, high-friction interfaces and various forms of condescension. As I have pointed out to many senior telco stake-holders, were they to treat customers the same way they treated developers (who are a customer for the platform), they would be out of business and probably get fired. For example, imagine charging customers to enter a shop, or asking them

[7] Before it went legal of course.

[8] Have designed sync products myself. I can tell you that I'd rather stick pins in my eyes than ever try it again.

[9] When I talk of stack, I mean the actual software that underpins the Groove application. The "platform" of Groove is not just the stack, but the stack combined with the users and a means to add additional services on top of the stack, making them available to all the users.

to wear a tie before entry. Imagine ignoring them when they got in and then asking them to take one of those tickets you see at the deli-counter lines. But this is how developers have been treated by operators for years. It's no wonder that many of them hate operators with a passion and can even get religious about supporting any technology that makes telcos irrelevant. I should add that whenever I've worked with or for operators, it has only ever been in the role of "developer," trying to fix the problem from the other side – trying to convert operators into applications and developer-friendly companies. As one of the pioneers in creating mobile applications in the 1990s, I suffered too many headaches bashing my head against the wall of operator ineptitude towards developers.

With platforms built on the Web, there is often a tension between a services play and a platform play, similar to what I described with Groove. Sometimes the venture can't decide if it's really an application or a platform, resulting in poor attention to the developers, the **only** folk who might be able to make it work as a platform. Many of these ventures suffer the same fate, which is aggressive disruption from a copycat who gets the platform play right. Joel Spolsky documents a number of such examples in his article about platforms from which I quote:

> The best way to kill a platform is to make it hard for developers to build on it. Most of the time, this happens because platform companies either don't know that they have a platform . . . or they get greedy (they want all the revenue for themselves.)

One reason for the tension between application and platform is the need for the venture to promote its own platform with an application that will attract enough users to gain critical mass in the first place. After all, without a user base, developers might not be interested in developing for the platform. This is not always true, as I said during my presentation of this topic to Mobile World Congress in 2010. I pointed out that a platform very often needs "cool power" in order to attract developers, which itself promotes the platform to end users and to other developers keen to join the action and fun. Twitter and iPhone are two examples where the shear coolness of the platform had developers eager to build apps.

When I'm asked what does cool mean, another of those thorny "C words," I usually say, not entirely in jest, "if you have to ask, then you won't get it." My answer seeks to expose the fact that software development is a very broad spectrum of activity and people, and not everyone is going to "get" cool. Anyone can open a computer and start coding. It's mostly free and doesn't really require permission. This is just like playing a musical instrument and we can draw an interesting analogy here. Music can be both a playful, artistic, creative endeavour and an industry making billions. Composer John Williams makes vast amounts of money out of writing scores for films. However, much of the innovation in music, like the invention of Blues guitar, came from individuals who had the tools and inclination to experiment. Similarly, coding software is just as much a creative pursuit as it is a commercial endeavour. At the creative end of the spectrum, new and shiny tools will attract a lot of attention and generate a lot of innovation. This innovation is then funnelled by various entrepreneurial processes into money-making industries. So, very often we can substitute "cool" with "new." In the case of a telco, we might suggest to give developers something they just haven't seen before or can't be done anywhere else. Sending a text message hardly qualifies as new and exciting. However, location-finding, had it been exposed to developers way back in the early 00s, would have been a very interesting platform enabler to offer developers.[10]

As I have shown earlier in the opco models for the future, becoming a platform player is going to be key for some operators. The challenge here is understanding exactly what makes a successful platform. Operators tend to focus, quite proudly, on their assets: "we have this, that and the other." Frankly, nobody cares unless the assets are easily usable, like Lego bricks. If ever there was an example of high-friction, it's probably working with an operator. Even attempts to lower the friction have mostly failed. This is usually because operators try to run their platform efforts by following a regular telco playbook. It's not enough for the usual suspects (e.g. Marketing) to run a platform business just like any other. The people running it have to understand first-hand what running a platform is all about. Developers are generally very smart people with certain expectations from technology that operators often fail to grasp. It's a bit like trying to run a wine-tasting event for expert wine tasters without knowing anything about wine. Who would dream of it? Yet this happens frequently with operators and their developer efforts. This has the unfortunate effect of causing failure that is interpreted by senior stakeholders as being a sign that platforms don't work.

Most successful software platforms have exploited direct and indirect network effects between applications and users – more applications attract more users, and more users attract more applications. We are beginning to see this with the iPhone and iPad, probably more so with the iPad which is marketed and purchased for its diverse number of applications, which add up to significant perceived value in the eyes of potential consumers. Nurturing both sides of the market helped Microsoft attract thousands of applications and hundreds of millions of users for its Windows platform. During early telco developer programs, I pointed out the stark contrast between how Microsoft courted developers and how operators did it, even taking into account their inexperience with developers. Frankly speaking, the contrasts were embarrassing. Sure, Microsoft has years and years of start on the telcos, but that isn't a developer's problem. If I were to launch a new car into the market, I can't ask consumers to overlook basic shortfalls, like missing window-wipers, because Ford has a head start on me. Again, such excuses are part of a culture that really seems to say that operators don't really want to be in the platforms business. I believe that the luxury of such a stance is rapidly disappearing.

The same business strategy of exploiting indirect network effect worked for Sony PlayStation in games and Palm in personal digital devices (before they lost their way[11]). But some software platform vendors have done little to help application developers make a success of the platform. The canonical example is IBM, who neglected developers with their mainframe operating system. The same is true today for many ventures whose businesses are built on software platforms in the technical sense, but for dedicated devices such as ATM machines – and operator networks – which remain closed and underexploited as commercial platforms. Indeed, as we all know, operators did their hardest to keep their IT stacks from becoming platforms, keeping them firmly closed behind "walled gardens."

I will explore various platform opportunities for operators in some depth throughout the book (e.g. see Section 9.1 Opportunity? Network as a Service). Meanwhile, in this section I

---

[10] Location finding has been possible on networks for a long time but was never exposed because operators couldn't find a business case, or, as was the pursuit back then, a "killer" application.

[11] Platform businesses are just as susceptible as any other business to the rigours of the open market and the usual challenges of sustaining a business lead.

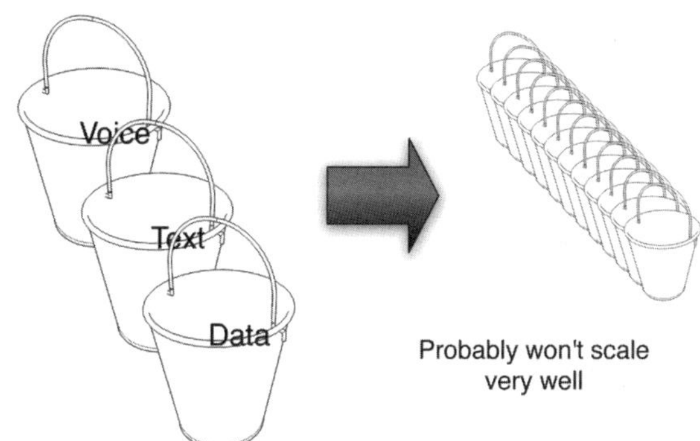

Figure 1.9 Three big buckets replaced by lots of smaller buckets.

will briefly explain the basic configurations for platforms, so that the key concepts are clear before you read the rest of this book. Let's begin by thinking about some of the rationale for platforms. Apart from extension through acquisition and merger, which is an entirely separate type of innovation beyond this book, operators will be forced to find new sources of revenue. The current sources are the big three: voice, messaging and now data. These are three very large revenue buckets that are heavily seeded by the licensed oligopolistic nature of the telco industry. As the license advantage wanes due to technological and regulatory disruption of one form or another, operators, just like every other business, are being forced to find new revenues through innovation. However, the new sources of revenue, at least until "the next big thing" is discovered, are likely to be a range of service innovations that generate revenue in smaller buckets than the big three sources, as shown in Figure 1.9.

The challenge here is that operators are heavily structured and aligned to support the big three buckets. Creating new services quickly enough is a massive challenge, especially if each service is treated with the same mindset as the big three, requiring similar business processes that are hardly built for the speed and flexibility that new services often require. Ask any senior stake-holder how he or she is getting on with new business revenues and they will almost certainly bemoan high revenue aspirations with poor delivery thus far. Little wonder really. Without a "new service factory" pattern, how can an operator be expected to develop substantially new services at all, never mind ones that are profitable? An alternative view of the problem might be shown by Figure 1.10.

This is a new approach. It says that rather than focus on building new businesses per se, focus on building new capabilities that can be turned into businesses both by the operator directly and via a platform play wherever possible, especially if both can be done with similar efficiencies and processes, almost like a "two-for-one" deal.

1.5.1 Platform Patterns for Telcos

Let's boil the background business problem into its component parts in order to understand the potential advantages of platforms before reviewing the different patterns to address the problem.

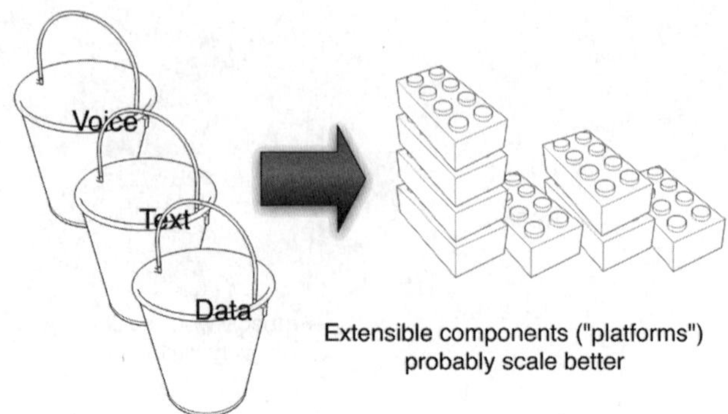

Figure 1.10 Three big buckets becomes lots of building blocks.

- **Problem 1:** How to try out, build and deploy "connected services" utilizing the low-cost and high-speed efficiencies of Web.
- **Problem 2:** How to create services that easily combine Web 2.0 services with telco services.
- **Problem 3:** How to continually evolve a live service to keep fresh and relevant.
- **Solution:** A connected services platform play.

What does a connected services platform look like then? There are a number of key architectural patterns for Web platforms, so let's summarize them here. We will get into the details throughout the book. The first principle of platforms is to evolve from a closed technology stack used to provide proprietary services towards a more open stack that can be used by other parties, namely developers.

1.5.2 Marketplace and Service Platforms

On the left-hand side of Figure 1.11, I'm showing the typical operator pattern used to deploy services today. It is tried and tested and can't be argued with from a historical perspective as it has served its masters well, generating billions in revenues around the core big-three services: voice, messaging and data. At the bottom of the diagram we have the business, which is all the stakeholders and their cohorts who go about the enterprise of bringing ideas to market. They do so via projects that get instantiated on the existing IT and Network infrastructure, which I'm calling the Technology Stack. Typically, any project requires an iteration to the stack. A new service can only run on the stack with a degree of iteration to the stack's IT components, ranging from configuration changes to wholesale sourcing and installation of new solutions (invariably called platforms[12] from an internal IT perspective, e.g. "Voicemail platform"). These iterations take time and money, from little to a lot. The exact amounts aren't

[12] I thought that I should mention this because the word platform is obviously widely used within a variety of contexts. Some telco IT folk reading this book might think that they're already delivering platforms. Please read on to understand the important distinctions.

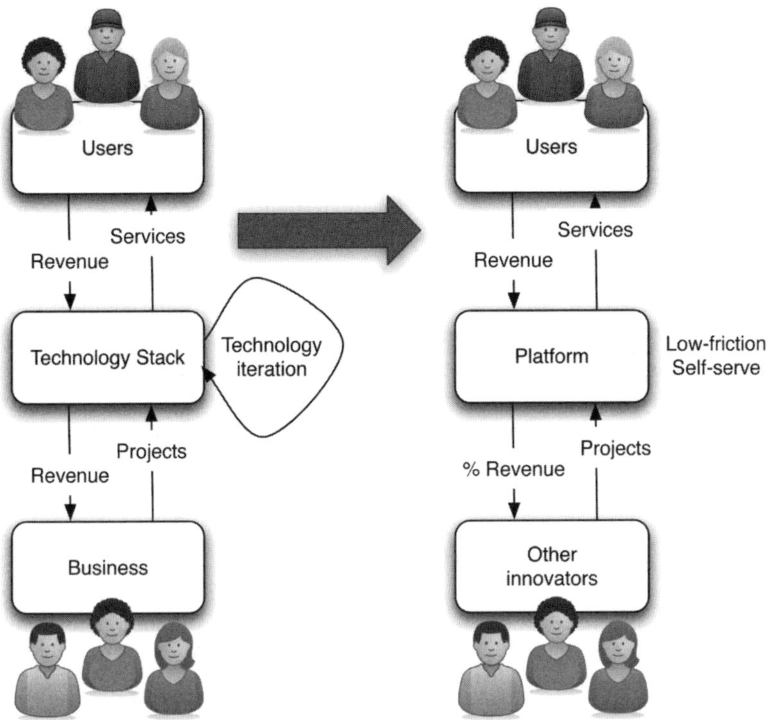

Figure 1.11 From closed technology stack to open platform.

of interest, just the principle of iteration to the stack, which is not without a lot of friction of all kinds (organizational, political, financial, methodological and so on).

The IT division will spend a good deal of its resources in just managing the iteration process. However, it is not without rewards of course. Eventually, when the first iteration is complete, a new or modified service is ready to run on the stack and can be delivered as a service to the users. Of course, all kinds of other business wraps might be required, such as marketing and retailing, but these too will have impact on the stack, requiring their own feature outputs from the iteration process. If the service is successful, the users will pay for it and pass revenues back to the platform to be enjoyed by the business, including new iterations and profit margin.

Moving over to the configuration on the right hand side of Figure 1.11, what if we can alter the approach a little. Take the IT stack and modify it in a way that some of the key enabling components, in whatever configuration makes sense, expose themselves to the outside world via a method that is easily consumed by external software systems. A standard pattern here is the use of Application Programming Interface (API) approach that was first seen inside of operating systems in order to achieve the same goal, which was to empower external developers to build new services on top of the core operating system.

The key is to allow external innovators to use the platform with as little friction as possible for both parties. The platform owner should not have to iterate the internal IT stack to allow an external service to access the platform APIs. The developer should not have to request or wait for any changes to the stack in order to use it as a platform. As much as possible, the goal is to achieve self-service. Apart from entering some legal agreement between the two parties, also in standard format, the innovators should be free to consume the platform services within the bounds of the API specification (including terms and conditions). The mantra of platforms is "low-friction." This is a theme that I will emphasize often.

With this platform configuration, external innovators are free to run their own projects, pursuing their own ideas, which may or may not be congruent with the platform owner's business ideas. They write software that consumes platform services via open APIs and deliver services using a combination of their software and the platform-capabilities. In many cases, the platform itself is also the vehicle of delivery. In other words, users of the service are also users of the platform. In the case of a telco, this would mean that only Telco-X users can use the services built by external companies via the Telco-X APIs. However, we shall see that this isn't the only configuration. Users could well be indirect users of the platform, in the way that Google Maps are often consumed by a large number of users via channels outside of Google's mapping website. This "Mash-up" configuration isn't unique (it was around before the Web) but is increasingly prevalent in Web 2.0 for reasons that I shall explain and explore later.

In the platform configuration shown, external innovators provide services that are consumed by the users of the platform for a fee, which could be collected by either party depending on the channels to market. However, the most common business model is some kind of revenue share between the platform owner and the external innovators. For example, if a telco provides an API for call control of some kind, perhaps allowing mobile users to call a party-line, then revenues from the minutes could be shared with the service integrator consuming the call-control API.

Examples of this kind of platform are:

1. Amazon marketplace
2. ITunes app store
3. Games consoles
4. Operator app stores

1.5.3 Data and Mash-Up Platforms

Figure 1.12 shows a different platform combination. On the left hand side, the telco is exposing data via APIs. External innovators on the right are using the APIs to request and consume data from the platform. The data could be anything, but is usually related to either the users of the platform (e.g. Customer data) or one of the key enablers of the platform, closely aligned to the core business, such as weather forecasting data from a weather forecasting venture. In the case of operators, a good example would be the location of customers.

Again, the telco goes about its business as usual. There is no special configuration or iteration of the core IT stack. Data is consumed without operator intervention, notwithstanding any data governance enforced by the API infrastructure put in place to expose the APIs in the first place. The innovators on the right hand side of the diagram have their own business, which doesn't have to be aligned at all with the core platform business. This is quite different to a platform play like games consoles where the innovators are narrowly confined to providing games per

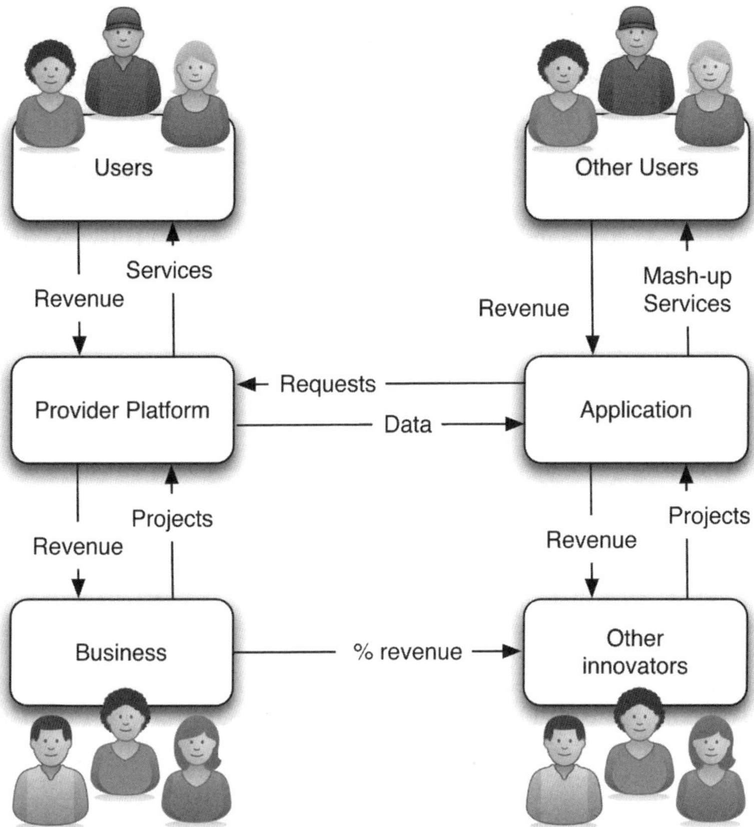

Figure 1.12 Mash-up platform configuration.

the business objectives of the platform provider (e.g. Sony). In this configuration, the platform could be something related to security services, for example, totally without alignment to the communications services of the platform provider.

The innovators build an application (or have an existing application) that requests data from the platform APIs and consumes it into the application itself, using it to deliver services or service features that add value to the application users. Note that the application users might not be platform users, although this is often not the case. The data consumed via the APIs is mashed with the data from the application itself, hence the term "Mash-Ups," which was a term coined to described the appearance of Google Map interface elements inside of other websites outside of the Google Website properties. However, there is no reason for interface mash-ups unless this is the only way that the API generates data (i.e. as a finished map). It might be that the API generates raw data that is consumed and then integrated into the application. The data might never be displayed at all, used only to control the business logic inside of the application.

There are various business models for these sorts of API configurations, but they generally fall into either direct or indirect revenue increments for the platform provider. In the case of direct increments, it is likely that usage of the APIs is monetizable – the platform provider charges for usage of the API. There is no single business model for this, but low-friction is still important, so we expect to see the "freemium" model used where initial (or low) usage of the APIs is free with a charge levied later on, most likely for substantial usage of the API. This is a common model across the Web.

With indirect revenues, we expect to see more revenue generated from users of the platform as a result of the API being used. Google Ads is a great example. The more that ads are consumed via the API, the higher the chance that more users will click the ads, which causes the platform to levy charges to the users. These revenues are then shared back with the API consumer.

Examples of this platform approach are:

1. Google Ads
2. Amazon affiliates
3. Navteq Maps
4. O2 #Blue Service

1.5.4 Platform as a Service

The final pattern is shown in Figure 1.13.

This pattern is the logical conclusion of the platform idea, in that the platform itself becomes the service. Platforms are no longer an extension or strategic side-dish for the provider – platforms become the core business within their platform play. For the innovators using the platform, it becomes their exclusive means of delivering services to the end users, causing the innovators to host their applications on the platform.

In this model, the platform owner charges a fee for hosting the application, wrapping the hosting with a number of features that make the process of delivering services as easy and powerful as possible. The idea behind the Platform-as-a-Service (PaaS) model is that the application developer is freed of the various infrastructural overheads of delivering a service to users, allowing the innovators to focus on the central proposition that adds value to users (whereas the infrastructural components are largely invisible to the users).

The innovators bring their projects to life through the platform. The innovators construct the applications using their own tools and processes, or using tools provided by the platform. Some platform configurations might embed enough tools to design and deliver the service entirely, such as through the use of visual programming paradigms accessible via a Web browser. I have piloted such platform designs myself, allowing developers to invoke applications on the platform and then create the service by integrating a number of modules that call various APIs both on the platform and out on the Web. The platform orchestrates the tasks carried out by the various modules to deliver an on-demand service to the users.

An example would be a service that combines data from an online diary with location data in order to build an online booking service for a solo merchant, such as a plumber. The platform here also includes the various infrastructural components to allow for optimal delivery to a range of handsets and to integrate additional services like text messaging and click-to-call, should the merchant wish to provide the widest set of tools for booking new appointments with his or her customers.

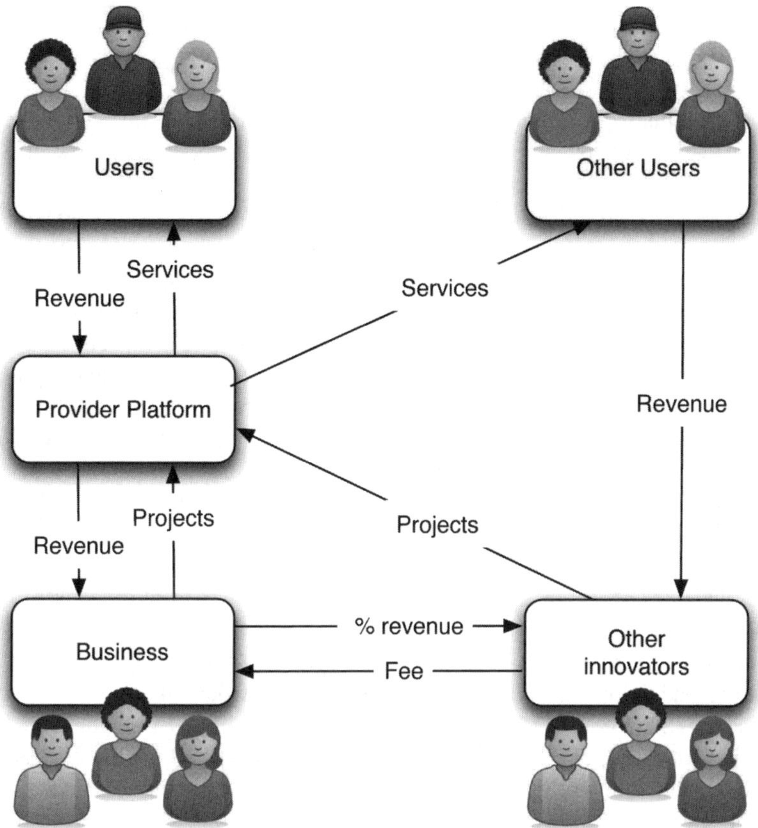

Figure 1.13 Platform as a service.

A typical business model for PaaS is to charge the innovators for the service. The model would include a monthly fee that scaled according to usage, which has a variety of possible quanta, including:

1. Number of applications hosted by the platform.
2. Number of users signed up to the applications.
3. Amount of usage of the applications (measured in all manner of ways).
4. Amount of infrastructural services being consumed by the applications – with various charges per infrastructural element.

The various chargeable infrastructural components might include:

1. Hosting and delivery of the service.
2. Charging and billing mechanisms.
3. Support mechanisms.
4. Analytics.

5. Personalization services.
6. Communications and messaging services.

Of course, the platform provider has the means to be flexible with the business model. An alternative is to share revenue with the innovators, which depends on how the revenue is collected.

In the case of a telco providing the platform, the dynamics of the business model will depend a lot on how the platform services assist directly with the core business versus being an entirely separate business operation and source of revenue. As the diagram shows, the platform can deliver services both to the platform owner's users and to the innovator's users. It might be that the platform owner sees an external service as being particularly attractive to its users and complimentary to its core business, in which case it might choose to enter a revenue share agreement if this enables a dynamic that leads to better service provision and uptake. In any case, the beauty of the platform approach is the elimination or reduction of any upfront capital costs, moving the costs to operational expenditure only. Moreover, the platform owner will usually provide a scalable pricing model with an attractive starting fee, thereby enabling a low-friction entry point, which is critical for platform adoption. As the service gains traction with users, usage goes up and the platform can deliver scalable infrastructure that grows with usage. The dynamics of scalable platform pricing are a crucial part of the PaaS story that has attracted so many application developers to various PaaS solutions.

As of the time of writing this book, there are signs of a new wave of PaaS solution that seem to be enabled by the ubiquity of underlying PaaS providers like Amazon Web Services (AWS). What AWS has done is to provide a very low friction platform on which innovators can build their own platform services, causing some commentators to think of AWS as something more like Infrastructure as a Service (IaaS), although such terms are not standardized. AWS has spawned an entirely new service industry where innovators have packaged common application components into their own PaaS offerings. This is a model that I have supported myself via the O2 Incubator program that I designed and ran for O2. One of the start-ups built a platform that aggregates and filters various information feeds across the Web (in particular social-network feeds). The platform gathers the channels into streams and then presents these back out via APIs. The advantage of the platform is that it removes a considerable processing and design burden from the application developer wishing to consume data from the social Web. Metaphorically, I imagined the service as a "Web switch" that switched social-Web conversations rather than voice conversations handled by a conventional telephony switch.

1.5.5 Do Platforms Work?

The answer is a resounding yes. The concept is not new. As Andrei Hagiu asserts in the title of his book, platforms have been the *invisible engines* powering vast business networks – ecosystems – that grow up around the platform, sucking in more users and innovators via network effects. The platform concept is not new to operators either, such as those who have ventured into the realms of enabling MVNOs. However, clearly operator business models are single-sided, dominated by a collection of revenues from the end users. All operators are aware of the vast potential hidden away in their IT stacks. The enticing notion of monetizing this potential has been discussed for some time and finally seems to be making its way into the daylight, thanks mostly to the Web, which in the broadest sense has become the ultimate platform of platforms. This trend will only continue as the Web becomes more powerful, enabling new forms of data economics, many of which I shall be exploring throughout this book.

Whether or not platforms will be a success story for operators, it is still too early to tell, although I am strongly of the opinion that platforms need to become an integral part of operator futures. Many senior stake-holders remain sceptical of this, which is understandable. Stake-holders in telcos are from a different background and mindset than the Web-savvy entrepreneurs building new empires on the Web. I have been here before, many times. When I first set up a business in 1996 to offer Web applications to the telco industry, the Web was viewed with much suspicion and as a fad. Stake-holders have habitually repeated such reactions to all manner of Web-born ventures and patterns, like social networks and instant messaging and now APIs. On each occasion, they have lost out to disruptors who have nothing to lose in trying to build new platforms on the Web.

However, this time around I would caution telcos to take platforms more seriously than previous emergent Web patterns. This is because the platform model has accelerated so fast that it has almost become a hygiene factor in the delivery of any modern digital service. If it is ignored by telcos, they will find that non-traditional competitors eat their lunch in the emerging world of mash-ups and highly personalized digital services and connected experiences.

1.6 From Platforms to Ecosystems

It's not enough these days just to talk about platforms. With the acceleration of Web framework technologies and the open source movement, combined with the vast numbers of fixed and mobile internet users, it is increasingly easy to start a platform play. In fact, degrees of "platform-ness" have crept into many Web ventures – it is almost standard to offer some kind of API on which developers can extend the basic offering.

With so many platforms vying for attention, the name of the game has changed to ecosystems. You can think of this as supporting an entirely new industry around a platform, including developers, entrepreneurs, investors, evangelists, fans and all kinds of niche outcrops from the base, as shown in Figure 1.14.

Twitter is a great example. The platform has already attracted over 50,000 applications with all kinds of commercial and non-commercial intentions. In turn, these ventures have attracted venture capital into the mix, chasing lots of opportunities with an opportunity to scale. Many niche categories of application have emerged, such as social-marketing, sentiment analysis and so on. Chances of success might be remote in most cases, but too tempting to ignore overall. Once established, ecosystems can grow rapidly because too many players don't want to be left out.

Bloggers, journalists, even stars, have been attracted to the Twitter platform and its various offshoots. Many businesses have adopted Twitter to participate in the conversation about their brand and services. It has become a defacto way to follow what brands have to say about their wares.

Digital ecosystems, like natural ones, might become successful "by accident," but the various inflows leading to the tipping point can be encouraged and primed. Hence, it is important to have a sense of ecosystem in the strategic plan for some of the future operator models. This means paying attention to a much wider set of ingredients than just developers, service-enablers

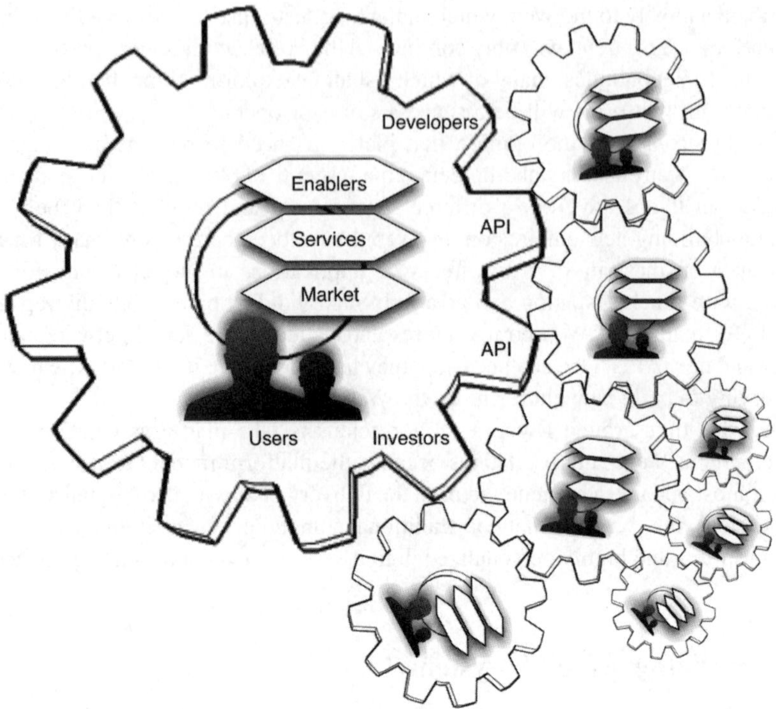

Figure 1.14 Ecosystems – a mutually supportive and competitive environment.

and APIs. You need evangelists, investors, start-ups, bloggers and a lot more besides to get that all important "echo chamber" effect that we see often in Silicon Valley. A platform is a lot bigger than just APIs and an ecosystem is even bigger still.

1.7 Where's the Value?

I'm not going to answer the "how do we make money from the Web?" question here. The title of this section is more a recognition that this question almost always gets asked inside telcos. I don't have the answer, nor is this book about giving the answer, at least not directly. But, it is important to know the right and wrong responses to this question from the outset, which is why I tackle it right now in the opening chapter.

In my experience, although this question is clearly important, the wrong conclusions are reached all too quickly, which I can explain within the context of what I have covered so far in this chapter, particularly the future models for operators (see Section 1.3 Six Models for Potential Operator Futures).

> **Lesson 1 – Money is where the users are –** That seems obvious, because it is. Yet, in this context it means that if we can't build platforms, services or applications that attract lots of users in interesting and new ways that address their problems, then we can't ask them for money. This is generally agreed. However, what many

folk in telcos still don't get is that delivering the initial service for free, or at least part of it, is often the cost of building a platform on the Web. There simply aren't that many examples of Web platforms that charged from day one and made a good profit, except in certain SME and enterprise categories. However, there are arguments both ways and I will return to the issue of charging from day one when I explore the patterns of start-ups (see How Web Start-ups Work).

Lesson 2 – Low-friction services are key – If it is possible to gets users to benefit from a service with minimal fuss, and drop an extra (small) charge on the bill in return, they might well go for it. The moment the service friction increases, to even a slight pinch of pain, all bets are off. The friction-value sensitivity curve has a very sharp inflexion.

Lesson 3 – Web is hygiene – To remain a relevant service in a world of increasing convergence and digitization, service delivery via the Web is essential and all that it entails in a post Web 2.0 world, which includes being socially-connected, context aware, open etc. In other words, the idea of becoming an "online business" is simply wrong-headed. Online is hygiene. Now the challenge is that born-on-the-Web companies generally do it much better. This needs to be fixed regardless of the money question.

Lesson 4 – Value is in the data – We have reached a point where the cost of storing and processing vast quantities of data has created new economies of scale. It pays to have a strategy that involves exploiting vast swathes of data. It pays to have a strategy that exploits potentially deep and complex number-crunching. Don't forget – it might be hard for you or I to contemplate complexity, but it isn't for a cluster of computers. There are entirely new opportunities to make money from handling large volumes of data.

Lesson 5 – The answer is in the method – This is probably the most important lesson, which is that we should accept from the outset that we don't yet have a business model, but we will set up our approach such that we maximize the chances of finding a viable business model. In other words, in Web start-up ventures (whether by a brand new company or an established one), the objective is to search for the business model as part of the initial phase of the venture. We suspend judgement about the model until we've actually built something that resonates with users and provides the answer to the money question. The secret to this approach is in the agility of the venture. The quicker and leaner we can operate to iterate through various service ideas, the better our chances of finding the business model before we either run out of money or patience. Being agile is key!

1.8 What Should We Build? It's Still About the Experience!

I still get asked this question time and time again – "what should I build?" Partly, it is that desire we all have to unearth a gem – to unlock the killer app. I always respond with the same advice, which has two parts:

1. Just build something and put it out there.
2. Focus on a compelling user experience.

Let's boil these down a little.

Firstly, it is essential to grasp, per the Eric Ries Lean Start-up Methodology (see Chapter 10) that the Web is still so young and its canvas so vast that we are constantly operating in the domain of the unknown. We don't know what will work, what won't, in terms of *new* ideas. This book will certainly outline a lot of the technological patterns that are behind things that work or appear to be working, so you're in the right place. However, one characteristic of these technologies and the Web frontier is that they move quickly. In the domain of the unknown, agility is king. You have to move fast, try things out, tune, adapt and keep moving, trying to find resonance with users before running out of cash.

But one could still be agile at delivering rubbish, so the focus must continue to be on the user experience. This message can't be said enough times because it is repeatedly ignored, even though it is startlingly obvious. We can talk about user experience (UX) in depth. It's a big topic. UX is almost always confused with user interface (UI), but let me be clear about this:

$$\text{UX} \neq \text{UI}$$

They are not the same thing. UI contributes to UX, but does not define it. I believe that many of my colleagues who are self-proclaimed UX experts also don't understand this topic in totality because UX is almost always seen as synonymous with visual design. It is not. UX is synonymous with utility.

BEFORE you start emailing me on this one, which I hope you do – I don't want to get into the semantics here. I'm not a UX expert, per se, although I have always seen my job, as someone who designs products and evangelizes various technologies, as generally being in the "user experience business." I just want to make a few simple points about this topic that seem to trip many of us up, especially in the telco world where they have historically seen UX as synonymous with handset usability.

Let's start with its most basic synonym, which is benefit. If the service has little or no benefit, then it doesn't have utility and it will be a poor user experience, no matter how much time and effort is invested in the design. This is bag-of-hammers-dumb obvious. Yet, the mistake made time and again is to define benefit in the mind of the product owner, or, more typically in large companies, inside a room of "intelligent people," which is usually a bunch of marketing guys armed with customer data.

In the Lean Start-Up methodology, the notion is to test the hypothesis with real users and real feedback, not conjecture and insistence on an idea because its owner has attachment. In the field of UX, it's often better to either discount one's own experiences, or at least to validate them. Unless you happen to be a product guru with a genuine feel for your users – and such people are rare – your tastes and product expressions are unlikely to resonate with the bulk of your users. Wherever possible, be guided by the users, not your instincts. Better still, be guided by the data obtained from real feedback and experience with your product.

It is also vital to appreciate that a lot of products and services can gain a foothold, even become wildly popular, despite a relatively poor design. This occurs when the product does

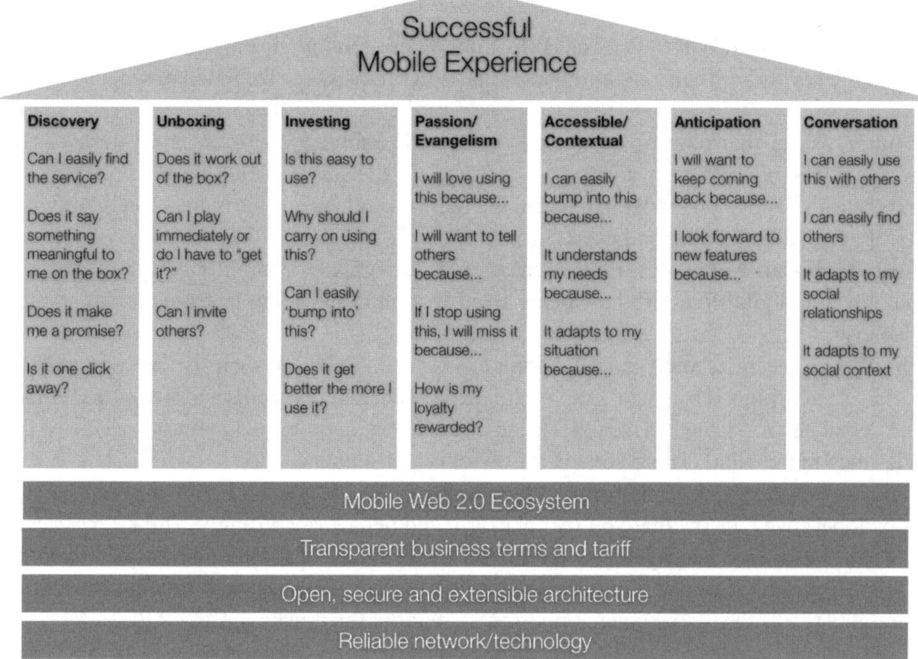

Figure 1.15 Components for a successful user experience.

an existing job but at a much lower price. When the price is lower, users will be more tolerant of product limitations, so long as – and especially if – they gain access to a new solution that was previously out of reach. Again, this is about utility. Because the user can do something previously unreachable or impossible, the experience will be satisfying.

The other failing of thinking of UX as UI, is that it misses overall context. Firstly, the UI is only one aspect of how a user experiences a product. For example, when one of your users tells another person, say a non-user, about your product, this is part of their experience of the product. When a user tries to discover the product, this is part of the experience. For these reasons, I came up with an example diagram to illustrate the various ways that UX can be approached in the total product experience, as shown in Figure 1.15.

The above diagram is only illustrative of the kinds of experiences that a user might encounter in using a product or service. It is deliberately in the first-person because of the way I use the diagram in workshops. However, in a wider setting, it would be matched with different user personas that we might envisage. We should not forget that in a platform model, there are two sets of experiences and personas to consider: the users and the developers. They are both important. UX should always be the starting point in our product, service and platform strategies. It has an important influence on the overall success regardless of underlying technologies and business models.

1.9 Summary

What this book will reveal to you is a collection of key technological patterns. I have outlined some of the key themes early in this chapter that will set the scene for much of what's to come:

- "Http://' is the dial tone of connected services. Combined with our ability to access IP from anywhere, Web software is "every-ware." In the telco world, we tend to think that transactions happen on the Web in IP, but the real vocabulary of Web innovation is software – widely available, flexible and powerful.
- Operator models will evolve, most of them to include greater Web-intimacy, if only because that's where digital lives are now being lived with more personal investment and attention than ever.
- The Web is the InterWeb – a collection of ecosystems. One of them is the "Social Web," which is underpinned by social graphs and their associated platforms and software elements. Big social networks have driven this trend and are now defining it. Operators need to regain relevance in this field. There are still plenty of opportunities.
- The name of the game is platforms. These are frequently hard to execute and need a lot of "developer love." Know that you're a platform and behave accordingly. Think low-friction.
- Beyond platforms are ecosystems – a much wider collection of actors and resources that make a platform successful. The best platforms will spawn into ecosystems, often by accident, but the rights seeds need to be sown – platforms are more than just exposing APIs.
- It's no longer viable to use the money questions as an excuse not to become an "online business," meaningless though that term is rapidly becoming, with so much of our world already online. In the fast-moving world of online, where business models aren't always as certain as in the telco world, a key competence is agility.
- Even with the best platform strategies and grasp of all the technologies in this book, we can never neglect the overall user experience. It's the starting point of everything that follows.

2

The Web 2.0 Services Ecosystem, How It Works and Why

```
<h1>The Platform is...</h1>
<ul>
<li>Web 2.0</li>
<li>LAMP</li>
<li>Open APIs</li>
<li>And lots more</li>
</ul>
```

- Low-cost and easily programmed software is at the heart of Web 2.0, especially the LAMP stack and its derivatives.
- A common pattern for using the LAMP stack is the Model-View-Controller pattern, which orchestrates the creation of Web pages with the underlying database fuelling the content of those pages.
- Open systems are a major theme on the Web. Being open can offer a number of strategic advantages.
- The browser is still one of the most important platforms for the Web, and, increasingly, for connected services of all kinds. HTML5 is an exciting development in browser platform evolution.
- The Web has evolved from being informational to being social.
- Platform capabilities are exposed via APIs, allowing new and unimagined services to emerge.
- Internet-centric mobile devices make it possible for the user to do other stuff besides talking. The mobile device is becoming a major platform for connected services, at the nexus of Web and telco. New combinations and possibilities are emerging under the banner of Mobile 2.0.

2.1 Introduction

Many of you will be somewhat familiar with Web 2.0. No doubt, sites like Facebook, Twitter and YouTube immediately come to mind as lighthouse examples, which they are, but can you explain why? We all know the brands and icons of the modern Web era, but we often fail to

Connected Services: A Guide to the Internet Technologies Shaping the Future of Mobile Services and Operators, First Edition. Paul Golding.
© 2011 John Wiley & Sons, Ltd. Published 2011 by John Wiley & Sons, Ltd.

explain exactly which of the various Web 2.0 attributions are actually all that important and have beaten other ideas in the Web-evolution story.

In this chapter, I want to make sure that the key ideas underpinning Web 2.0 are well understood, as they form the foundation for the current inflection of the Web timeline into a new era of "Right-time Web," which I think is a more important emergent trend in our connected services story, especially as it places mobile centre stage.

We are currently at the point of Web 2.0 maturity where many of its influential geek participants and architects, both from the technical and entrepreneurial domains, view the current part of the journey as being about building a "Web operating system," that will host many key services for the next decade, spanning all connected devices, including mobiles – especially mobiles. We should have no doubt that the very future of digital services is being debated, built and battled for right now on the Web. However, the fight isn't happening in the labs of telcos or even in the labs of many of their suppliers. It's happening in the minds and pioneering antics of born-on-the-Web ventures and hackers.[1]

Although somewhat dated already, it's still worth a read of the seminal Tim O'Reilly article that launched the 2.0 meme. I also suggest consulting the book *Web 2.0 – A Strategy Guide*, by my deeply respected associate Amy Shuen. For an update of Tim O'Reilly's original thesis, he published a follow-up article, with John Battelle, which they delivered at Web 2.0 Expo. If you don't have time to read these references, don't worry, as what I intend to do in this chapter is to outline some of the key technological and methodological underpinnings of the Web 2.0 services ecosystem.

If you work in a telco, then you probably already appreciate much of what Web 2.0 is about, but in my experience with many telcos, the actual understanding of Web 2.0 is too shallow, blocking any truly creative and strategic thinking amongst middle and senior ranking stakeholders in key telco projects. As one board member in a leading operator bravely said during a review of Web-related strategies: "if this company were formed more recently than it was, guys like us [who don't 'get' Web] wouldn't be sitting in this board room."

It takes a while to sort out the hype (and there is much of it) from the really important trends that have the potential to be transformative, especially in pursuing our various telco futures. I hope to act as a filter for you in the rest of this chapter. Let's begin with a simple point, which is that Web 2.0 is an ecosystem of parts beyond just technologies and key categories of usage, like blogs. This is a point still missed by many folk who locate the Web 2.0 idea in their minds by pinning it to one or two key trends, such as the rise of user participation and the rise of YouTube and social networks. Web 2.0 is about the people, technologies, platforms, collaborative work patterns, emerging business models and cultural expressions. It's ALL those things. It is unique. It is also quite different to the enterprise IT world that many in the telco world will be familiar and comfortable with. Of course, they overlap, but the differences are important and worth knowing.

2.2 Beneath the Hood of Web 2.0: CRUD, MVC and REST

In my previous two books, I explained in detail how the world of IP works, allowing devices to connect and exchange data. The golden protocol is HTTP, which allows a Web browser, or

[1] Again, a hacker is a pioneer constructor of new ideas and their expressions, mostly in software, but also in hardware. A hacker is not a destroyer of IT security perimeters.

a mobile app, to say to a web server – "Hey, GET me some data," which is invariably a Web page, but could be the co-ordinates of a friend. However, the genius of the HTTP protocol was the inclusion of another command – "Hey, let me POST some data to you," which is often triggered by clicking a submit button on a form, but could be the operation that lies beneath uploading a picture from a smartphone to an online gallery, like Flickr.

With the ability to push and pull data using the POST and GET verbs of HTTP, interactive services are possible, which are what underpin all of Web 2.0. HTTP is the lingua franca of most services that want to connect via the Web. "http://" is the dial tone of connected services.

The HTTP protocol has four main verbs: GET, POST, DELETE and PUT, except that most browsers don't really implement DELETE or PUT, which are intended for deleting and modifying pages, or datasets ("resources") on the Web server. Such actions weren't considered that useful. After all, we don't want users deleting stuff from our servers, do we?

Well, yes we do. Not pages, but records in a database. Example: delete an item from your shopping cart on Amazon! Let's explore this a bit further. Most web applications are a combination of templates, which guide the visual layout of the page elements, such as where the title block goes, where the top menu goes, the footer, the sidebars and so on.

But these templates aren't what make the page interesting. It's the data that gets squirted into the placeholders within the templates that brings the page to life, as shown in Figure 2.1. The data comes from the data storage layer, which is commonly a relational database like MySQL, PostGRES or Oracle, which you might have heard of. Alternatively, other data stores are increasingly being used, like those that come under the so-called "No SQL" banner, like MongoDB or CouchDB, both of which I will explain later in the book (see "No SQL" and some of its flavours).

The data is divided into two sets. Firstly, and perhaps most importantly, there's the application data itself, such as your account data on a shopping site, the shopping items in the catalogue, those in your cart, their prices, descriptions and so on. And then there's the state information, which tells the web server whereabouts the user is in the myriad pathways through the application. For example, it keeps track of the fact that you are at the checkout page and not the home page. This state tracking is vital to web applications.

All this data gets stored in database records which can be Created, Read, Updated or Deleted, hence the term "CRUD." It is the mapping of these four data-manipulation verbs to the HTTP verbs, as shown in Figure 2.2, that provides the backbone for how Web applications work across the browser-server divide.

It takes a bit of getting used to, even as a Web programmer, but all Web applications boil down to the judicious combination of the CRUD verbs to guide a user through a compelling Web experience. Of course, the user is oblivious to all this CRUD nonsense, but it's the fundamental vocabulary of a Web developer.

We can think of each software action on the Web server as carrying out one of the CRUD verbs in response to whatever the user is clicking or submitting in the browser combined with what these actions mean within the context (state) of where the user is in the application. Now it turns out that there's a way of organizing chunks of Web software to carry out the CRUD verbs that allows the programmer to architect the solution in an efficient and maintainable fashion. This organization is called the Model-View-Controller (MVC) pattern, which I talked about extensively in my previous books and shown in Figure 2.3.

The Controller, as its name suggests, controls and co-ordinates any requests from a Web browser. It's like the brains behind the whole operation and is responsible for co-ordinating

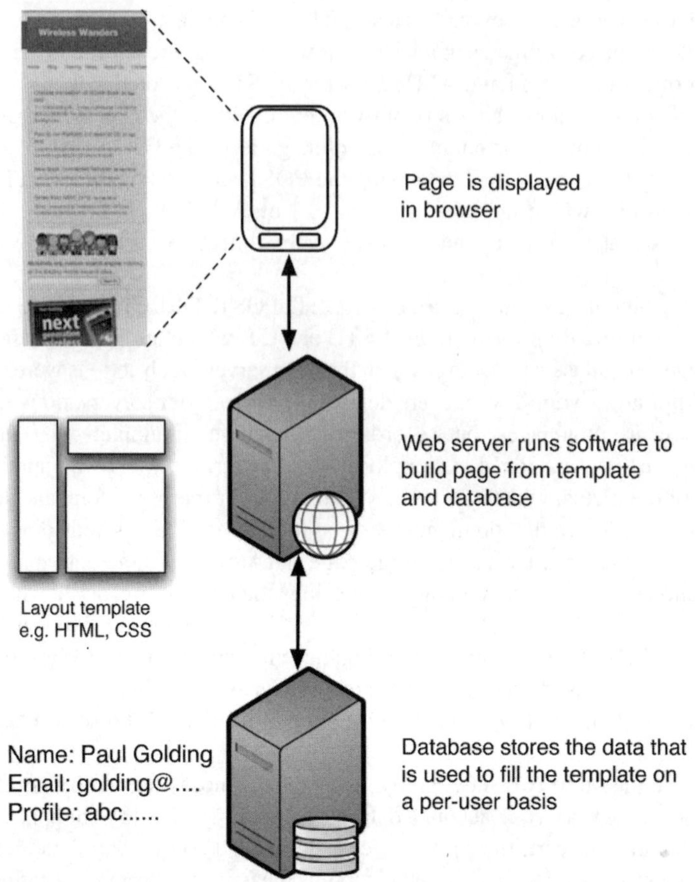
Page is displayed
in browser
Web server runs software to
build page from template
and database
Layout template
e.g. HTML, CSS
Name: Paul Golding
Email: golding@....
Profile: abc......
Database stores the data that
is used to fill the template on
a per-user basis

Figure 2.1 Database and Web servers collaborate to deliver page to browser.

all other actions that take place on the Web server to fulfil a Web request. It recognizes from the HTTP request what type of action is required. Let's assume that it's a GET request to view the contents of the user's shopping cart on an e-commerce site.

Eventually, the content of the shopping cart is going to be made available for displaying in a Web page in the browser. It is the job of the View block of code to construct the view, which is does by combining a pre-programmed template of what a cart looks like with the actual cart data (e.g. products, descriptions, prices, quantities), which it gets from the database, as shown in Figure 2.1.

However, the view-code doesn't actually interact with the database directly. It does so via another chunk of code called a Model. The model is what speaks the CRUD verbs directly to the database. However, databases are just tables with columns and rows of data, not unlike a basic spreadsheet layout, although far more scalable. Rows and tables are an artefact of database design, not of the real world. For example, a shopping cart has items in it, not "rows" per se. Items have descriptions, prices, offers, dimensions, specifications, and so on, not "columns" per se.

HTTP Action	Intended Database Action	SQL Mapping
POST	Create	INSERT
GET	Read	SELECT
DELETE	Update	UPDATE
PUT	Delete	DELETE

Figure 2.2 Mapping of HTTP operations (verbs) to database actions.

A model then is a piece of code that appears to the rest of the Web application as a representation of the actual object being modelled, a shopping cart in this case, rather than rows and columns in a database. There will be a shopping cart model that has attributes like "cart customer" (that's you or me, or the customer name), "items," "total price," etc. The other software components, such as the view code, can talk directly to a model by sending it a message like "get me the total price." Or, in the case of initializing a brand new model for a new customer, it might be a message like "set the owner to be Paul G."

It's the model's job to get or set these attributes and to keep track of them by storing them in the database. Internally, the model knows about rows, records, tables and so on. It knows the dirty intimate details of how to talk to the database, which is usually via a special language called Structured Query Language, or SQL, a language that's messy and can get complicated very quickly. Databases only know about data, they don't know anything about the types of

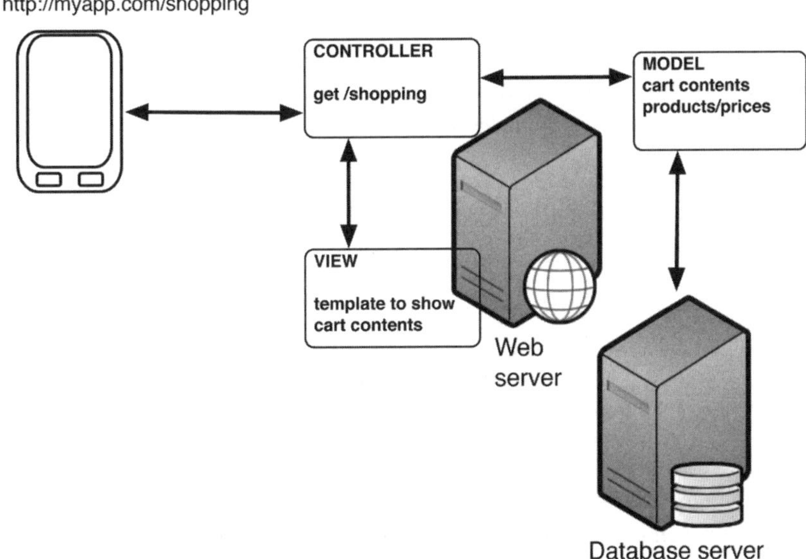

Figure 2.3 Model-view-controller pattern.

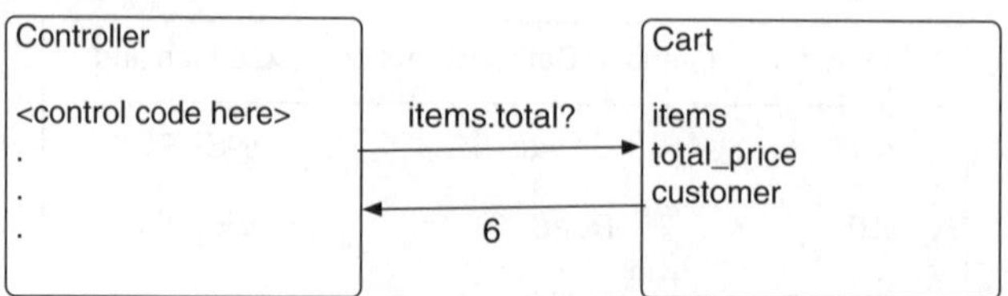

Figure 2.4 Calling a model object.

objects being stored in the data, which could be shopping cart items, phone numbers, patient records or anything. Models are meaningful to the rest of the Web application. It's quite clear what a "Shopping Cart" model is. Its interface will also be very obvious and understandable: things like "add an item," "get an item," "get the total price," and so on.

This notion of models provides what software folk call a level of abstraction. We can overlay the world of rows, columns, tables and SQL with a meaningful world of shopping carts, shopping items, tax, prices, shipping address, etc. This allows the programmer, when building his views, to refer to items when populating the shopping cart part of a Web page, rather than table rows, as shown in Figure 2.4. Here we see that the Cart model has attributes like "items," "total_price" and "customer." When a controller object wants to know how many items there are in the cart, it can send a message like "items.total?" and get a response indicating the number of items in the cart. The programmer doesn't have to know anything about how this data is stored in the database, nor how to read the data from the database. In this fashion, it's easier to program and allows software to be created much more efficiently and quickly. It's also a lot more maintainable.

Finally, we return to the controllers. These are the pieces of software that fire up first when a command comes in from a Web browser via the user entering an address (http://something.com/somepath/subpath), clicking on a link or button, or submitting a form. When the user attempts to carry out an action, the Web application first needs to interpret what the user is trying to do. It then co-ordinates all the server-side activity to deliver the result, which means co-ordinating models (e.g. "Let's set up a shopping cart model because the user is about to add something to a cart") and then handing results back off to views (e.g. "Go display the shopping cart using the model just created").

There might be a number of controllers in the application: a "shopping controller" to take care of the shopping cart stuff, an "account controller" to take care of setting up an account, a "products controller" to take care of viewing the product catalogue, and so on. Typically, a Web application will divide into various groups of related tasks – like shopping, account management and catalogue perusing – each warranting a dedicated controller. This divide-and-conquer (decoupled) approach is generally the best way to write any kind of software, making it easier to maintain. It is perfectly feasible that one controller could do everything, but it would be very unwieldy, difficult to write and maintain, never mind debug (because software never works first time).

So how does the Web server know which controller to run? Moreover, how does each controller know which sub-task it should be running (e.g. "Add to cart," "Subtract from cart," "update items in cart," and so on)? The secret to unlocking the answer is in the URL used to initiate the action in the first place. In fact, this is the only way that a Web browser can tell a server what to do.[2] And there's an additional complexity here. The HTTP protocol itself doesn't maintain any state information. In other words, as far as the server is concerned, each request from the browser is independent of any previous or next requests. The HTTP protocol has no way of saying that a particular request follows on from a previous one.

However, the server can track state, as it must. For example, if this is the third item I'm about to add to my shopping cart, I shouldn't have to tell the server the previous two items again – it should remember them. It does this via the database of course. However, as far as the browser is concerned, it is just executing an "add to cart" operation. It doesn't know if this was the second, third or tenth item. It leaves the server to keep track of that stuff. If HTTP didn't operate in this way – that is, stateless – it would be a much more difficult protocol to design and implement. Its stateless design is in many ways responsible for its success, especially because it allows users to jump from one server to another (via page links) without those servers having to keep track of where the user has already been, which would be complicated.

So let's take a quick look at how this "add to cart" example might work. Let's say we have a site called myshoppingstuff.com, the URL for adding an item might look like this:

```
Http://myshoppingstuff.com/cart/13764
```

The myshoppingstuff.cart bit is what enables the browser to route the request to the right website in the first place. After all, we wouldn't want the /cart/13764 going to book-me-a-dentist.com where the Web application probably has no concept of carts.

The second bit (the path), which is "/cart" tells the application that this request should be handled by the controller called "cart." This is how the right controller gets selected. And that final bit "/13764," which looks a bit cryptic, is actually a unique ID for an item from the shopping catalogue, such as a pair of socks. But how does the server know that this request is to CREATE a shopping item in the cart and not, say, READ the item. Well, this is taken care of by the use of the right browser (HTTP) verb, which in this case, to add an item, is POST.

This kind of URL pattern matching and HTTP verb selection to fire the appropriate response at the server is extended to all controllers. Here are a few URL examples, along with the HTTP verbs, to indicate what's required:

GET Http://myshoppingstuff.com/catalogue/13764 – fetches information about shopping item 13764 from the catalogue (handled by the "catalogue" controller).

GET Http://myshoppingstuff.com/account/324asdve134 – fetches the account information for shopper with the ID 324asdve134 (which could be Paul G in the database).

POST Http://myshoppingstuff.com/account/324asdve134 – update the account information for shopper with the ID 324asdve134.

In this case, where would the information to update the account come from? It would come from a form in the Web page (see Figure 2.5).

[2] This isn't quite true. It's also possible to pass data in the actual HTTP body itself, but it is usually the URL that drives the invocation of controllers in one way or another.

Figure 2.5 Example of a web form.

Note that in all cases, it is the job of the server to change state (which happens via the models) and to reflect any changes accordingly in a new view presented back to the user (e.g. adding a requested item to the shopping cart view).

The technique of mapping URLs directly onto actionable CRUD requests that update state in the Web server (and not shared state between the client and server) is an example of a software paradigm called Representational State Transfer, or REST. Just to emphasize this point further, the "command" to add an item to the cart – for example:

```
POST Http://myshoppingstuff.com/cart/13764
```

is always the same.[3] It doesn't matter if this is the first, second or 99th time that this item has been added to the cart, it is always the same URL. There isn't a series of URLs like this:

```
Http://myshoppingstuff.com/cart/add-first-item/13764
Http://myshoppingstuff.com/cart/add-second-item/13764
Http://myshoppingstuff.com/cart/add-third-item/13764
```

It's always the same URL. It's the job of the server to remember that this is the first, second or third time that this item has been added to the cart.

It turns out that this "stateless" approach to the architecture makes the Web very scalable and easier to program for. It is relatively easy for a server to maintain state. For example, we could have a row in a database called "item counter" and just keep incrementing the number stored in it to show how many items are in the cart. This is a lot easier than tracking and including this information in the interchange between the browser and the server, requiring the browser to keep track of what's going on. You can imagine how difficult it would be to write code like that and how easy it would be to get it wrong, using /add-third-item instead of /add-tenth-item, for example.

This tidy way – the REST way – of accessing data via a Web server is called RESTful and have been extended to enable data to be accessed in the raw, without a Web page. These are

[3] Note that this is just an example. The number 13764 has no meaning. It could be a customer number, or, more likely, a cart number used to uniquely identify to the server what cart is being updated.

called APIs and are driving a lot of exciting activity on the Web right now. Here's an example of calling the following URL from Facebook's Social API:

```
GET http://graph.facebook.com/559089368
```

This returns:

```
{
        "id": "559089368",
        "name": "Paul Golding",
        "first_name": "Paul",
        "last_name": "Golding",
        "link": "http://www.facebook.com/paul.a.golding"
}
```

This block of data between curly brackets and formatted as "string:value" pairs is an example of a common format used to suck data from Web services, called JSON (Javascript Object Notation).

So now you know a little about how modern Web software works, let's move on to explore some of the software that does all this stuff.

2.3 LAMP and Beyond: Web Frameworks and Middleware

2.3.1 Introducing LAMP

Lots of Web applications have been built using the so-called "LAMP stack," which in many ways has become the basic building block of the Web. LAMP and its variants are now the undisputed champion software stacks for building Web applications. LAMP stands for Linux, Apache, MySQL and PHP. It's a set of software applications that stack on top of each other to form a complete environment for creating software applications, as shown in Figure 2.6.

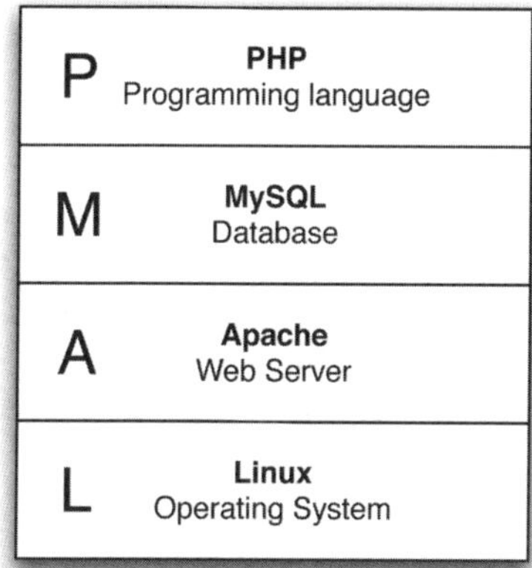

Figure 2.6 The LAMP stack.

Linux is the underlying operating system, created by Linus Torvalds in 1991. It is one of the best and most successful examples of free and open source software collaboration. Not only is it up to the job technically – a fully fledged advanced operating system – but it's free. Anyone can use Linux, free of charge. This is why it was picked up by the early Web community as an interesting and enticing platform. The Web was born very much in a time when commercial use of the Internet was either discouraged or unheard of. This spirit of collaborative working and sharing in a non-commercial context has played a significant role in the spawning of ideas like open source, which is a subject I shall explore later in the book (see Section 2.4 Open by Default: Open source, open APIs and open innovation).

Continuing with the stack, next we have Apache, which is still the most prominent Web server on the Web, powering the majority of websites. Apache is the component that responds to the GET and POST verbs and knows how to talk HTTP with Web browsers. Apache was instrumental to the early growth of the Web and remains important. It is a well-written, thoroughly tested and reliable piece of software. One of its key features is virtual hosting. This allows a single instance of Apache to support many different website domains, such as www.domain1.com, www.domain2.com, www.domain3.com and so on, thus enabling Web-server sharing between sites. This enables cost-effective and scalable Web hosting, which has been instrumental to the widespread growth of the Web.

MySQL (pronounced "my sequel") is a relational database management system (RDBMS) that enables many databases to be hosted at once, thus aligning with the multi-hosting model of Apache. Just like Linux and Apache, MySQL is also free and underpins a good number of popular Web platforms, such as the Wordpress blog solution. It is also used in high-profile projects like Wikipedia and is used by most of the born-on-the-Web ventures, including Google, Facebook and Twitter. You won't be surprised to learn that it is also open source.

At the top of the stack, we have PHP[4], which is a programming language that runs on Linux and knows how to talk with MySQL. PHP files can also be called from Apache, thus enabling HTTP verbs to be converted to CRUD verbs via an MVC pattern (see Section 2.2 Beneath the Hood of Web 2.0: CRUD, MVC and REST). It's a relatively easy language to learn and is widely used, including by such big names as Facebook. It is also a very forgiving language. Programmers of all abilities can quickly get a website up and running in PHP. The language was designed for creating websites, so it is no surprise that it has lots of friendly features that support this purpose.

So there you have it – the LAMP stack, which in many ways is the real engine of the Web, powering millions of websites. As fantastic and versatile as it is, the problem with the LAMP stack is that it doesn't really obey the DRY[5] principle that I discussed earlier. PHP is just a language rather than a set of higher constructs that act like building blocks to make website construction easier. By analogy, imagine writing a weekly progress report and each time having to write everything from scratch, including the headings, even though you've done the same thing many times before. What you really need is some kind of template to make life easier.

[4] PHP stands for Personal Home Page.

[5] Don't Repeat Yourself, which means that, where possible, a software environment should follow a set of patterns that reuse common components rather than forcing the developer to create them from scratch each time.

The same idea applies to website creation. There are a lot of common operations and components for each website, including the need to create MVC components, the ability to support CRUD verbs by mapping them to model objects, and so on. We don't want our developers to keep telling each model how it should connect to a database on MySQL and that the database is called "Shopping," for example. Once we have the code to connect models with the database, surely we can reuse it. Recognizing this need for a common set of software components to make website coding easier, some developers took it upon themselves to create generic Web frameworks. These are collections of useful software that take care of most of the common tasks in creating a website. They also provide a means for developers to template the construction of a website, accelerating the development, leaving them to "fill the gaps" with the particulars of their actual application.

2.3.2 Web Frameworks

Experienced Web hackers who build websites time and again don't just use LAMP. They will add another layer above the PHP, called a Web framework. In the PHP world, some of the most popular Web frameworks are CakePHP or Code Igniter. These are examples of frameworks that contain software modules to make it easy to program with CRUD and MVC patterns. The frameworks also contain other code libraries to support a number of common and important tasks like testing, page template ("view") construction and so on. Let's take a brief look at the typical ingredients for a Web framework:

Object Relational Mappers (ORMs) – These are software components that enable database tables to be represented as models in the code (from our MVC pattern). When used properly, most of the time an ORM will automatically and transparently "build" the code required to convert a model into a database table, taking care of constructing the SQL queries. For example, if a model is created to model a shopping cart, it might have an attribute like "Items," which is a list of all the items in the cart. These items are probably stored in their own table in the database. However, the developer doesn't need to know this. All of the code to convert operations on the attribute, like "Add item X to the items list," to an action applied to the database (which will be a SQL query) is created by the framework. The developer doesn't have to write the database code or even see it!

Mark-Up Template Systems – As I mentioned in the previous section (see Beneath the Hood of Web 2.0: CRUD, MVC and REST), Web programmers write software (e.g. in PHP) that generates Web pages (output in HTML) dynamically. This can get messy quite quickly because the developer is often forced to embed HTML inside the PHP code, even though this is what PHP was designed to do. The beauty of HTML is that the HTML is usually quite readable and follows a flow that largely matches the expected layout in the Web browser. However, PHP follows the logic of the code, not the layout of the page, so it becomes difficult for the programmer to see what's going on the page level versus code level. A solution to this problem is a template system that allows HTML to be mixed with PHP, attempting to get the best of both worlds. As the following code snippet shows, we can mix HTML tags (like `<body>`) with PHP code (like `echo "Hello`

World";) by embedding the PHP inside the HTML between special tags – <? and ?>.

```
<html>
<body>
<?php echo "Hello World"; ?>
</body>
</html>
```

Scaffolding – This is a way to produce lots of skeletal code quickly for a typical Web application, like a template for a whole app rather than just a page. Scaffolding came to prominence with the introduction of a modern framework called Rails that used Ruby rather than the PHP language. One of the Rails' inventors would turn up at conferences and proceed to demonstrate building a blog in about 15 minutes from scratch! OK, I exaggerate. It was about 40 minutes in total, but the bare bones – the scaffold – was laid down in minutes. Some programmers complain that scaffolding is a gimmick, as most of the scaffold code gets thrown away by the time the real app is finished. But isn't that true of real scaffolding? In any case, the principle of providing a set of files with some skeletal code that the developer expands is a common theme in many Web frameworks.

The point of scaffolding is to get the programmer going, not to complete the job. On the other hand, a Web app usually has two sides to it. There's the side that the user sees and then the side that the administrator sees, the bit where accounts can get deleted, product catalogues updated and so on. Scaffolding is often a great way to build an administration interface that doesn't have to look all that pretty or function with as much elegance and efficiency as the user interface.

Testing Tools – Frameworks will usually include support for automated testing, which is important when building any sizeable Web application. As we shall see, building for the Web is often about frequent iterations to add new features, fix bugs etc. It's important to know that each iteration is enhancing the site without adding new bugs or breaking parts of the existing app. A comprehensive and automated test regime is vital. Most frameworks will provide tools to make this easy. There are two common strategies for testing: unit testing and behavioural testing. Developers tend to use a combination of both.

A unit test is where a test is written to confirm that a piece of software does what it's expected to do, such as add a new item to the shopping cart. The test code will initiate the sequence of operations to create an item and then also include additional code to check that the new item was indeed created and can be found inside the database.

Behavioural testing is a different approach. It requires that the developer write tests first that indicate the required behaviour of the code that needs to be written to implement a particular feature in the website. Returning to our shopping cart example, the developer would first write a test to initiate construction of a cart and confirm that the cart existed. Further tests would be written to add and remove items, and so on. However, none of these tests will pass when run because the code has yet to be written to carry out the behaviours being tested for (or specified by the tests). The indication of failing tests is often colour-coded as red in the testing tools, which is why developers talk about the cycle of RED, GREEN, REFACTOR. This means to create the tests and then watch them fail in RED (as they ought to because the code hasn't been written yet).

As the next step, the developer then goes about writing the code to make the tests pass, such as writing some software to construct a shopping cart in order to make the "test that a cart is constructed" test pass. Once the code is written, and works properly, the test should pass, which is often indicated by the colour green in the test tool. However, getting the test to go from red to green is only the first step. In getting to green, the developer is more concerned with just getting the test to pass rather than the elegance of the code. The next step is to refine the code, which is a process called refactoring. Now that the developer has a test and a basis for passing the test, it is much easier to refine the code. Any mistakes in the refactoring will immediately get picked up by the test, as it will turn red again.

Helpers – These are ways to create snippets of code that can be re-used easily across the entire Web application. For example, we might want to check on each page that the user is authorized to access the site. This authorization check is a process that is repeated on each page, so it's better to write the code for it once and then call that piece of code as a helper, thus following the DRY principle.

There are so many frameworks now, that it's hard to keep track. There are frameworks for most of the common and popular languages used to construct websites. As mentioned above, Ruby on Rails has become a popular and fashionable framework. (What? You think geeks don't follow fashion?) Until Rails, much development was done on the LAMP stack, which meant using PHP, or even Python, which was already a popular scripting language to carry out various programming tasks on top of Linux.

Rails was written using a language from Japan called Ruby. In fact, the creation of Rails is what sparked a big interest in Ruby, which turns out to be a highly expressive language that enables a lot of functionality to be expressed in very few lines of code (compared with other languages) and in a format that many developers find more natural, somewhat akin to writing in English.[6]

Thanks to Rails, in addition to the wonderful and ubiquitous LAMP stack, we can now add the LAMR stack to our toolkit, which is becoming increasingly popular (although LAMP remains the most popular by far). The same principles of free and open source still apply. Rails is an open source project with an increasing number of contributors to the code. It is also a maturing framework with a large number of extensions provided by a large community of Rails developers. There are even hosting sites dedicated to hosting Rails websites, taking away much of the pain of deploying and scaling a Rails application. Later on, I will describe a case study of one such site, called Heroku (see Section 8.4 On-demand: Platform as a Service).

[6] Although this is far from the case, should you be expecting to be able to code in English.

There are other variations to the stack. Some websites run on Windows Server, often because it's an easier platform to integrate with other Microsoft components. Some sites use other relational databases, like PostgresSQL, another free database, or those coming from enterprise origins, like Oracle RDBMS and SQL Server (Microsoft).

In case you're wondering, which I hope you are, let me declare now that it's impossible, and often fruitless, to claim that one language, framework and stack is better than another. The old adage seems to apply: "horses for courses." Various coders will debate the pros and cons, but this is usually on the basis of personal experience and "what works for me," rather than any usefully objective comparison criteria, especially as it is difficult to create a standardized benchmark across all frameworks. And that's where the problem occurs – without a single use case to drive the comparison, many of the framework and stack debates become meaningless pretty quickly. My advice to you, in your environment and with your projects, is don't be put off by lengthy debates, nor the temptation to carry out an N-month analysis. Just pick a popular framework/stack and start coding! Be agile . . .

2.3.3 Agile – Coding at the Speed of Thought

Besides those listed above, there are various other components that the various frameworks might include, all of which substantially increase the productivity of the Web programmer, moving us to a remarkable place where hackers can often code as fast as "the speed of thought!" This is an exaggeration of course, but the general improvements in software productivity are significant compared with website construction even a few years ago. It's like moving from the typewriter to the Word Processor. In my experience, many of the technical project managers in telcos seem to have fallen behind in understanding what's possible with modern Web methods (which is one of the reasons that I decided to write this book).

It is certainly possible to build prototypes in days, not weeks or months. It is possible to circumvent lengthy specification stages by simply stating requirements as small feature stories that immediately get interpreted into code, thanks to the speed of developing when using frameworks and the growing ecosystem of code and support that surround them. This is the essence of the agile method for Web programming. The product owner(s) comes up with a set of stories, which are often one-line statements like "As a user, I can view the shopping catalogue."

All of these stories are added to a list of potential features. Then, each week (or other short time period), the product owner meets with the development team to decide which stories get placed into the backlog of things to do (i.e. they become real requirements rather than just wish list) and then which stories from the backlog will be picked up and coded that week, or however long the team decides a single sprint should take.[7]

There's nothing particularly clever or magical about this so-called "Agile" approach to software. Breaking the tasks down into manageable chunks and then iterating around those chunks is a lot more manageable and has long been a wisdom of effective coding. However, it mostly works because coding has reached the stage where it is possible to produce results quickly at relatively low cost and in a controllable fashion. Therefore, there's no point in

[7] I don't intend to describe the various agile coding methods in this book, but it's worth mentioning that the unit of activity, often called a sprint, doesn't have to be 1 week, which seems a common misconception.

spending too much time specifying functions and requirements in ways that don't lead to the production of software.

If the specification can be decomposed into small chunks and then rapidly interpreted in code, then this provides something real to ship to the customer (internal or external). This is far better for getting real feedback, insights and experience with the problem – and its solutions – than writing lots of words in an unwieldy specification that might get reviewed and debated forever. However, this approach only works when the whole team is involved in the sprints to set goals, review targets and confirm results. The whole team must communicate with each other on a regular basis to expose all the issues. By whole team, I mean all the people needed to ship the product. If a key stakeholder is missing, then the sprints are in danger of missing a key element for success of the product in the absence of any guidance and feedback from the stakeholder.

There are other tools and patterns that support the agile approach. Firstly, it is relatively easy to store the code into a version control system to keep track of changes. SVN and Git are examples of popular revision-tracking tools. These can be extended across work groups so that a team can easily share coding, even if working in different physical locations. In fact, coding in pairs (one of the popular ideas to come from the Extreme Programming agile method) is still a popular way to enhance productivity (mostly by eliminating mistakes during the code authoring).

Just to be clear, pair-programming is two people working on the same piece of code at the same time, like two people with the same Word document open, writing the same document together – that is, taking it in turns, one person writes while the other watches, pointing out obvious typos etc. As with all software production, the sooner bugs are spotted, the better.

Pair programming needn't happen at the same desk. There are plenty of tools to enable live screen sharing, which mean that pair programming can take place virtually. Indeed, the ability to work virtually has been a great boon to many projects because it enables the work to "flow" to the best people, irrespective of their location. This is often easier, cheaper and more effective than bringing the people to the work. That said, physical co-location also has a number of benefits, not to be dismissed too easily. Those regular sprint-planning sessions (sometimes called "Scrums") need to be effective. If this means putting all the team in one place, then so be it. However, this is often difficult. Therefore, effective shared-working tools become essential.

One thing that Web programmers have excelled at is creating their own tools to speed up the process. Many of these tools then get shared back with the Web community, either as free tools or as paid-for services. It is an industry that breeds its own success by feeding back the best ideas from successful projects and hard-earned project experience, often in the form of code – and advice – free of charge.

Here's a list of useful and interesting (Web-based) tools used by many Web hackers:

- Pivotal Tracker – a tool to manage the backlog of stories.
- Basecamp – a lightweight tool for shared project management.
- Campfire – an online chat application with persistent memory of the conversations.
- Zendesk – an online support ("ticketing") tool for customer support issues.
- Jira – a bug tracking tool.
- Skype – a free-to-talk telephony application for group conferences.
- GetSatisfaction – a tool to collect user feedback about your site.

The list of tools goes on and on and on – it's possible to get tool fatigue. Developers really are spoilt for choice. These aren't crappy tools just because they're free (and the best ones usually aren't anyway). Many of them are labours of love by a number of passionate software collaborators. Many are very cool tools with a serious monetization model behind them – and developers will and do pay to use good tools.

2.3.4 *Summary – "Why Frameworks Work"*

If it isn't clear already, the widespread availability of open Web frameworks dramatically speeds up website development and helps to improve efficiencies and lowers costs – much of this stuff is free or affordable. Another key point, often overlooked by observers, is that these frameworks also have very active communities supporting them, adding to the code, improving the framework, helping each other out and helping the framework to become very robust very quickly.

Conferences and events are held just for coders to gather and talk about their favourite framework, share common practises, new ideas, advanced methods and so on. And, like with many of these communities, there's an active job market too – people posting jobs, contractors responding, teams gathering and, yes, even start-ups forming. The web framework world isn't just a collection of software, it's a community too – it's an ecosystem. It's also worth noting that many developers use more than one framework. This is more of a reflection of how much easier it is becoming to use the tools on offer.

2.4 Open by Default: Open Source, Open APIs and Open Innovation

2.4.1 *The Different Types of Open*

A lot of the Web infrastructure supporting connected services is open. However, this is a controversial word that scares telcos. It's a word that comes with all kinds of baggage and prejudices, some of them justified, but most of them stemming from ignorance. In my experience, there are too many misunderstandings of the word and its connotations, so I thought that a section of this book should be dedicated to exploring the topic in more depth. What I intend to do here is to review the common uses of the word open in the software world and on the Web. Many of the topics below will be expanded in greater depth throughout the rest of the book.

2.4.1.1 Open Standards

I start with these because open standards are the most important foundation of the Web and its ongoing success. Standards like HTTP are open, which means that the entire operation and specification of the protocol is openly documented, available to anyone who wants to read it and implement it, free of charge. It is not a hidden secret. This is the first aspect of openness worth clarifying. The HTTP standard is maintained and published by the World-Wide-Web Consortium (abbreviated as W3C). You can find the specification published on their site – http://www.w3.org/Protocols/rfc2616/rfc2616.html.

The strange labelling of the specification – "RFC 2616" – is a reflection of the template used to write and publish the standard, which is the template called "Request for Comment," a method of documenting Internet-related standards. This method is proposed and maintained by another non-profit standards body, called the Internet Engineering Task Force (IETF). The W3C use the IETF method for standardizing Internet-related protocols because it was already in existence, so no need to reinvent the wheel. However, all protocols and standards relating specifically to the Web are initiated and managed via the W3C. When a standard begins to take shape, involving a protocol, it is converted into an RFC.

More importantly, the standard is not only open to read, but is also open for anyone who wants to implement it in a product, like the Apache Web server. There are no licensing fees preventing anyone from including the HTTP protocol inside their product or software. In fact, there aren't any compliancy tests either. There's no body to approve HTTP implementations. Obviously, implementations of the protocol that are not compliant won't make it very far, especially on the "World Wide" Web that is brutal – and effective – in exposing technical flaws that crop up in its global network of servers.

As with most of the Internet protocols, there is a documented method for adding extensions in order to make the protocol as flexible as possible. The actual use and meaning of any extensions will then be application or domain specific, but still following the "HTTP way," so to speak. In the example of HTTP, it is possible to add custom headers to the HTTP messages that go back and forth between browsers and servers. For example, I might want a browser to inform the server about a particular (and possibly proprietary) security method required to protect the HTTP messages. This method can be declared and negotiated via the use of customer headers. Many of the Internet and Web standards are extensible in this fashion so that the core standard can be widely adopted, but without hindering edge-case innovations.

2.4.1.2 Open Source

This is where the code (called the source code) used to produce an application is then made available in its original native (source) form for others to download, use and often extend. Just to be clear about what this means. When you or I download a software program from the Internet, like a photo editing package, we are downloading what developers call "the binary," which is the finished program presented and accessible via its user interface. We don't get to see the code behind the interface and we don't get to see how the program was constructed inside. With open source, we would be able to do exactly that – we get to download the code inside the photo editing package. We could then re-create the finished binary ourselves, or, perhaps more interestingly, we could make changes to the source code in order to create a new binary. For example, we might add support for a particular photo format that the original creators of the source code missed out.

Release of source code might happen after the product has been fully developed in a closed environment, which is what Symbian did with its operating system. This was a tactical move to gain more developer support for the Symbian platform. It might also happen from the outset in order to allow co-creation of a product from idea to deliverable. Examples of open source projects are Linux, the operating system, and Webkit, the Web browser. You may not have heard of Webkit, but it's the engine beneath many successful Web browsers like Apple's Safari. It has become a standard Web browser engine for many mobile smartphone platforms, including Nokia's Symbian, Google's Android and Palm's WebOS. An important aspect of open source projects is how the use of the code and extensions to it are managed by the governing body, whether that's a person, a company or a non-profit. The governance model is usually stated in an open license.

2.4.1.3 Open License

The open license states the terms of conditions for using the open source material, which is usually source code. There are a variety of open license models that have been adopted and adapted by the open source community. Examples include the GNU and the MIT license models. The license is a legal document and legally states the extent of the rights to use the code by those who accept the terms of the license. For example, the rights might state whether or not the source code can be used in a commercial project and what the licensee's rights are in that case. In some instances, any use of the open source code, including modifications, must be openly published so that anyone could potentially access the modified code. However, even when users are free to modify and extend (or fork) the code base, they might not have their code contributions accepted back into the main code base. This is one method used to maintain an open source stance while keeping strict controls over the actual code itself, carefully governing what stays in and what stays out, which is a means to exert strategic and commercial control over an open product. Google use this approach with Android.

2.4.1.4 Open API

These have been one of the key ingredients of the Web 2.0 explosion of creativity. An API (Application Programming Interface) is a way for one software application to access the functions of another. For example, Flickr, the popular online photo storage and sharing website, has APIs that enable software developers to access the photo database programmatically using the RESTful techniques described earlier (see Beneath the Hood of Web 2.0: CRUD, MVC and REST). This feature would be useful to an online service for creating calendars of birthday cards from photos, allowing users of that service to import photos from their Flickr gallery, as shown in Figure 2.7.

When talking of open APIs, open here means that the APIs are accessible via the Web, potentially accessible from any other website, rather than only available to the developers of Flickr. The website wishing to use the API will typically have to sign-up for a special access token (called an API key) and agree to the terms and conditions (legal document) for using the API. There's usually some signing up process to get the API key (which is the API equivalent of a username and password), but this is almost always self-serve – there isn't usually a corporate guy sitting there to approve applications (except for certain cases).

Think of all the myriad things that a child builds with Lego building blocks and magnify that over and over. Developers love APIs: "Oh look, something new I can play with – let me play!" And play they do, creating lots of new ideas and services. And that's often how the

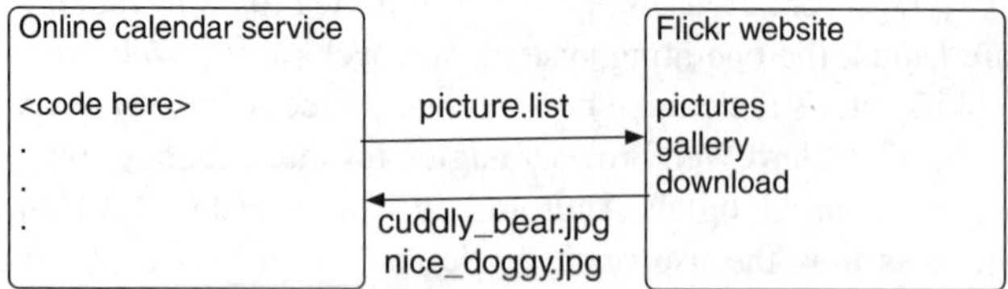

Figure 2.7 Online calendar service accessing photos from Flickr to create a calendar.

revolution begins, like Twitter. It's safe to say, as widely acknowledged by its creators, that the unplanned success of the Twitter ecosystem has been largely due to its easy-to-use open APIs.

Only a few years after launch, there were already 50,000 apps created by third parties, including desktop apps like Twitteriffic and Tweetdeck. I mention these because the founders of Twitter did not have desktop access in mind when they launched Twitter, nor when the desktop clients appeared. The whole 140 character rationale was motivated by mobile-centric access via texting. It was a distinctly mobile play. But they opened a bunch of APIs to allow any developer to submit and consume tweets in any way they felt like. Along came these surprising desktop apps and suddenly a whole bunch of people found ways to use Twitter that enabled them to make sense of the service. This created a large following very quickly.

It's worth mentioning here that "Open API" (and "Open Platform," – see next definition) does not mean free of commercial control or the absence of a business model. Open does not mean free for all, like the lawless Wild West. It isn't like an open protocol that typically has nearly zero commercial or legal restrictions on its usage, besides making it work as intended by the specification. With Twitter, the protocols for formatting, submitting, consuming messages all belong to Twitter. The company can change them whenever they like, such as in ways that might spoil certain types of application built on the APIs, if they wanted. The Twitter folk own the infrastructure and are unbound by any commitment, legal or otherwise, to unfettered access, or to regulation of usage (Twitter can block users at any time). Also, if a developer wants to attach to the Twitter APIs and consume the entire feed of tweets, known as the Fire Hose, then Twitter will charge for this. After all, that feed is worth money and also requires a lot of computing resources to maintain. As we shall see later, APIs can often be monetized, even when some access is free.

2.4.1.5 Open Platform

This is what happens when a business opens enough APIs that a substantial amount of the business usage occurs via the APIs, which become central, rather than incidental, to the business strategy. Twitter is a great example, with about 75% of the usage coming via the APIs rather than their own site. As a business strategy, "Open Platform" is deliberate. It exists in order to encourage and exploit direct network effect – more users doing more stuff with each other. It works because it encourages new use cases that the platform designers didn't consider or didn't have the resources to consider, often bringing in new users from adjacent communities, thereby exploiting the indirect network effect.

Many start-up ventures deliberately attempt to become platform plays, but it is an exceptionally hard nut to crack. Open platforms only work when there's enough users to generate a productive network effect. We should not be so naive as to think that open platforms will simply gain adoption because they are free and available. The rules of promotion, discoverability and marketing apply equally to open platforms as to any other product. However, the key is the absence of friction in using the platform. If the APIs are very simple to access and use, then they're more likely to get used. Sounds simple and obvious, but seems to escape many platform attempts, particularly telco ones.

The Twitter APIs are well known for their extremely low-friction implementation. It's possible to access the APIs and start consuming/producing Tweets within minutes. This has made it easy for developers of all abilities to produce Twitter applications. The number of different apps and services that have arisen around the Twitter platform is quite remarkable, helping to propel the business to 140 million users before the Twitter team had even reached 100 people. The Twitter APIs are so complete that it is possible to create an entire user interface that supplants the official Twitter one. Indeed, most Twitter users do not use the Twitter website to interface with the service.

With open platform plays, the venture is letting go of the "command and control" influence over the user experience. Of course, there is a central set of functions and premises around which the venture operates, but how these are interpreted and consumed by developers and users is left to the community of innovators that are attracted to the platform.

2.4.1.6 Open Stack

This is the concept of an entire set of open source products that combine to produce a more powerful workhorse for other applications, often via open standard protocols and data formats. We've already explored the LAMP example earlier in this chapter. Another example is what some call the Open Social stack, which is a collection of open source software, open protocols and open APIs that combine to allow any Web-connected software service to become socially aware, capable of extending its services along the social graphs exposed to it, often via external social-networking sites (e.g. Facebook). I explore the Open Social stack in a lot of detail later in the book – see Section 3.4 Future Web: "People OS?"

2.4.2 Open, Open, Open!

The LAMP stack and many Web frameworks, tools and Terabytes of related code samples (contributed by the community) are all open source, freely available. It's hard to find another industry with such low friction as the Web apps industry, a wonderful and productive mix of people, ideas and their software. It's just so easy to start a venture or project on the Web. It requires almost no money and, increasingly, fewer skills to get more results than before.

Being open is often a default approach on the Web, and one that many traditional companies, especially telcos, find hard to accommodate beyond an intellectual appreciation coloured by certain anecdotes that fail to penetrate the real scope and meaning of open. There is still a view that such ideas are just the trapping of a "net culture" that is somewhat "unrealistic," anarchic and faddish. This is not the case. With traditional retail and grocery companies like Sears in the US (over 100 years old) and Tescos in the UK now offering open APIs to access their wares, we might safely say that the days of "Web skepticism" are over. It is important to appreciate that business is being done on the Web in a manner heavily influenced by the "Web way" that has been largely driven by proponents of ideas like open access. It turns out that "open," in its various forms, can be a very powerful part of business strategy, even though it's probably fair to say that the idea has been largely informed by a certain mindset we find with many Web pioneers, which is often the willingness to take something and change it, almost to destruction. We call this hacking (which has nothing to do with breaking into people's websites, as in the film War Games.)

Openness has been at the core of the Internet for decades, providing a strong network effect for knowledge sharing and innovation. Such is the nature of code, that smart guys can produce work that can be easily and openly shared, re-used in interesting and meaningful ways, picked up and extended by others – hacked. This behaviour isn't just restricted to lone hackers or students with time to kill. Many of the born-on-the-web ventures solve a problem using open-source components and then share the improved solution back to the community. In fact, the terms of many of the open source licenses that govern open source projects insist that any contributions to the codebase need to be made open too.

There are myriad open source licensing arrangements, which have become an area of confusion and controversy – no one is quite sure what "open" really means, or should be. The term is frequently abused. For example, some ventures use the term to mean "widespread," and open for anyone to use. As my very good friend and leading industry analyst Andreas Constantinou told me in a podcast interview: "Open is the new closed." What he was getting at was the difference between open source code and open source products. When the governance of certain open source products is scrutinized, the arrangements aren't so open. In fact, they're closed. The concept of open is just being used for marketing purposes. Sure, the code might be open, but that doesn't mean that the platform is, especially if the governing body decides who to accept code contributions from, or not, using business principles identical to those seen in traditionally "closed" ventures.

This "open but closed," approach is clearly a means to retain commercial control and advantage. However, it isn't always about being duplicitous with the word open. Making the code open source still has its advantages – it encourages innovation, which the product owner often gets for free. Putting it crudely, it works like this:

- The venture releases open source code.
- The innovators play with the code and publish (openly) their ideas.
- The venture benefits from the ideas without ceding any commercial power.

Let's move our attention back to open APIs. As I've noted above, these can fuel platform growth and even push a platform into becoming an ecosystem play. This was certainly the case with Twitter, which serves us well as an example. I attended Chirp (in 2010), the first ever Twitter developer conference. There wasn't any doubt that developers had contributed hugely to the success of Twitter as a platform, thanks to the way that Twitter had created a compelling set of APIs to more-or-less give open access to its entire platform.

Twitter follows a pattern of "eating their own dog food," which is to say that they build their own services atop of the same APIs that they expose to the developer community. This is quite remarkable in many ways and worth reflecting upon. It would be like a telco installing a core network and then giving outsiders the same access to the core as it has to build its own services. Can you imagine?

No. Probably not.

This degree of openness is quite unique. In Twitter's case, the API strategy was in alignment with the core idea of the business, which was to allow these follow-me discussions to happen in public. Don't forget that the entire conversation of any Twitter user is, by default, shared in the public stream. The aggregate of all Tweets flowing through Twitter is available via the public timeline. Users can elect to make their updates private, but this is the exception, not the norm.

So, in a world where everything is public – "open" – anyway (by user consent), there's no harm in exposing that information via programmatic interfaces for other applications to consume and make use of. At the Chirp conference, we were told that there are already 50,000 apps registered with Twitter. The way to think about this as a telco is access to unprecedented amounts of innovation. Instead of just your marketing department dreaming up yesterday's ideas, you get 50,000 heads on the job (or more). This is one of the major advantages of running an open platform, as I will explore further in Section 8.4 On-demand: Platform as a Service.

An ecosystem has arisen around Twitter with so many developers, investors, entrepreneurs and Tweeters flocking to the platform. Open APIs enable open innovation. This is undoubtedly an important source of ideas and future value. Again, the challenge here, at least for telcos, is in being prepared to unleash the unknown. No one at Twitter, at least so they claimed at Chirp, was expecting a desktop client for Twitter to be important. But, along they came, starting with Twitteriffic. In turn, these sparked a much more avid following of Tweets by users who were suddenly able to expose a lot more value from the Tweet streams. This should be food-for-thought for telcos.

2.4.3 Summary ("Why Open Works . . .")

As with open frameworks discussed above, the widespread availability of open technologies also speeds up website development. Not just that, it speeds up innovation and adoption. The open approach takes advantage of the network effect. It does so by significantly lowering friction, lowering the barriers to entry and adoption.

2.5 One App Fits All? HTML5 and the Modern Browser

Early on in its development, the Web browser was referred to as "the universal client." A brief reminder of software history will highlight the wisdom of that name. If I had financial data, then I needed a financial app to use it, like a spreadsheet or book-keeping package. If I had staff data, then I needed a "Staff app" to access it, like a human-resources database application. If I had office correspondence, then I needed a Word Processor. On and on the list went – one app for each type of data or task. Along came the Web, capable of exposing all kinds of data via HTML over HTTP, that it seemed all we would need is the one app to use all data – the Web browser. It promised to be the universal client for all applications.

It didn't quite work out that way:

"Can I show this graph from my spreadsheet?"

"Sorry. No. Browsers don't do graphs."

"Oh boy, what about this data widgety-thingy. . .?"

"Sorry, HTML not very good at widegty-thingys just yet."

"And this database?"

"Sure, we can probably do that, but might be a bit slow. . ."

Early browsers sucked at anything besides displaying text with blue hyper-links on grey backgrounds. Remember those days? The main issues preventing the Web browser actually becoming the universal client were:

1. Poor graphics – Limited ability to support rich graphics of the type users were familiar with in Windows and Mac apps.

2. Limited local processing – Early browsers could only display the data fetched from the Web server. They weren't able to process it locally, such as to show an alternative view of a graph.
3. Sluggish interaction – The downside of HTTP's otherwise useful simplicity was that it dictated the need to fetch a new page from the server each time an update was required to the page. Early links between browsers and servers were slow, painfully slow.

An early solution to some of the more challenging UI limitations of browsers was the invention of the plug-in by those clever guys at Netscape, which led to Adobe's creation of Flash, bringing a whole raft of interactive multimedia capabilities to the browser. This worked, to an extent, but broke the whole "Web platform" experience for developers. Flash uses its own language and tools to create Flash apps, like a mini-web within the Web. Great for multimedia creative stuff, but not so useful for nuts-and-bolts programming via the CRUD and RESTful patterns.[8]

Several technologies have come together to enable exceptional performance from browsers that will really make them contenders for the universal client ("platform") for connected services running on mobiles, set-top boxes, tablet devices, handheld games machines, and even cars. These are:

1. Powerful Javascript engines and libraries
2. AJAX
3. Web sockets
4. Offline storage
5. Widgets
6. HTML5

I'll be returning to these technologies in depth later in the book (see Section 6.5 Key Platform: The Mobile Web), but I'll give a brief summary here:

Powerful Javascript engines and libraries – It is now possible to do lots of powerful processing within the browser itself, rather than rely on going back to the server. Gmail is a good example. When opening the webpage with its email contents, all of the email messages are pre-loaded into the Javascript processing engine. When the user clicks on an email heading, the message is opened immediately from memory, rather than fetched from the server. This results in better responsiveness and a much more user-friendly experience.

AJAX – This technology uses a local Javascript function to fetch data from the server without having to fetch a complete page. Think of this as like a mini-browser within a browser, updating a partial piece of the page. An example is the way that Google search can suggest search terms before the user hits the submit button to fetch the search results. In the background, a Javascript function is sending the characters, one by one, back to the server where a RESTful API is returning

[8] This isn't strictly true. It's possible to create RESTful applications from a Flash client, but in a way that is often disjointed from the wider web-programming environment.

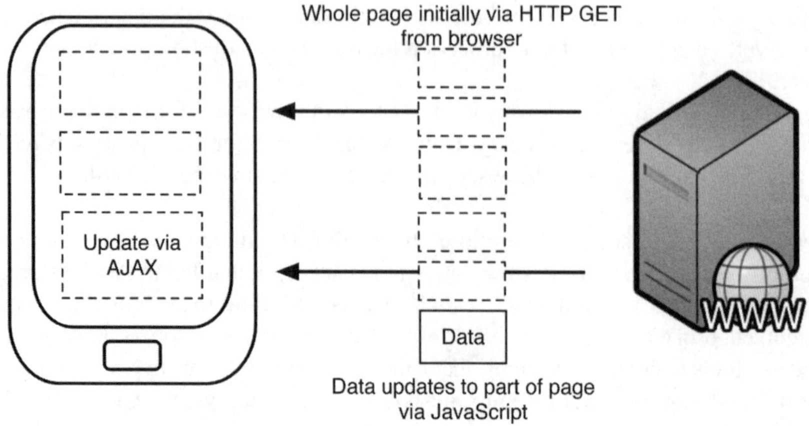

Figure 2.8 Simplified view of AJAX updates in a Web browser.

possible word-completion matches based on the user's search history and similar search-terms ranked by popularity. Figure 2.8 shows how the whole page is initially loaded via a browser request (HTTP GET method). Subsequent updates to the data in the lower part of the page are pulled across from the JavaScript running in the page, only updating the lower portion and avoiding the need to pull the whole page again.

Web sockets – This is a method to push information directly to a Web page, or, more accurately, to a Javascript function running in the browser that is associated with the current page, as shown in Figure 2.9. Think of this as like AJAX in reverse – "Push AJAX" – where data is being updated in the client, but without the browser requesting it. An example would be pushing Twitter status updates, or stock alerts, to a display area in the page. In fact, Web sockets is a two-way process allowing real-time information exchange between a browser and a Web server without the need for the HTTP GET and POST methods. This technology will become increasingly important as the Web becomes more real-time in nature, which I cover in detail in Section 5.1 – Real-time Web and Twitter.

Offline storage – This includes technologies like Google Gears (now call just Gears, which is an open source method for storing data (i.e. Web pages and associated Javascripts and data) on the client side rather than back at the server. Google originally invented Gears to allow the editing of Google Docs to continue even when the browser was without a connection back to the server (i.e. when disconnected from the Web, like on an airplane.) There is now a W3C standard, called Web Storage (originally part of HTML5, but later removed). I will explore this in detail in Section 6.5 Key Platform: The Mobile Web.

Widgets – There are all kinds of widget technologies, but the idea is for a "Web app" to be stored with the browser (i.e. in the mobile phone memory, desktop box memory etc.) rather than being fetched from a Web server. Of course, to

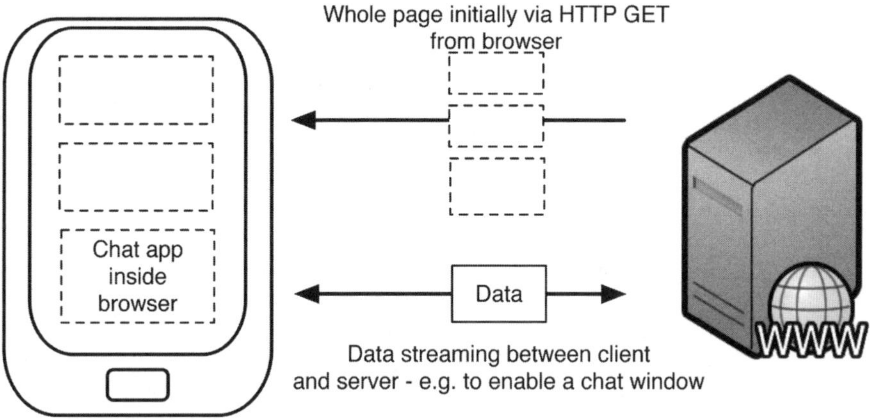

Figure 2.9 Web sockets allowing bi-directional flow of data initiated by a Web app.

be useful, the widget will interact with a Web server, including updating itself, but the advantage of widgets is the immediacy of response. A user can press a widget icon on their mobile and immediately see the app spring into life via the browser, even without a connection. This enables Web apps to appear like apps that are native to the phone, at least in terms of fast start-up and response. Widgets use offline storage techniques, as shown in Figure 2.10. The W3C has a widgets standard, which I explore along with a review of widgets technologies in Section 6.5 Key Platform: The Mobile Web, along with a cross-industry widgets initiative by telcos, called Wholesale Applications Community.

HTML5 (and friends) – This is the huge body of work to define a new browser standard, which includes a large feature set (although much of the original scope has since been relegated to other associated standards – hence "friends"). HTML5

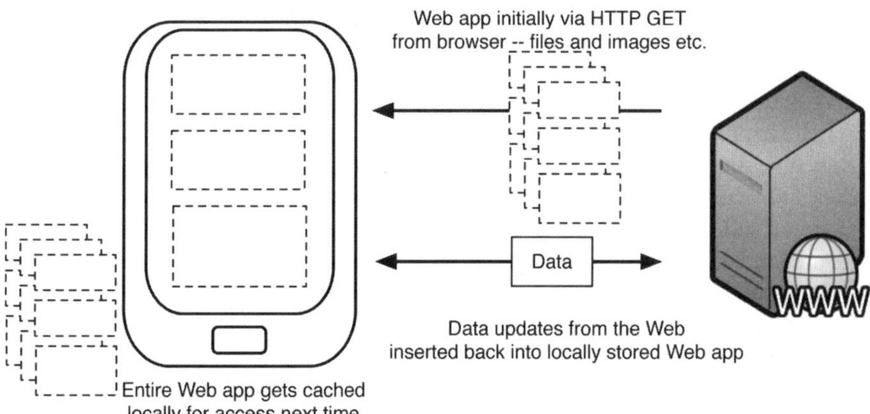

Figure 2.10 Offline storage of Web applications.

is a huge topic, enough to fill a whole book, but we can think of it here as being everything needed in a modern browser to make rich, responsive and powerful Web applications. This includes the ambitious Canvas API, which is an attempt to implement many of the rich multimedia capabilities of plugins like Adobe's Flash directly in the browser (i.e. without plugins). This doesn't mean embedding the Flash player into the browser. It means providing a whole set of browser functions to support rich vector-based graphics using an open standard. This also includes video. YouTube already runs a beta version of their site to support native HTML5 video streaming without the Flash player. I will explore HTML5 in a lot more depth in Section 6.5 Key Platform: The Mobile Web.

2.5.1 Summary ("Why the Browser Works")

The Web browser, which I explore in detail later (see Section 6.5 Key Platform: The Mobile Web) is in many ways the ultimate in application user interface (UI) because it has the ability to deliver a UI for any type of application. This whole paradigm has attracted vast amounts of research, investment, mindshare and hackers to building stuff that works in the browser.

The paradigm has only got better because, with HTML5 and all its friends, there are increasingly fewer reasons to leave the browser for something else. Rich graphics? Sure, stay inside the browser. Video? Yep – stay inside. Running apps without a live HTTP connection? Sure, stay inside again (with limits).

The current exceptions are really anything that requires very demanding graphics performance, such as gaming. Even here, though, efforts are being made. Palm's WebOS allows browser applications to access the underlying GPU[9] to get blisteringly fast graphics performance within the browser. The other key exception to using the browser and going native is to access the incredibly rich APIs available in modern smartphone operating systems, which I explore in some detail in Section 6.5 Key Platforms: iOS and Key platforms: Android.

2.6 It's all About People: Social Computing

2.6.1 Exploiting Relationships – The Social Graph

With the explosion of social networks has come the realization that people increasingly want to interact with their friends within and across a number of different applications and services online. For example, it might be useful within a music application to pull in a social graph of friends and start interacting musically between friends, such as sharing "liked" items, favourites and so on. A social graph is simply a model for representing human relationships in software, as shown in Figure 2.11.

There are a number of ways to create services that utilize the social graph, but the emergence of various socially-aware APIs, frameworks, stacks and services has led to a socially-connected infrastructure that makes the Web much more social in its nature than before, evolving from document-centric to people-centric. It is now relatively easy to create socially-aware apps. Many of these will ride on the top of a social graph that exists within a social network,

[9] Graphics Processing Unit – like the CPU, except it just does graphics, very fast!

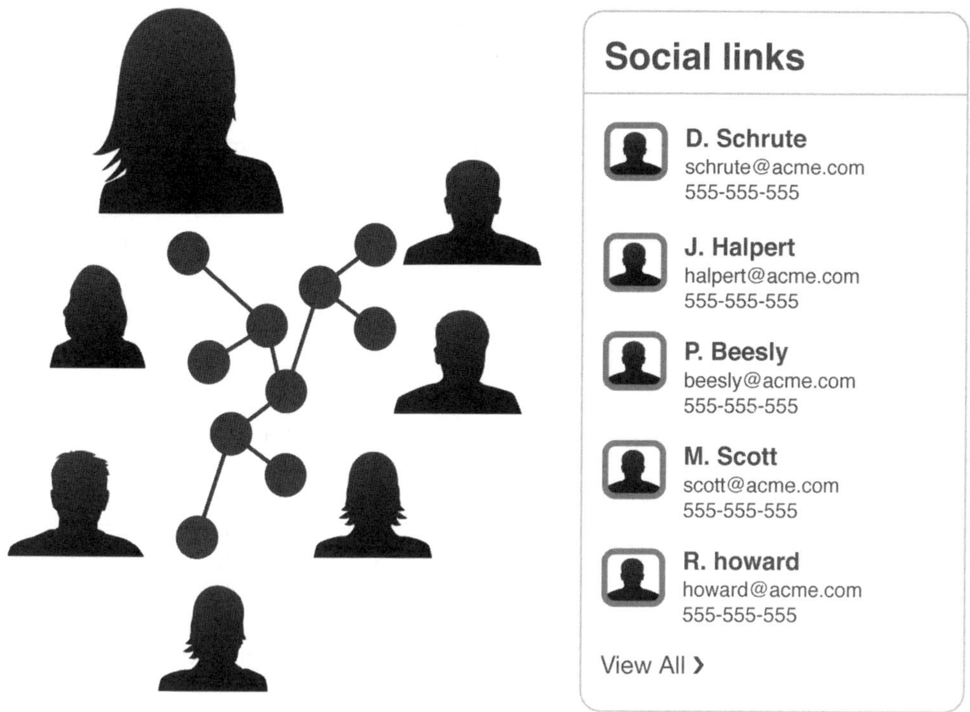

Figure 2.11 Social graph maps people's social connections.

Facebook being the obvious and most dominant one. Facebook has an API just for gaining access to a user's social graph.

Operator networks already have social graphs of course – millions of them, buried away in call records and text messaging logs and elsewhere. It's just that these aren't exposed to the world of software services on the Web and so don't really figure much in the various platform plays taking place on the Web. A challenge for operators is that much of the interesting and potentially useful data in their networks is buried in OSS/BSS enterprise IT systems that weren't born on the Web, and so struggle to release their inner data and functions to external systems. This is compounded by a predominant culture that is generally reluctant to adopt open strategies, even though, somewhat ironically, operator successes have largely been due to the same direct and indirect network effects that exposing the social graph could also exploit.

2.6.2 Exploiting Interests – Context Awareness

Social computing isn't just about the social graph. It includes harnessing personal data to improve the online experience. It is also ironic that the concept of personalization was an early fixation for mobile services designers who wanted to save mobile users the pain of navigating (slowly) through various WAP pages just to find the information they wanted. Better to dish up a set of personalized pages and eliminate the need to present generic data. The phrase "Contextualized Services" was actually coined by operators, not Web hackers.

After such enthusiasm for context, operators mostly failed to capitalize on the idea. Meanwhile, and still not fully appreciated by many telco managers, the Web has excelled in generating not just personalized services, but lots of software infrastructure to support them: frameworks, APIs etc.

Given that much of the Web is monetized by the delivery of adverts, money is made in getting people to click on those ads. That process works a lot better with personalization. Similarly, the sale of products online, such as via Amazon's store, is highly personalized. Online stores are all about conversion rates – converting visitors into buyers. The more personalized the experience, the higher the conversion rate. It's the same with search. It works best when the search engine somehow exploits what the particular user intends by his or her search term, rather than the generic (and usually vast) interpretations of the term.

Now, here's the difference between the telco world and online. All of these interesting Web challenges, like ad-targeting, visitor conversion and search relevance, have attracted lots of "Web Sourcing" power and attention. This is how the Web can source a solution to a problem very quickly and cost effectively due to the various positive network effects of its open mechanisms and low-friction potential for solving problems. Hence, the Web has become the ultimate personalized environment and is still headed in that direction with great speed, aided by social computing. Meanwhile, most telco services remain rather generic in nature and are only personalized to a crude level using the more conventional (and un-automated) approach of marketing's "customer segmentation." This needs to be addressed by telcos with some urgency if they want to stand a chance of relevance in the connected services world blossoming on the Web. It is not too late.

2.6.3 Portable Data

Personalized computing requires personalized data – the more, the better. If we know that a person likes to eat out, that's great. If we know that she likes Thai food, that's even better. If we know that she's also a vegetarian – now we're cooking. But wait! We can do better.

What if we know how much she earns and when she gets paid? Where she lives? Who she likes to eat with, and why? What if we know everything she's actually eaten while eating out, say for the last year, maybe two? I'm sure you get the picture. The more data, the more ways we can query the data and move along various vectors of personalization, as shown in Figure 2.12.

But where does all this data sit? At the moment, unlike the diagram above, our personalization data sits all over the place, smeared across the Web. That's not too great if I'm running the restaurant booking service and the data for my customers is sitting elsewhere. I need access to the data. But, there are a number of technologies and trends that are already addressing this problem:

Data access points – for example, APIs, which is the ability for a service to make its functions and data available to another service on the Web rather than via the user interface.

User-controlled data-sharing mechanisms – for example, OAuth, which is an open standard for a mechanism that allows a user to control access to his or her data via open APIs. Think of it as giving a username and password to a third party

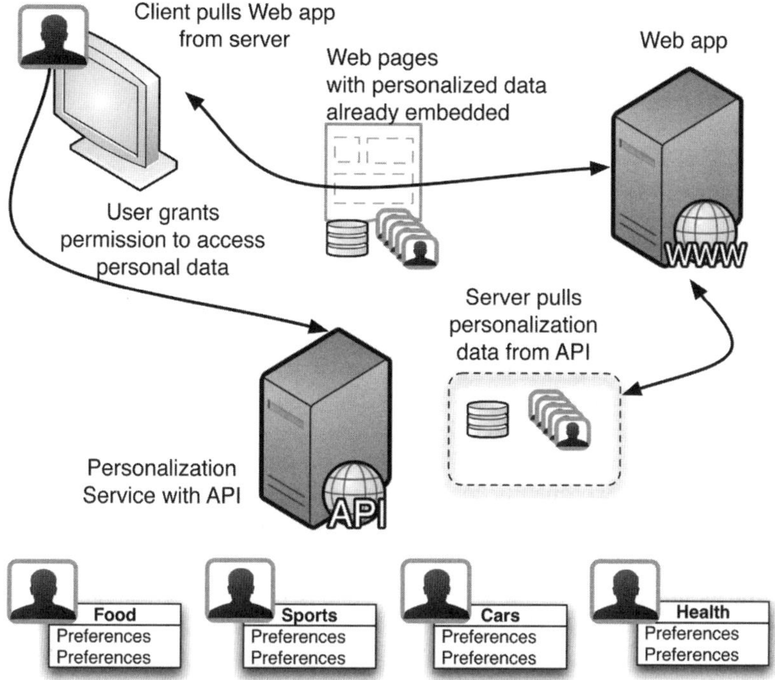

Figure 2.12 Personalization aggregation via the Web.

service to gain access to the user's account, although without actually having to share the password (a token[10] is used instead, which is associated with the account).

Portable data formats – for example, Micro-formats, which are open specifications for how certain types of commonly used data should be packaged, such as contact information (hCard), calendar information and so on. By sticking to a single format, applications can share data more easily. Micro-formats are designed to be embedded into Web pages so that browsers can detect the information too and then process it accordingly, such as downloading someone's contact information and storing it into a local address book.

In a Web page, my hCard might appear like this:

```
Paul Golding
Wireless Wanders
Swindon UK
123-1234-1234
```

[10] A token is the same as a password, except that it is longer and only used for this account. If, later, the user decides to revoke permissions, then the user can mark the token as invalid with invalidating her password.

Whereas the Web browser understands it as a piece of code, like this:

```
<div id="hcard-Paul-Golding" class="vcard">
 <span class="fn">Paul Golding</span>
 <div class="org">Wireless Wanders</div>
 <div class="adr">
  <span class="locality">Swindon</span>
  <span class="country-name">UK</span>
 </div>
 <div class="tel">123-1234-1234</div>
</div>
```

This snippet of code is a micro-format, recognizable by any software programmed to look for hCards. Micro formats are being developed all the time. There are ones for contacts, events, lists, licenses and even recipes (as in cooking).

The idea of a license micro-format might seem a bit odd, but in the context of data sharing it's important. Users of a website like Flickr can specify the types of permission they want to give for sharing of their photos, such as one of the flavours of the Creative Commons license.

> **Standards for describing data** – for example, RDF, which stands for Resource Description Framework. This is a method for describing digital objects, but in a way that can be understood by software programs, not just humans. In other words, it is a means to add meta-data, which means data about the data.
>
> When looking at a Web page, it might be perfectly obvious to you or I that a particular block of content is a set of instructions for building a shed. However, a software program, such as a Web browser, just sees data, like you or I staring at hieroglyphics. It might be a list of instructions for making ice cream, or just a shopping list. There's no way of telling unless we describe the data in a standard way that other software can recognize.
>
> This is the point of RDF. It provides a standard and software-recognizable way of describing data. The use of RDF is steadily increasing as we find it increasingly necessary and useful to allow one program to understand the output of another without any previously intended (or programmed) commonality between them.
>
> **Open data policies** – for example, data export mechanisms, which means simply the idea that data can be exported from one program to another. This is not a new idea in computing, but a relatively new idea in Web programming. During early Web, hackers just built sites with only their sites in mind. If you wanted to log your running times, then here's a site to do it. The idea of taking the data elsewhere wasn't considered. However, in the Web 2.0 era, moving data from one app to another is increasingly common. Not just that, but in some cases there is a growing consumer sentiment towards open data: "If I can't get my data out, I'm not gonna use this service!"

As we've already noted, the open approach can bring a number of benefits. In particular, by allowing users to take their data out, it increases the likelihood of service adoption. The idea of being locked-in hasn't just fallen out of favour in the corporate IT world. Consumers are cottoning on fast. Moreover, various Web ventures have found that providing data export via APIs increases network effect. If another venture sees that data can be exported from a particular platform, then it might be more willing to build its service on top of that platform. It's a safer bet. In turn, this attracts more users to the platform and the network effect comes into play.

2.6.4 Mobile is THE Social Device

What could be more natural than using the mobile as the key device for organizing social data? It is a socially-aware device from the outset, designed and built to make social connections. Some of us always knew this and understood its significance in the Web age. In my previous books I wrote extensively about the many ways that mobile address books could be seamlessly merged with Web services to create a powerful social computing platform. This realization has now become mainstream. Many hackers and ventures now see the mobile as the predominant social platform – the main entry and control point for using socially aware applications. Indeed, with Facebook typically being the number one site accessed via operator portals, there is no longer any doubt about the importance of keeping in touch via the Web on a mobile device.

The mobile device, especially via the smartphone format, presents the ultimate possibility for a hyper-personalized social computing experience. We've lived through the age of the PC – personal computer – which meant a world in which we each had a computer, rather than sharing the company mainframe, if we happened to have access to one. Now, we've entered the age of the true PC – the *personalized* computer – a device that is increasingly ours in every way imaginable.

2.6.5 Summary ("Why Social Computing Works")

In the beginning, there was data, as in really hard, yucky data, like scientific data, numeric data, and financial data. This made computer scientists think a lot about data in a strict numeric sense. Then there was business data – stuff like documents, files, folders, and cabinets. This made computer scientists think a lot about data in document sense. We ended up with a lot of things called files and folders on our computers. The analogy with physical objects made those same computer scientists think about visual metaphors, like files and folders as icons. We ended up with Windows.

The Web was originally a publishing technology – more documents again. After the advent of social networks, things changed. And, what's more, we don't need computer scientists so much – Web hackers (more people than scientists) can build stuff for us, easily, cheaply, well-crafted. With lots of people on the Web doing people stuff, the Web has become a lot more social in all respects. This works because ...

No Surprises! The Web is used by people.

2.7 User Participation, Co-Creation and Analytics

2.7.1 User Participation

User participation was part of the original vision for the Web. It was intended as a means for writers to publish data directly to an audience who could interact with the data by clicking on hyperlinks. The readers participated directly in the network – the Web. Moreover, anyone was free to link to published data and, albeit indirectly, comment upon it.

Nonetheless, the early Web emerged as a publishing medium. We eventually arrive, via blogs, YouTube and faster upload speeds, to a point where users are regularly involved in the production of content without the need to be in any way technically involved with the Web – these are ordinary users getting involved, not Web programmers.

Places like Facebook are worlds in which the landscape is entirely filled with content created and donated by its inhabitants. Users construct the narratives for themselves. This is a far cry from the origins of the Web. The producers and consumers have merged or blurred. The audience has a voice – a very vocal one. In the case of Facebook, when it first started posting status updates onto friends' home pages, there was a rebellious uproar from some 800,000+ users who formed an anti-fan club to tell Facebook where to go stuff their new design. Facebook ignored them – or did they? I shall return to this story in just a moment to reveal a possibly surprising conclusion.

2.7.2 Co-Creation

Giving the user a voice has become a predominant feature of Web design and Web venture management. Users are given every chance to participate, as this is often what creates a sticky site that users will, well, stick with. Let's not forget that with other sites only ever a click away, stickiness is as important as ever. But in terms of getting to a useful level of stickiness, the users can help here too. By letting users give feedback, such as recommendation of new features, and then – critically – rapidly responding to the feedback, Web ventures have a better chance of navigating their way to sticky success and, hopefully, monetization. The name of the game is not to create something in isolation of the users, or guided only by a fixed set of initial insights and ideas, but to create something in concert with the users – which is what co-creation is all about.

Returning to the Facebook story, how does a Web venture know if it's being successful with its co-creation strategy. For example, should it ask users to vote on their ideas and then implement the most popular? Should it try a few and then ask users for more feedback? They will surely give it, but this isn't always the answer, as Facebook showed us when it ignored a mob of protestors who threatened to leave the network.

Facebook persisted with the changes that were upsetting so many users. And they did so for good reason – usage of their website, especially of the new feature, was on the increase. They knew this by monitoring the usage stats. Users (some of them) were saying one thing – "stop this we hate it!" Other users (most of them) were "saying" another – "keep this, we love it!" At least that's what their usage of the site was saying. Welcome to the world of analytics.

2.7.3 Analytics

This is the detailed measurement and scrutiny of all manner of site usage statistics. We're not talking here about those early "web hits" stats – page counters and so forth – "You are visitor 10,101." Web analytics is a very refined and involved science, requiring swathes of experts to implement it properly. Actually, that's not true. Implementation of the various widgets to measure site usage, feature by feature, isn't all that hard. Analysis of the data and having the discipline to create and maintain an analytics-driven culture is a lot harder. So many sites get it wrong. The same old problems apply. Most people know a little about statistics, but very few can understand them.

A key point to grasp with Web analytics and co-creation is that this stuff really works because a website can be updated easily and often and at relatively low cost. We are dealing with very malleable technologies and agile delivery methods. This is a totally different world to what most telcos are used to in terms of product and service refinement.

Typically, in most telcos, in-life management of a service is executed poorly, often neglected entirely. Millions are spent bringing a new service to market, with lots of additional marketing budget, only to find that it doesn't get the response hoped for. Instead of tuning the service, product managers get frustrated by some perceived design flaw, mostly informed by a personal bias – "hunch" – about what's wrong. Marketeers scratch their heads and start scrambling around to review their original market insights, wondering what went wrong. Eventually, they blame it on some missing features that the technical department didn't have time to implement, conveniently forgetting that said features were clipped from the program (by marketing) to save time and money ("tactical advantage"). The service is deemed a failure without ever knowing why, not only spoiling any chances of success with the service, but sometimes with an entire category of services: "What? You want to build an online doo-dat service? Didn't the last doo-dat fail?"

In the online world, service optimization is a breeze. Moreover, it's essential. It's the only way to get towards success, which is an ever-moving target. We shall return to this topic in some depth in "How web start-ups work," when we explore the various online venture strategies in some depth, especially the idea of "user voice," which is the collection of all these themes rolled into one term that really says it all. Remember that old adage – "customer is king." Turns out to be true.

2.7.4 *Summary ("Why User-Voice Works")*

Hey! Isn't it obvious? You'd think so, but so often it's totally ignored. The idea of involving the user is really essential to Web ventures. It's an integral part of the equation, yet in some ways difficult to do. The obstacles are usually cultural. People (i.e. product managers and HIPPOS[11]) have a habit of thinking that they know best. That's mistake number one. The second mistake is to rely too heavily on a single set of insights, often obtained in isolation of the actual venture. It's one thing to postulate a solution to a problem based on good market insights, but quite another to test the hypothesis in real software with real people. Things seldom work as predicted! Remember this! It's why user-voice works. Users, if you listen, will tell you – either directly (feedback) or indirectly (analytics) – what's up with your venture. It's down to you to act!

2.8 Standing on the Shoulders of Giants: APIs and Mash-Ups

When Isaac Newton coined the phrase "Standing on the shoulders of giants," he was being cheeky, using it as a put-down to dismiss Robert Hook, a man of relatively small stature,

[11] Highest Paid Person's Opinions.

for his "derived" works. But there isn't a scientist in the land who doesn't derive his or her theories without building on the works of others. It's a fools strategy to rely totally on original work. Most of us aren't that original. Let go of the notion that your ideas are precious. This is one of the greatest cultural barriers to success in life generally, but in telcos particularly.

Here's how it goes:

"Hey – we have a calendar idea."

"Oh great, but didn't They-Did-It-Already-Dot-Com do that?"

"Yeah, but ours is [unique | better | *our* brand]."[12]

"Really?"

"Yes. We have great customer insights that say they're like our one."

In Web 2.0, using other people's stuff is often an option. The concept of mash-ups have become widespread, first popularized by early hacks built on top of Google Maps. These days, when we see a link to a place, we almost expect it to open someone else's map, nothing to do with the venture whose site we're currently visiting. Exposing some of the features of a site to other developers to include in their sites has become a standard in many Web ventures. It's common to find a "Developers" link that leads to API documentation.

To be clear, I'm not saying not to bother building something because someone else has already. Perhaps you really do have the best calendar since, well, the last one. But, here are a few ways to think of APIs and mash-ups:

1. Gravity – If other sites already have a lot of user pull ("gravity"), such as various calendar sites might, then you might be better off leaving the users to enjoy their favourite calendar site in a way that they can also use your venture. Create a mash-up of your service with their calendar service. This is assuming you really aren't in the calendars business. Maybe you're providing a service for groups to organize mini-league soccer. You focus on the soccer bit, or the team communications, and re-use someone else's calendar bit.
2. Gravity (again) – If you want to pull users to your site – always a good idea – then providing an API is a good step. For every Web venture, there are two on-ramps for users: the user interface and the APIs. Wait! I can feel the doubts. If we let users come via some other site to our site (API), then don't we lose control?

   Yes. And no. You gain users, which is, ultimately, the real control point. If controlling the entire user experience is your thing, which it might be (and in a telco, typically will be), then the mash-up strategy is going to hurt. But, it's only a mild headache compared with the quite likely alternative, which is no control at all because of that more thorny problem of no users.

   So, the name of the game is to create a platform that pulls in the users and then to try as hard as possible to create an ecosystem around it. Constant "user-voice" optimization and constant innovation are essential. But, guess what? The API strategy can do a lot to help with an ecosystem play. All those unanticipated innovations built on your platform will attract all kinds of activity to the platform.

[12] Delete as appropriate.

Figure 2.13 From talking to doing.

2.8.1 Summary ("Why Mash-Ups Work")

As discussed in Section 1.6 From Platforms to Ecosystems, most of Twitter's usage takes place via mash-up applications outside of its own user interface. It has attracted over 105 million users, climbing rapidly.[13] Twitter still has control over these users because Twitter governs, technically and legally, the use of its platform. When innovators are building services using your stuff, then you have an opportunity to monetize.

For the innovators building their services on top of Twitter, and other services, the logic is fairly simple. If users, say, are using services X, Y and Z, and I create a new service that works with X, Y and Z, then I can potentially attract those users to my service. It's a case of indirect network effect. It often works, certainly in terms of attracting users. Keeping them and then adding enough value to support a business model is another matter. But then these two problems exist no matter what you build, so why not take the route of lower friction – build a mash-up, even if it means conceding (initially, or forever) some ground to another player who will be both the friend (need them for the mash-up) and the enemy (they might eat my lunch) – welcome to the world of "frenemies." No one said this was easy.

2.9 Mobile 2.0 – It's Really a Developer Thing!

2.9.1 Mobile 2.0

Mobile 2.0 – Isn't that just about making all this Web 2.0 stuff work on the mobile? Mostly, yes, but with some unique and interesting twists. But let's focus on the core features of mobile Web and work outwards. With the advent of smartphones, particularly post iPhone, the Web is seriously usable on the phone. We have moved from Internet-capable mobiles (early WAP phones and now today's feature phones) to Internet-centric mobiles, packed with computing power.

[13] As of May 2010, announced at Chirp conference.

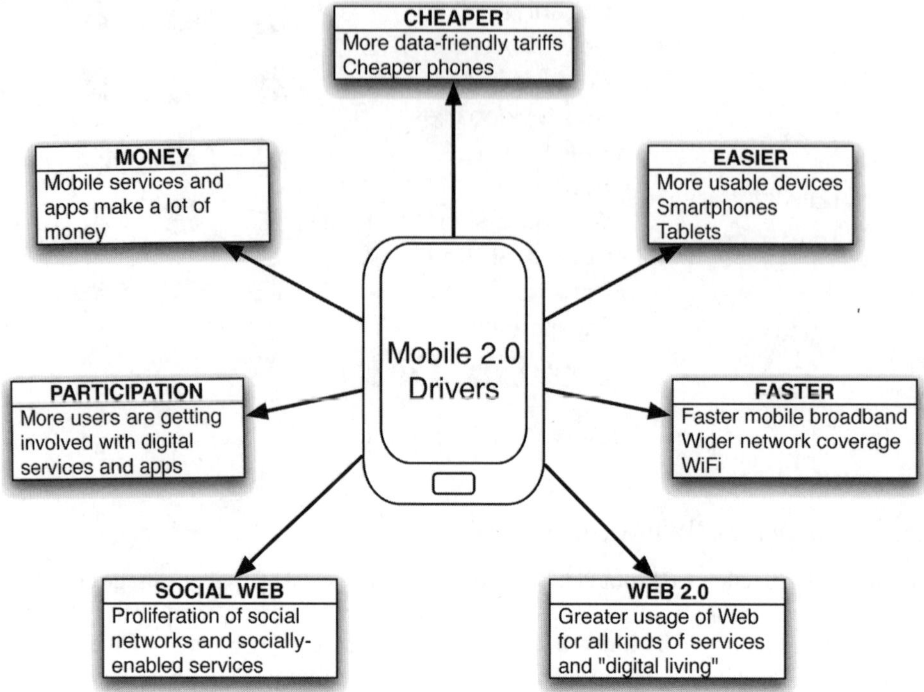

Figure 2.14 Mobile 2.0 drivers.

Internet-centric devices make it possible for the user to do other stuff besides talking, as shown in Figure 2.13, and this is another dimension of the Mobile 2.0 generation.

As shown in Figure 2.14, Mobile 2.0 is driven by smartphone adoption, although the more important driver is the availability of transparent "data-friendly" tariffs.

With Internet-centric devices, we are able to engage more meaningfully with the increasing number of online services that we interact with in our daily lives. For many users, the mobile device has already become the central device for engaging with the digital world. It contains and controls our digital alter-ego. It has become *the* computing platform that matters.

2.9.2 Mobile as THE Platform (Again)

Personalized social computing is very much the future of computing, at least from the user perspective. In many ways, we are headed towards a new stage in both Web and computing evolution. Historically, there has been much attention paid to the enduring "operating system wars," which began with the Windows versus everything else saga. For a while, this moved to the "Browser wars," with Internet Explorer versus everything else, but this particular battle has largely subsided with the recognition by the original software warriors that the browser isn't really a fulcrum of differentiation – innovation in the browser wasn't leading to any monetizable business model, despite billions of dollars thrown at the chimerical opportunity. Of course, that hasn't stopped niche players, like Mozilla's Firefox, from gaining traction.

Now we have the mobile operating system, or platform, wars. This didn't used to be an area of much interest. The dominant player in the handset world was Nokia, but no one really cared much about their operating system. Most other phones used an embedded OS from one vendor or another. Users, in particular, didn't care – and mostly still don't. Widespread consumer interest and awareness in mobile operating systems per se is practically zero.

However, developer interest *does* matter. The iPhone "app-for-that" boom has only happened because developers have taken an interest. Why are developers important? Well, it's nothing new. With Windows, once developers started making apps, the ecosystem flourished. It began with the spreadsheet and blossomed from there. Ditto the Web. Once developers started building apps, it blossomed too. Developers drive innovation, which drives adoption, one way or another. Ever since the industrial age, technology has played an important part in commerce. However, what's different about the Web is that the tech guys really do lead. It's hard to find any example of a successful Web venture that wasn't started by a coder, or some kind of tech guy as one of the founders. We don't find this in any other industry, where the captains are usually people with money and power.

The current thinking – and reality – is that the developers are the kings of the hill. Whatever they pay attention to matters. Put simply, developers have begun to pay attention to mobile, so now it matters. This is what Google's Eric Schmidt speech encapsulated when he said "mobile first" at the Mobile World Congress in 2010, the same conference at which I explained (to telcos) how to empower developers to pay attention to your wares.

Operators are not stupid of course. They realized the importance of developers many years ago when they attempted to create their own platform, which was called MIDP[14] – the standard for running Java on mobiles. However, this failed due to many issues that I won't explore here (but will revise when we look at mobile device platforms – see Chapter 6. Modern Device Platforms). We can put it all down to poor timing and execution, not forgetting that many technological successes are precisely due to the opposite – great timing and execution – but *only in hindsight.*

Following the success of the iPhone/iTunes platform and ecosystem, developers have flocked towards the mobile. This is not a new breed of developers. In the past, it seemed that there was a certain clique of developers who favoured the pain of developing for mobile. I was one of them. Back then, developing for mobile was 99 per cent mobile-specific pain and only 1 per cent application freedom to develop something with a vague chance of success beyond an interesting lab experiment. Today, things have reversed.

The power of Web and mobile platform technologies, along with the other Mobile 2.0 drivers (cheap data and smartphones especially), make it possible to focus 99 per cent on the application possibilities, spending only 1 per cent of the time on the mobile pain. However, this is still a relatively niche play, with many developers still focused on building Web apps.

This is changing rapidly though. With the smartphone adoption rate and the likely success of a few gorilla mobile platforms (iPhone OS, Android and HTML5), which is the usual market dynamics, the world of Mobile 2.0 is set to explode for developers, as in the world of doing other stuff besides talking for users. Now that we've arrived at this opportunity, for real this time, the question is how to exploit it. The remaining chapters of this book explore the key emergent technological themes of the Web and how these are going to shape the future of mobile.

[14] Mobile Internet Device Profile – a "standard" flavour of the Java language for mobile devices.

3

The Web Operating System – The Future (Mobile) Services Platform

`http://the.web.os/is/mostly/about/data`

- A Web Operating System allows the developer of Web and connected services to "hand-off" much of the underlying plumbing to a set of existing services that provide key common functions.
- The emergence of a dominant set of common Web Platform services, like social networks, will create a Web Operating System by default.
- The Web Operating System doesn't run on one machine, but across various server farms in different ventures, accessible via APIs.
- The Web has moved from a network of linked pages to a Web of linked data. The data is generated by all kinds of objects and represents various things in the world, both concrete and abstract.
- Data itself is important and central to connected services. However, the meaning of data ("Semantics") is increasingly important. The vision of the Semantic Web, a world in which all the data on the Web is usefully annotated, is still some way off, but rapidly approaching.
- Telcos need to recognize the strategic importance of the Web Operating System and act quickly, before their rightful place in the ecosystem of consumer and business related data is lost to Web ventures.

3.1 Why is the Concept of a Web OS Important?

In this section, I want to explore with you the concept of a Web Operating System – an all-embracing platform upon which we run connected services that enable us to participate and collaborate. Since Web 2.0, the Web has become more than just a place to consume content passively. It has become a place of mutual interaction and participation. Therefore, when designing any digital or connected service, it's important to think about the *design and architecture* of systems that fully exploit all of the Web 2.0 potentialities.

Connected Services: A Guide to the Internet Technologies Shaping the Future of Mobile Services and Operators, First Edition. Paul Golding.
© 2011 John Wiley & Sons, Ltd. Published 2011 by John Wiley & Sons, Ltd.

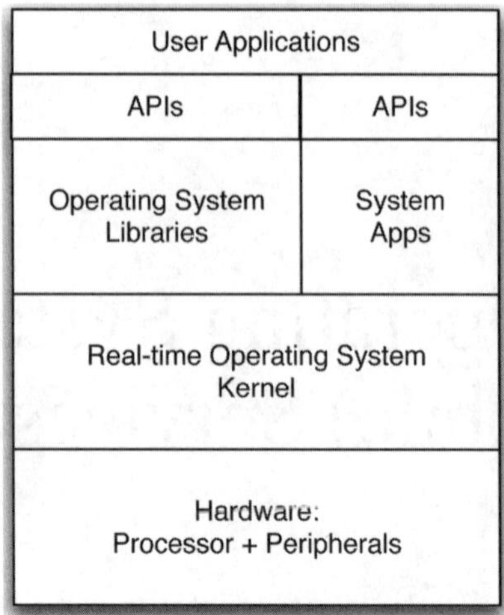

Figure 3.1 Simplified Mobile OS.

I should begin with a caution: *there is no such thing as the Web Operating System.* At least not officially. You won't find a product, service or technology called the Web Operating System.[1] The Web Operating System is a concept that gets talked about by various thought leaders on the Internet, such as Tim O'Reilly. It is especially important for this book because it has a potentially profound impact on the future of connected services. To understand why, let's briefly review what an operating system is and then explain why the emergence of an Internet version poses a challenge – and an opportunity – for telcos.

In my previous book, I described in some detail the workings of a mobile operating system (OS).[2] It is a set of software functions that enables applications to run on the hardware of the device. As shown in Figure 3.1, the OS is a set of software functions. The lowest layer is called the kernel, which takes care of managing the lower level hardware, such as sending and receiving data from the wireless networking chipset. It also takes care of making sure every software process on the handset can run in a timely manner, including support for real-time behaviour. For example, if the device is busy running a game whilst a phone call arrives, the kernel has to be capable of responding quickly to the incoming call, suspending the game process and then running a new process to handle the incoming call.

This process to handle the call is part of a library of operating system functions that run on top of the kernel. All of these functions are exposed via Application Programming Interfaces (APIs). These APIs allow third party applications to access the inner power of the device. For example, if a developer wants to write an application to filter incoming calls, then the app will need access to the mechanisms that control incoming calls, which are OS level mechanisms. This can be done via APIs.

With this configuration of software on the device, an application developer can focus on building only the functionality needed by his or her application, leaving all of the lower level

[1] Except Palm's name for their mobile OS, which they call WebOS. This is not the same thing.

[2] See Chapter 10 – "Mobile Devices" – of *Next Generation Wireless Applications*, 2nd edition.

device management and operation to the OS. With modern operating systems, such as Linux, which is popular on Internet servers, and Android, increasingly popular in mobile devices,[3] quite a few higher level functions are included in the OS too, which I have shown as "System Apps" on Figure 3.1. For example, the Android OS comes with an extensive set of software libraries that are closer to what we think of as applications, such as the address book. In addition to providing the application, the OS also exposes the application's inner functions via a set of APIs. This includes functions such as "Add Contact," "Delete Contact," and so on, allowing developers to write their own applications that can interact with the address book.

In computer-science language, the OS provides a *layer of abstraction* between the underlying hardware and the application itself. Otherwise, programming would be extremely complex indeed. As far as the application developer is concerned, the OS provides a set of functions that can be glued together to form an application. The developer doesn't really know, or care (most of the time), whether or not a particular function is implemented in hardware or software. For example, a method to compress images could well exist in hardware (via a specialized graphics processing chip) or as an operating system software library. The developer doesn't really need to know. A developer wants to include a reference to "http://somelink.com" in his or her code and for it to simply connect to that URL. She doesn't want to know how to drive the mobile wireless chipset to make this connection happen.

Taking a step back to think about the industry of producing software, the challenge with operating systems has always been their diversity. Windows is not the same as Linux, which is not the same as Mac OS X. Therefore, when writing an application, the developer can't expect the same code to run on different OS platforms because the interfaces (APIs) to them are different. The method to add a contact on one OS is not the same as on another. The micro version of Java (J2ME) attempted to solve this problem for handsets by building a common facade of OS interfaces that were then mapped to the underlying ones on different platforms.

I discuss this at length in my previous book.[4] Without wishing to sound like an anti-telco fascist developer, J2ME ultimately became a somewhat grandiloquent and eventually vacuous attempt by mobile operators to provide a common programming environment.[5] It was somewhat undermined by device manufacturers constantly adding new APIs in order to differentiate their offerings. The age old problem of standardization versus customization reared its ugly head and spoiled the affair. However, probably the real issue was the lack of a commercial ecosystem around J2ME to incentivize developers to create J2ME apps and to incentivize the various technology providers to make J2ME successful as an industry standard.

Now that we've reviewed what an OS is in general, we can turn to the idea of a Web Operating System, which is a very similar concept, but on a far grander scale. What developers are looking for, and what some Web ventures are attempting to provide, is a way to abstract many of the underlying and foundational *Web 2.0* mechanisms that are likely to be common to most Web applications, leaving the developer to focus on his or her application, not the underlying plumbing. This leads to the concept of a Web Operating System, as shown in Figure 3.2.

---

[3] Although as we shall see later, Android runs on top of a Linux kernel.

[4] See Chapter 11 – "Mobile Devices" – of *Next Generation Wireless Applications*, 2nd edition.

[5] This sentence uses a few big words simply because my kids challenged me to use them while I was in the middle of this paragraph. Apologies for the exuberance.

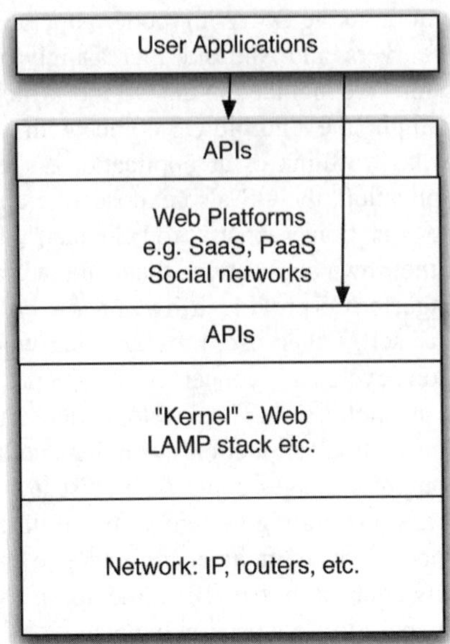

Figure 3.2 Web operating system concepts.

Let's briefly review the layers in the diagram:

1. **Network Layer –** This is the underlying Internet Protocol layer built on top of routers, firewalls and servers. It's the plumbing that keeps the data bits shifting around the net. Developers and service providers are seldom interested in this layer, except for telcos who provide data connectivity as a service. This layer tends to be very standards-driven. It is extremely difficult to differentiate network offerings.
2. **The "Kernel" layer –** This is what I have described earlier (Section 2.2 Beneath the Hood of Web 2.0: CRUD, MVC and REST) as the "software plumbing" for Web 2.0, such as Web servers, proxies, data storage services and so on. As with all operating systems, there is no single standard here, but the common LAMP stack and its variants tend to dominate, along with common implementation patterns, such as REST and CRUD. Much of the software in this layer is open source and is increasingly powerful. The Web 2.0 era has been underpinned by substantial commoditization of the software in this layer, which has lowered the barriers to entry significantly, especially cost and time to market. Most Web developers are dealing with APIs that hook directly into software frameworks designed to make building Web apps easy. Developers seldom have to think about the actual operating systems (e.g. Linux, Windows, OS X) that might be running these frameworks. Frameworks also come with vast arrays of software components to make certain tasks "plug and play," such as:

   A. User authentication
   B. File uploading
   C. Photo library management
   D. Data storage and indexing
   E. Search
   F. Batch processing
   G. Connectivity to charging gateways (e.g. for credit card payment)

H. Support and user forums
I. Blogs
J. Content management systems
K. Messaging (e.g. Email, SMS, IM)
L. And many more...

The list of components is very long, leaving developers to focus on the domain-specific functions needed for their applications. Without wishing to trivialize or underplay the skills required by Web developers, the whole industry of website production is increasingly a "building block" process, gluing bits together that not long ago had to be created from scratch. The open source movement has driven the commoditization process, but there is also a support ecosystem of developers building, testing and supporting more industrialized versions of these components.

3. **Web Platforms –** This is the most interesting and strategically significant layer of the Web OS stack. In some ways, it is akin to the "System Applications" layer in mobile devices, like the built-in address book, messaging and photo apps. However, there is one key difference. On the Web, anyone can provide software services at this level because the Web is an open platform in terms of allowing any component to work with any other component via the IP protocols, which are without boundaries on the Web. Web platforms include any *service*, as distinct from a mere software component, that can be consumed by another application via a Web API. These services tend to fall into two categories:

   A. B2B services that are only consumed by the application, such as Force.com, Amazon Web Services.
   B. B2B2C services that are also consumed by consumers directly, such as Facebook, Amazon Marketplace etc.

It turns out that this Web Platforms layer is quite important from a strategic point of view. Here's why:

1. **There are currently no standards for many Web Platforms, such as social networks.** Therefore, Web venture(s) which succeed in defining the platform and gaining adherents to its definition will have a great influence over the future of Web services, or *connected services.*
2. **Platforms will dictate how users engage with the Web.** For example, part of the plumbing might be a common way to login to services, which is intimately involved with user identification and control. Other examples are Web payments, social graphs (who my friends are), personal preferences (what I like) and so on. As you can imagine, many of these mechanisms are potential *control points* for how users engage with services. They are also control points that will figure strongly in various business models.
3. **From a telco perspective, most of the efforts to define a Web Operating System, whether formally or not, are taking place on the Web**, driven by Web ventures and entrepreneurs, not by telco standards bodies, except with a few exceptions, like the Wholesale Applications Community (see Section 6.5 Key Platform: The Mobile Web). In fact, given the strategic significance of Web Platforms, the absence of telcos from the game is notable. This suggests that, as things currently stand, telcos will have very little influence, and therefore role, in the future of Web platforms and therefore in the connected services that ride atop of these platforms. This will only accelerate the tendency to become a bit-pipe, stranded in the network layer of the OS. This may not be a problem to some operators, as

they might be happy with the bit-pipe model, content to leave service innovation to strategic partnership, acquisition, or just to others entirely. However, the main point to observe is that as an industry, the telcos are rapidly becoming irrelevant in the definition and operation of connected services, which was hitherto their rightful place to dominate. The pipe, devices and everything along the way to the core of the platform don't really count. To quote from Tim O'Reilly, when talking about search as an example of a core service:

> "The resources that are critical to this [search] operation are mostly somewhere else: in Google's massive server farms..."

Therefore, using O'Reilly's example, if "search the net" becomes a common component of the Web OS, to be widely used by Web applications, as well it might, then the power of the platform shifts towards Google. And this is the point about the strategic threat, or opportunity, of the Web OS. Unlike with conventional OS platforms (e.g. Linux), these underlying plumbing functions don't run on the machine. They run somewhere in "The Cloud" in somebody else's data centre or on somebody else's server farm, see Figure 3.3.

To put it plainly, a connected service running "on the Web" is actually running on a plethora of existing Web services. As is normal with most markets, there will inevitably be a gravitational pull to a small number of large underlying services. It is these who will become the Web OS providers and yield power on the Web. An alternative vision would be to influence the Web community to adopt a set of open standards that put the key parts of the Web OS

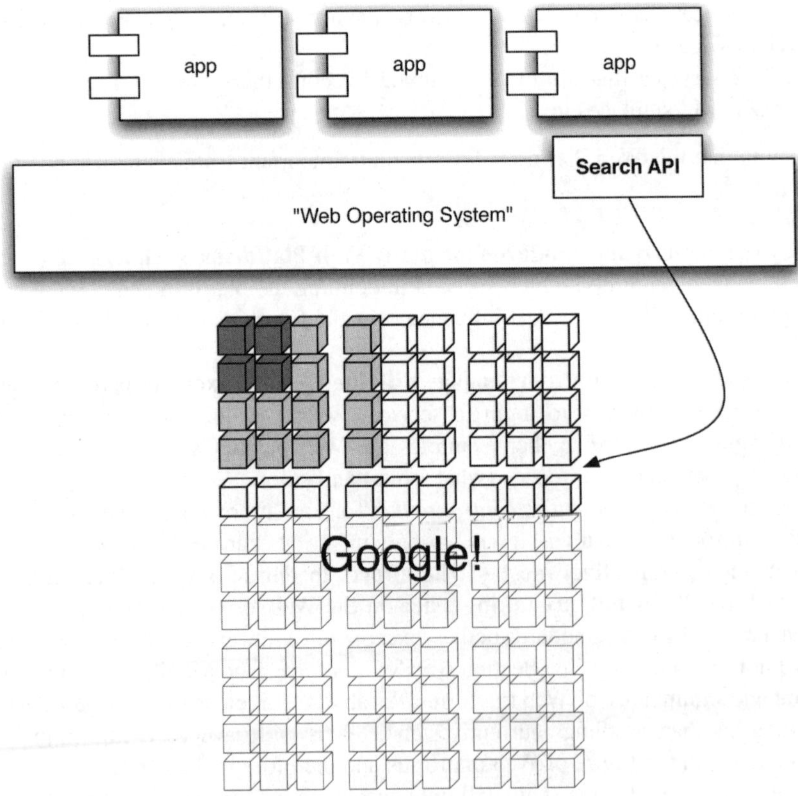

Figure 3.3 Demarcation of the Web OS.

into the public domain, not under any one organization's control. But then this is the ongoing standards versus proprietary dichotomy that is never easy to negotiate or navigate, left to market forces, regulatory bodies, market disruption and so on.

In the remainder of this chapter, and throughout this book, I shall explain the various ways that a Web OS is emerging and how it works. Along the way, I shall continue to point out the strategic issues of potential interest to telcos.

3.1.1 Summary

A Web Operating System allows the developer of Web and connected services to "hand-off" much of the underlying plumbing to a set of existing services that provide key common functions, such as identifying the user to login to the service. The emergence of a dominant set of common Web Platform services will create a Web Operating System. It won't run on one machine per se, but will run on various Web venture's server farms, accessed via APIs. These Web ventures will be in a position to exert significant influence over the digital economies that emerge on the Web, including connected services. They will have a major strategic opportunity to influence and control user experiences and the business model dynamics that interplay with those experiences.

3.2 Internet of Things

It's no longer current to talk about Web 2.0. Most of the useful and well-designed Web applications created since about 2006 have mostly followed the various Web 2.0 patterns, featuring the key hallmarks of modern Web programming and venture methods, as discussed in Chapter 2. The Web 2.0 Services Ecosystem, How it Works and Why.

Attention has now turned to the next evolution of the Web. We might say that much of the Internet thus far, especially post Web, has been about getting data onto the platform, kicked off by Berners-Lee's invention of the Web to "publish" scientific documents. The fulcrum of his invention was the hyperlink – a means to link from one piece of content to another, along with an open protocol/server paradigm that meant that links could point anywhere, allowing anyone to insert content into the open Web. Click it and you're away, over to the next page. There has since been a flood of content linked in this way: billions and billions of pages.

But what we've seen emerge with the evolution from pages to Web applications, using databases to build pages, is a transformation of the link from a page-pointer to a data-pointer. Returning to my example from Beneath the Hood of Web 2.0: CRUD, MVC and REST, I showed how the CRUD verbs of the RESTful Web can be used to manipulate data pointed to by a URI, which stands for Uniform Resource Indicator, which is the technical term for link. I gave the example of a link from Facebook:

`http://graph.facebook.com/559089368`[6]

This simple link, when activated by a HTTP GET command (which we don't really care about, as this takes place "behind the scenes") returns a set of data (a "resource"):


```
{
"id": "559089368",
"name": "Paul Golding",
```


[6] This link is real. You can try typing it in. You will see the same data returned as printed in the following paragraph.

```
"first_name": "Paul",
"last_name": "Golding",
"link": "http://www.facebook.com/paul.a.golding"
}
```

This is some basic high-level data about my Facebook account, formatted with JSON[7] (those curly brackets, speech marks and colons used to delimit the data fields). This is pure data. There's no page here. The idea is that a program (e.g. a connected service) pulls data from this link and consumes it for some purpose. Perhaps this is my dentist wanting to know details of my Facebook account in order to post an appointment reminder onto my Facebook wall. But we can use URIs (links) to point to any data we like, so long as there's some mechanism for getting the data onto the Web. Here's another possibility:

```
http://graph.telco.com/0751522128
```

Which is a link to a telco platform, appended by a mobile number.[8] This might return something like the following:

```
{
"id": "559089368",
"name": "Paul Golding",
"first_name": "Paul",
"last_name": "Golding",
"link": "http://www.telco.tel/paul.a.golding"
}
```

That name fields could be used to confirm that this is the person that the application expected and the link field to go find a page (or API) where my contact details can be downloaded in an hCard micro-format, as described in Section 2.6 It's All About People: Social Computing.

But, what about this link:

```
http://toyota.com/car/245534462342
```

This is a link to a car! Why not? A car has a computer in it, such as the engine management system. It might have a GPS device attached to it, tracking location. It might store a record of braking pressure, useful to tell how many times the driver brakes too hard.

```
{
"id": "559089368",
"name": "Paul Golding",
"first_name": "Paul",
"last_name": "Golding",
"engine_stats" : {"Brake_data" : [Array of braking data],
"Fuel_consumption" : [Array of fuel data], "etc" : [Array of
other data]}
}
```

[7] Stands for JavaScript Object Notation.

[8] In practise, using mobile numbers to access user data isn't very useful because the user might change his or her number. An alternative unique and unchanging ID is better.

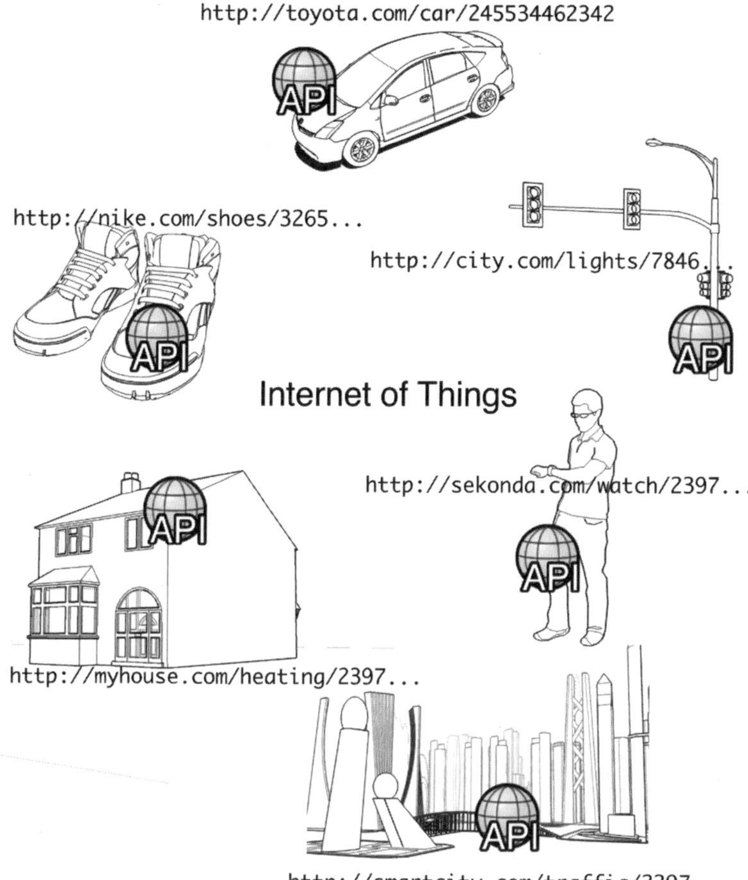

Figure 3.4 Links to things.

This humble thing we call a link suddenly has all kinds of power. It's not just a page at the end of links, it's data, as shown in Figure 3.4. Moreover, the data isn't just some amorphous record from a database, it's structured data that represents a thing: a person, a car, an engine, whatever we want. Welcome to the Internet of Things!

This Internet of Things is the basis for a really powerful platform. It's what the new Web is becoming. Within this context, the idea of a Web OS makes even more sense. Suddenly we have a plethora of objects to connect with, manage and organize online. If, for example, gaining access to car data becomes important, which it might for all kinds of reasons,[9] then we're likely to want a set of underlying software services that can take care of how we access car data in a repeatable and reliable fashion, abstracting car sensors and low-level interfaces (e.g. Bluetooth Low Energy) from high-level software functions (or API calls), as shown in Figure 3.5. Maybe car manufacturers provide such a platform, or possibly telcos, or even insurance companies. We're also likely to want a uniform way to connect car data with people.

[9] Such as security, pay-as-you-go insurance, safety monitoring, energy optimization, logistics, tolling, etc.

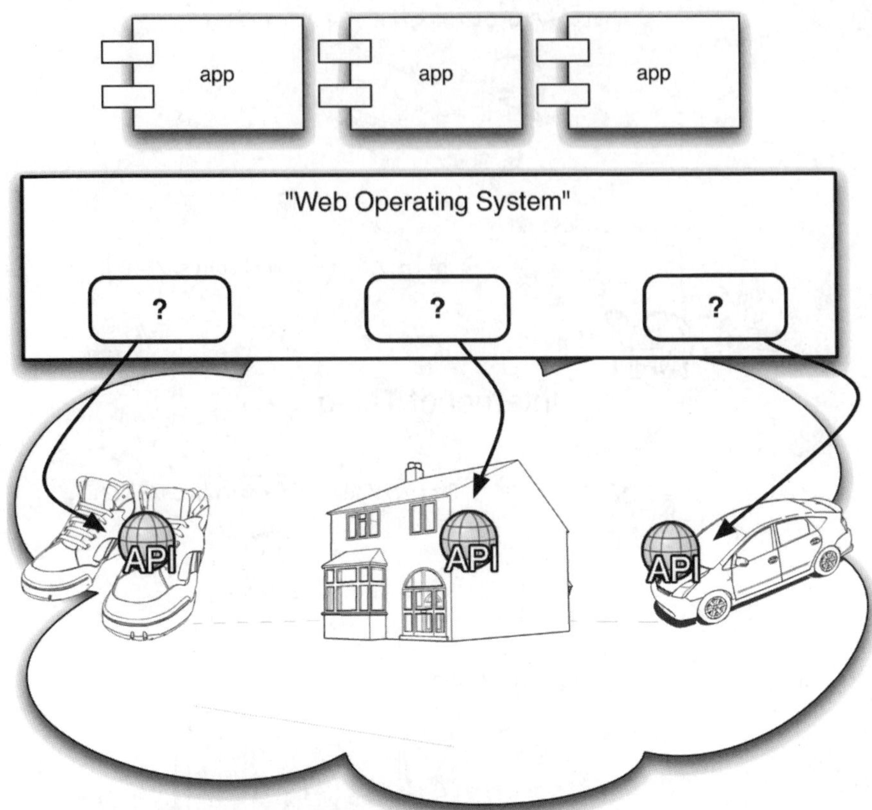

Figure 3.5 Web operating system for the Internet of things.

We might want a standard way to pay toll charges for cars, parking charges and fines etc. I hope that it's clear that in the Internet of Things, an underlying OS is even more important and even more likely to emerge.

There are many "utility" type services that might also prove to be important, such as control mechanisms for granting access to data. I might want to control who can access my braking data from my car, which means that security, authorization, authentication become important. These enablers are shown as the boxes with question marks in Figure 3.5, not because we necessarily can't define them, but because these are infrastructural components in the Web OS that are yet to emerge in terms of clear winners, commercially or technologically (including standards). If there are no clear winners, then this means that the opportunity still exists for telcos to occupy strategically important positions in the emergent battles to define and dominate the Web OS!

3.2.1 Summary

We've had Web 2.0, which has moved us from a Web of linked pages to a Web of linked data, of which a fraction gets turned directly into pages. The data is generated by all kinds of objects and represents various things in the world, both concrete and abstract. The growing complexity of a Web of linked data requires management and organization, which leads us even further towards the inevitability of a Web OS. Some thought-leaders are now talking about the Internet of Things as the next Web, or Web 3.0. But there are other perspectives, which I shall explore in the following sections. Please read on.

3.3 Making Sense of Data

3.3.1 Data Semantics

Returning to our example of linked data from the previous section, let's look at the issue of the various "things" that represent Paul Golding. We had a link to my Facebook account. We also had a link to my car and a link to my telco account. Conceivably, there might be lots of links to things that are related to me – Paul Golding, see Figure 3.6:

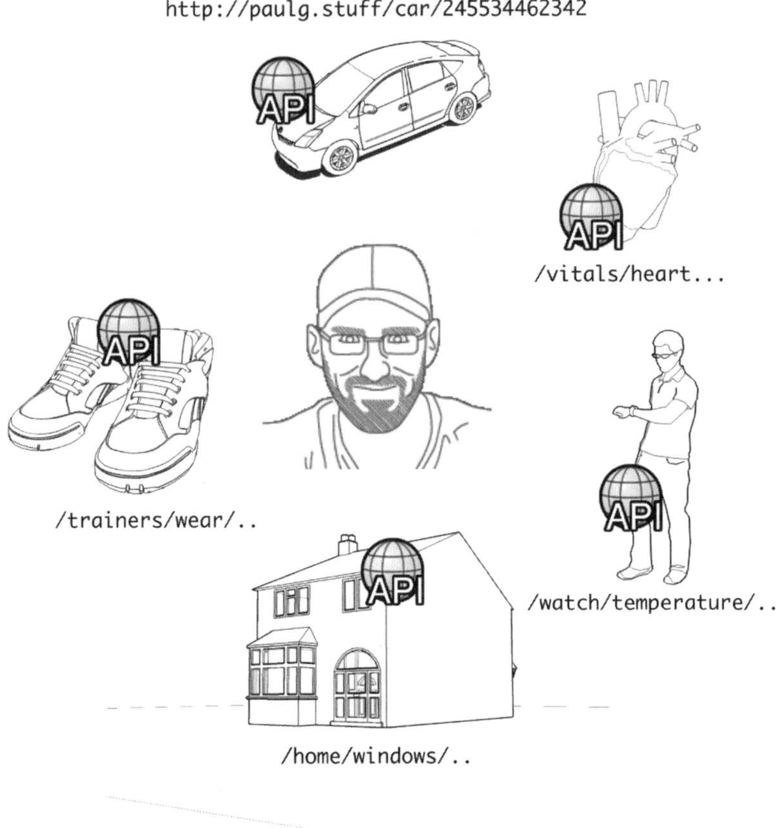

Figure 3.6 Paul Golding's linked data.

We said that all this data is accessible by links, which might look like:

```
http://someservice.com/app/ID
```

This is just an address. Looked at blindly by an application, it has no idea what to expect at the end of the link. It might be something like:


```
{
"id": "559089368",
"name": "Paul Golding",
"first_name": "Paul",
"last_name": "Golding",
"link": "http://www.telco.tel/paul.a.golding"
}
```


But then it might be something like:


```
{
"id": "8736918309",
"node": "Paul Golding",
"order": "WQ18329749",
"package": "Gecko",
"status" : "pending"
}
```


Or...


```
{
  "data": [
  {
  "id": 57,
  "title": "Woodsman, The (2004)"
  },
  {
  "id": 59,
  "title": "Wild Things (1998)"
  }
  // tons of movies here ...]
}
```


I'm pretty sure that you might recognize the first data set. It's from the last section and appears to be something related to my telco account, including where you might find my details (perhaps an hCard). What on Earth is the second set of data? It's hard to tell, although it has some words that look familiar, such as "order" and "package." Maybe this is something to do with an online ordering system. The {"status" : "pending"} field might tell us where my order has got to.

But look closer. What's the field "node" all about? The field name is ambiguous, but the data isn't, at least for our context. It's my name and perhaps this order has something to do with me. But then how do we know that the name is mine and not some other Paul Golding?

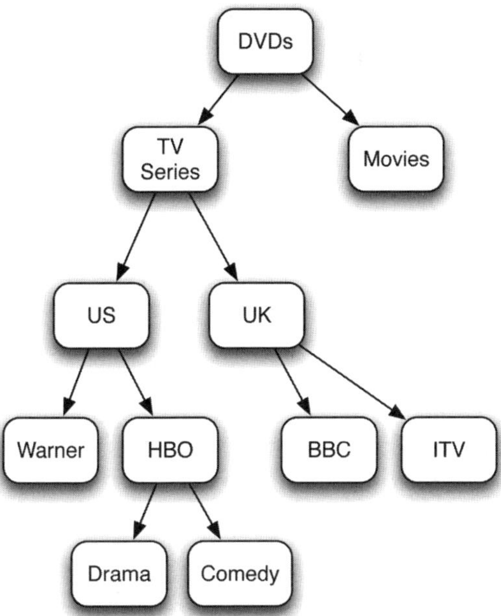

Figure 3.7 Tree structure of movie database via API.

What's the final set of data about? It seems to be about movies. It might be data that I've extracted from the IMDB[10] website via a link like:

```
imdb:/path/to/data/in/graph
```

The reason the path is to a graph is that we can well imagine that the movies we are querying are connected together by a network of relationships, see Figure 3.7.

Just looking at the data, it isn't clear what the fields are, as they just say "id" and "title." We happen to know that they are related to movies, but a field name of "title" could apply to all kinds of things.

What I am exploring here is the meaning of data, or what Web geeks called *semantics*. There have been various attempts to characterize Web 3.0 as the Semantic Web, a web in which applications can make sense of the linked data through the lens of annotations about the data, or what we call *meta-data*. I'm not going to elaborate here on the definition of Semantic Web, as there probably isn't a single definition that's useful, but the concept of meta-data and semantics is very much going to be part of the emerging Web Operating System.

3.3.2 Data Relationships

As shown in Figure 3.8, in the world of Web Operating Systems, the platform upon which applications run is very widely distributed across a number of loosely connected ventures and

[10] Internet Movie Database – http://imdb.com

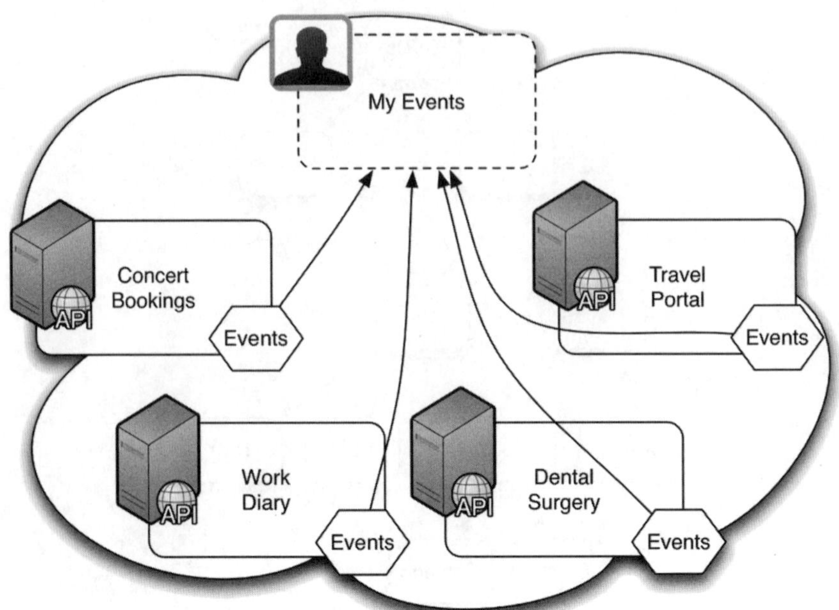

Figure 3.8 Event data smeared across the Web.

domains. It is also a Web in which there is often no single definitive source of data. Calendars and contacts are good examples. It is perfectly reasonable and common to have a number of services, each with their own calendar and contact management facilities. I have friends in LinkedIn, in Facebook, in Gmail, in various work-related project sites, in photo-sharing apps and on my mobile phones. I have event data in Facebook, in EventBrite, on my iPhone and all over the place, as shown in Figure 3.8. Data is very often smeared in this fashion, which is why the notion of "owning" the address book isn't necessarily the right way to think about the opportunity.

With smeared data, there is the matter of how to connect the data back up again for consumption by services that need a view across the whole data set. For example, is it enough just to scour links for any data that has "Paul Golding" and some kind of date entry in order to find out what I'm doing? Well, firstly we have to disambiguate (big Web-geek word) "Paul Golding" to ensure that we have the one single Paul Golding intended – that is, me. I wouldn't want to mix my data with someone else's, even with a cool name like that.

There are two challenges to joining the data. Firstly, the disambiguation part mentioned above. We need to have a mechanism for ensuring that certain types of data are authoritative, or the data we want. Secondly, we need to know something about the relationships, as shown in Figure 3.9. How do the various pieces of data labelled "Paul Golding" join up? It's possible that they don't join directly and that we have to go through intermediate steps. A classic example of this is the Friend-of-a-Friend (FOAF) means of describing personal relationships. This is a standard for describing social graphs. Using FOAF, it might be possible to connect "Paul Golding" with a FOAF who happens to have a solution to my problem, or who has a common interest.

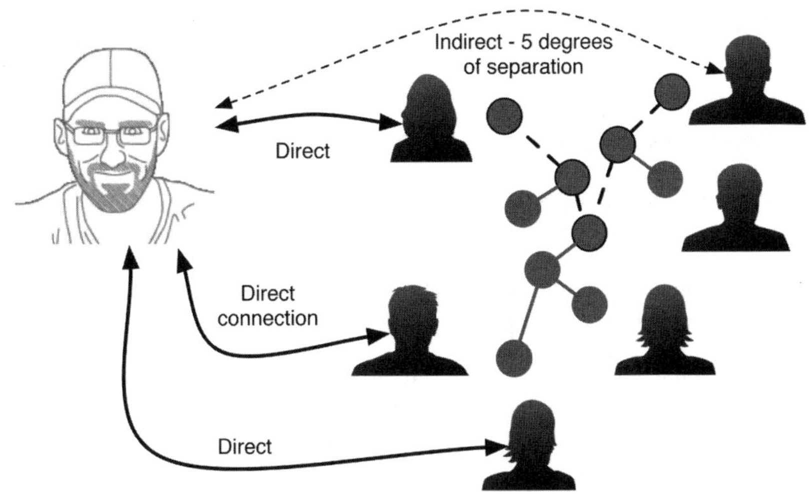

Figure 3.9 Degrees of separation across the social graph.

3.3.3 Meta-Data Tools: Ontologies, OWL, RDF

In our search for meta-data, we face the legacy problem – there is lots of data that isn't labelled and isn't about to be labelled any day soon. In looking for explicitly labelled data, we might have to wait a while. Meanwhile, for new data, which we're producing plenty of all the time, especially online, we have various ways of labelling the data.

I'll start with RDF, as this is the most difficult to understand and also the most formal in terms of various attempts to describe data on the Web. RDF stands for Resource Description Framework. In other words, it's a way to describe resources, which is the formal name for linked-data on the Web, noting that links are more properly called Uniform *Resource* Indicators – URI). We can think of RDF as a means to describe linked data.

In computer science, ontology is a formal representation of the knowledge about a set of concepts within a domain[11] and the relationships between those concepts. RDF allows ontologies to be constructed about linked data. It uses the same idea that we use with our descriptions in English. It might have been a while since you last thought about sentence structure, but sentences contain a minimum of two parts: a subject and a predicate.

Examples:

1. This field is a contact.
2. Paul@wirelesswanders.com is an email address.
3. Paul Golding is a person.
4. Jeff Daniels is a friend.

Item 3 seems obvious, but that's because we can recognize the words "Paul Golding" as a name of a person. To a computer, the words "Paul Golding" could easily be a brand, or the

[11] Here we mean domain more generally, as in an area of knowledge, not a domain name like www.domain.com

name of a breakfast cereal.[12] For even greater descriptive power, we can add an object to our statements:

1. This field is a contact of Paul Golding.
2. Paul@wirelesswanders.com is an email address of Paul Golding.

You can see that the predicate is useful for expressing relationships, enabling us to link the subject to the object. Not surprisingly, using a link (URI) is one way to express the connection:

1. This field is a contact of http://paulgolding/person
2. This field is an address of type http://hcard.com/rdf

The first statement enables the software to follow the link to a definition of Paul Golding as a person, whatever that means, but let's assume that there's some kind of standard way of describing a "person" and that it sits, for Paul Golding, at the end of the link. The second statement shows a link to an RDF file that explains what an hCard is. The idea with RDF is that any of the subject, predicate or object can be a URI. Here's another example:

Statement: "The author of http://wirelesswanders.com is Paul Golding".

The subject of the statement above is: http://wirelesswanders.com The predicate is: author The object is: Paul Golding

As you might expect, computers don't handle English too well. It's not a very well structured language, too full of ambiguities. Computers prefer unambiguous and highly structured data, which is why we turn to something like XML[13] to express statements in RDF:

```
<?xml version="1.0"?>
<RDF>
 <Description about="http://www.wirelesswanders.com">
  <Author>Paul Golding</author>
  <homepage>http://www.w3schools.com</homepage>
 </Description>
</RDF>
```

This format[14] expresses the statements in a structured and unambiguous way that a computer can read. RDF is also a serialization ("way of writing") of the Web Ontology Language, otherwise known as OWL,[15] which is a World Wide Web Consortium (W3C) standard.

OWL comes from a more concerted effort to enable ontologies to be described in a way that machines on the Web can better understand the meaning of data. Unlike RDF, which was somewhat designed for both humans and computers to read (per the intended advantage of XML), OWL is intended for consumption only by computers. It is a much more expressive language that overcomes some of the limitations of RDF.

[12] I'll let you know when I launch my own breakfast cereal brand ☺.

[13] For an in-depth explanation of XML, refer to section 8.3.1 of my book, *Next Generation Wireless Applications*, 2nd ed.

[14] Note that the sample shown here is somewhat simplified, omitting what we call namespaces. However, the idea is still the same.

[15] Yes, the acronym is out of sequence, which is somewhat ironic for a language aimed at accurate (i.e. ordered) expression of data.

XML and XML Schemas[16] are useful for passing data around the Web, but their lack of semantical expression prevents software from understanding new XML vocabularies automatically. For example, the same tag name can be used in different contexts with ambiguous results – software simply fails to cope. RDF and RDF Schema begin to approach this problem by allowing simple semantics to be associated with identifiers. However, RDF does not support enough semantic detail.

Taking the world of CRM as an example, telcos will know that the segmentation and characterization of customers isn't simple at the best of times. For example, it is common for a single person to purchase multiple phones and thereby appear to be some kind of "super user," whereas he or she has actually bought the phones for a family or small business. RDF will struggle to express the various semantics needed to define an ontology for CRM. This is where OWL becomes more useful.

OWL supports the W3C vision for a Semantic Web, which is a Web where software is able to navigate and consume linked data based on inferred and expressed meaning, quite separate to the human meaning that we are used to when reading Web pages. The same applies to linked data used in enterprise domains. We can well imagine how a variety of ontologies are required in a telco, covering CRM, operations, marketing functions etc. We end up with a Web of ontologies. What we don't want is to find that these ontologies become the equivalent of talking French, German and Chinese in the same organization. Actually, that's exactly how computers will see the world of disparate RDF schemas. OWL moves us a step closer to the possibility of automated translation between ontologies, so that, for example, a customer in the marketing IT systems can be meaningfully translated to a customer within the CRM systems.

Now imagine a world in which this internal Web of ontologies can connect with the external Web of ontologies sitting in blogs, social networks, Twitter and so forth. Whoa! The mind boggles, but the computer doesn't – it can suddenly make sense of these two worlds. It takes mash-ups to a whole new level of power and possibility.

3.3.4 Meta-Data Tools: Tagging and Folksonomies

RDF and OWL are all well and good, but they are the domain of computer programmers. We can hardly expect an ordinary user to start writing his or her emails and blog using RDF or OWL. Us ordinary folk have a hard enough time being understood by our contemporaries as it is, never mind by a Web crawler scrutinizing our prose.

But all hope is not lost, thanks to tagging. One of the first sites to popularize tagging was probably the photo-sharing site, Flickr. Users upload pictures and are then invited to tag them: "paul, beach, sun, California, family, kids." This free-form use of words to tag stuff has the advantage of being easy to use. Anyone can think of a tag, although very often it's an extra effort too far – never overestimate the willingness of the consumer to do more than the bare minimum to get where he or she wants to go. Other sites or Web applications that use tagging include:

1. Delicious – for tagging of bookmarks.
2. Wordpress – for (optionally) tagging blog posts.

[16] Which is a particular set of XML tag definitions.

3. Technoratti – for tagging of contributed content.
4. YouTube – for tagging uploaded video content.
5. Slideshare – for tagging uploaded slide shows.
6. Twitter (hash tags) – for tagging tweets.
7. Gmail – tagging of mail messages.

Tags are easy and tags are fun. The lack of rules works for and against the user. Without rules and structure, tags are easy to construct. If I want to tag: "beach," I can. If I want to tag: "gadget," I can. There are no rules. No one owns the definitions of tags. Folk use them how they like, which leads to the emergence of what Web geeks call "Folksonomies." An interesting mash-up of the words *folk* and *taxonomies*. We all know folk, but a taxonomy is simply a posh word for a bunch of terms used to classify stuff. That's exactly what we're doing with tags – creating classes of stuff. By labelling pictures with the "beach" tag, we end up with a class (or category) of pictures called "beach."

It is up to me how I use the tag "beach," although most of the time I, and most other folk, will tend to use the more obvious dictionary meaning. On Flickr, we would probably expect to see pictures of beaches if we searched with the tag "beach." And that's how tags are often used, to facilitate search, an important aspect of weaving the Web. Tags can improve search a lot, especially in areas where the meaning of content is particularly hard to unlock, such as binary-encoded images. Nothing in the bits and bytes of a beach JPEG say: "beach." This can only be inferred by recognizing patterns in the data, which is where we are headed. Meanwhile, tagging is used to give explicit meaning.

However, even with pattern detection, we are still stuck with a rather limited set of semantics. We might be able to detect a beach in an image, but where is it? Why was it taken? Who took it? What did it mean? Was it a holiday or a business trip? Is this a casual holiday snap or the work of a professional, available for sale? With these questions, you can begin to see how the field of semantics is quite challenging. Even with tags, it isn't obvious how we might answer all these relatively basic questions.

Folksonomies emerge in large communities where there's a lot of tagging going on. Over time, patterns in the tags will emerge, helping the taxonomy to take shape. In Twitter, the use of the #fail tag has become a convention for talking about any kind of service or product failure. However, tagging remains a weak form of ontological representation. Tags are too informal, too poorly structured and too ambiguous. When I use the tag "mobile," what does it mean? Does it mean that the picture contains an image of a mobile, or, as often the case, that it was taken using a mobile camera phone? Or, does it mean Mobile, the town in Alabama?

As with English words, tags can include the same tags with different meanings (homonyms) and multiple tags for the same idea (synonyms), which only adds to the ambiguity and poor resolution of meaning. This leads to inefficient searches, as easily demonstrated when looking for a particular image on Flickr. It is all too easy to spend a lot of time getting incorrect results due to ambiguity and a lack of any structure to guide the search. For example, if I'm searching for images relating to Orange, the telco, I probably want to narrow my search to photos relating to "mobile" or "business" and not "fruit." However, without any structure to declare Orange as a business in the first place, then we merely rely upon some judicious use of tagging to get where I want to go.

Even if I wanted to be unequivocal with the use of a tag, I couldn't. Unlike RDF and OWL, there is simply no means to formally define the meaning of a tag, at least not in most systems that support the simple free-form tagging in a tag field. If fields are named, allowing tags to be divided into classes, this might assist. But, one of the lessons of the consumer Web is that users are extremely reticent to expend any effort in adding information, or undertaking extra steps, just for some future gain, such as improved search or retrieval. There have been attempts to refine the tag structure slightly through the use of tag triples, which is a condensed form of subject:predicate:object expression, or think more of object:property:value. For example, "geo:long = 50.123456" is a representation of an object relating to geography (in this case the more narrow field of spatial geography), hence the "geo" namespace prepended to the tag. The "long" is then a longitude property, finally followed by its value "50.123456."

Tag triples are useful, but likely to remain niche. At the one end, basic tagging is likely to be increasingly adopted by users, as they become more familiar with the tagging habit and as tagging gets incorporated into more sites. At the other end of the spectrum, the fully blown RDF and OWL approach is likely to be increasingly adopted by developers to add even greater "platform power" to what they're building. Of course, both are possible. In fact, it is likely that tags can be incorporated into RDF serializations so that computers can make sense of tags, where it's possible to do so.

3.3.5 RDFa – Embedding Meta-Data Within Web Pages

There's a halfway house between tagging and RDF, at least for adding semantics to Web pages. It is called RDFa and is aimed at giving Web page developers a means to tag the HTML[17] mark-up used to generate Web pages in browsers.

HTML mark-up allows a Web developer to control how a page is structured and displayed by the browser. For example, when displaying my name on a blog article, I might code this using a couple of title tags, something like:

```
<h2>Why tagging is important</h2>
<h3>By Paul Golding</h3>
```

Another Web application (e.g. Search engine) consuming this page will only see the <h2> and <h3> tags, unable to glean anything about the content. You and I can tell, because of our well-trained minds, that this data represents the title and author of the blog post. So, without making life difficult for the developer, insisting on a whole new approach to Web page construction, we can simply extend the HTML tags, like so:

```
<h2 property="title">Why tagging is important</h2>
<h3 property="author">By Paul Golding</h3>
```

This scheme only works with the latest version of HTML, called XHTML, which means eXtensible, allowing us to pop in these 'property' extensions. With this approach, the browser can ignore (or not) the extensions and carry on displaying the header tags as usual. At the same time, software that recognizes the extensions can process them. For example, a browser

[17] HTML mark-up is explained extensively in Chapter 8 in my previous book *Next Generation Wireless Applications*, 2nd ed.

extension programmed to look out for 'author' tags might enable the user to look up the author's works in Amazon or an online library.

You might wonder how a software program is going to know what these extensions mean. After all, a 'title' could mean the deed to a piece of land, or an appellation like 'Mr' and 'Mrs.' Recall that our RDF schema allows a more detailed semantic field to be constructed, with OWL taking it a level further, if we wish. If we use RDF to construct fully a set of semantics for the domain we are working with, we can host this RDF file at some URI and then point to it as the schema to use for our property extensions. This is done using a simple extension to the code to define a namespace thus:

```
<div xmlns:dc="http://purl.org/dc/elements/1.1/">
   <h2 property="dc:title">Why tagging is important</h2>
   <h3 property="dc:creator">Paul Golding</h3>
   ...
</div>
```

That initial line of code sets up a moniker – 'dc' – that when used in front of any of our extensions – 'dc:title' – tells the software where to look for a definition of the 'title' property, which can be found in this example at http://purl.org/dc/elements/1.1/

3.3.6 Meta-Data Tools: Twitter and Annotations "Twannotations"

Twitter Annotations ("Twannotations") is an interesting example of one approach to the problem of semantics. It is useful to explore because it deals with a world where semantical expression is already extremely limited due to the terse nature of tweets – there isn't that much room in a tweet to say anything at all, never mind explain what it means.

The background to Twannotations is the emergence of Twitter as a kind of "messaging bus" for applications on the Web, as shown in Figure 3.10. People don't just tweet from a Twitter client anymore, they tweet from inside other apps, such as Spotify, the music streaming app, which allows music track info and playlist info to be tweeted out into the user's Twitter timeline. This app-messaging theme is very much part of Twitter's apparent strategic aims – to become *the* way to share real-time messages about anything and between anything, thereby occupying a potentially significant role in the Internet of Things. Of course, Twitter isn't an open standard or platform, as such, not like SMTP-mail and other Internet building blocks. Should a single system owned by a single company become a major plank in the future Web? That's another question, and one that I shall return to in later chapters.

With an increasingly varied number of uses for tweets, such as sharing of music interests and activities, it becomes more useful to identify what a tweet is about. For example, if it were possible for software to recognize certain tweets as being related to music, or film, or whatever, this would be interesting to the vast number of applications that consume tweets via the Twitter API. The possibilities are potentially endless and, as ever, left to the Twitter developer community to invent. However, a few ideas come to mind, just to give a flavour for the potential of Twannotations:

1. This [tweet] is about a *book* – such a tweet might be good for Amazon tweet-outs, enabling book fans to filter for book news, reviews, offers etc.

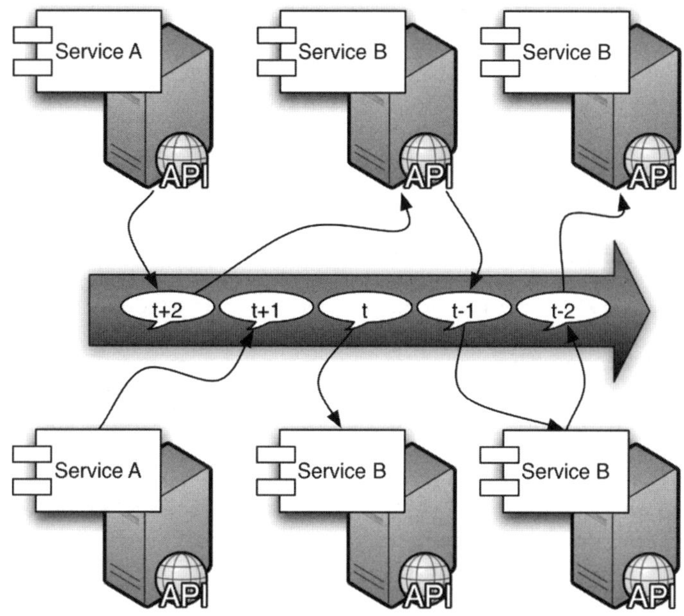

Figure 3.10 Twitter as a universal messaging bus for Web apps.

2. This is about a *product* – good for all kinds of services.
3. This is about a *stock* – "Buy! Sell! – go mad!"
4. This is a product or service *offer* – whoa! All kinds of interesting uses here – expect lots of innovation.
5. This is about a *song* – so many great use cases around the song theme!

You're probably wondering where all of these annotations are going to fit, given the 140 character limit. Well, here's where I can share an interesting fact, unknown by most Twitter users: tweets are much bigger than 140 characters! Much bigger! Here's what a tweet really looks like (in XML):

```
<status>
  <created_at>Tue Mar 16 17:31:46 +0000 2010</created_at>
  <id>10579490995</id>
  <text>[oauth-dancer] http://bit.ly/aCXf46 episod - Debug
more with GhostTrap</text>
  <source>&lt;a href = &quot;http://github.com&quot;
rel = &quot;nofollow&quot;&gt;GitHub&lt;/a&gt;</source>
  <truncated>false</truncated>
  <in_reply_to_status_id></in_reply_to_status_id>
  <in_reply_to_user_id></in_reply_to_user_id>
  <favorited>false</favorited>
  <in_reply_to_screen_name></in_reply_to_screen_name>
  <user>
```

```
    <id>119476949</id>
    <name>OAuth Dancer</name>
    <screen_name>oauth_dancer</screen_name>
    <location>San Francisco, CA</location>
    <description></description>
<profile_image_url>http://a3.twimg.com/profile_images/730275
945/oauth-dancer_normal.jpg</profile_image_url>
    <url>http://bit.ly/oauth-dancer</url>
    <protected>false</protected>
    <followers_count>7</followers_count>
<profile_background_color>C0DEED</profile_background_color>
    <profile_text_color>333333</profile_text_color>
    <profile_link_color>0084B4</profile_link_color>
<profile_sidebar_fill_color>DDEEF6</profile_sidebar_fill_color>
<profile_sidebar_border_color>C0DEED</profile_sidebar_border
_color>
    <friends_count>3</friends_count>
    <created_at>Wed Mar 03 19:37:35 +0000 2010</created_at>
    <favourites_count>0</favourites_count>
    <utc_offset></utc_offset>
    <time_zone></time_zone>
<profile_background_image_url>http://a3.twimg.com/profile_ba
ckground_images/80151733/oauth-
dance.png</profile_background_image_url>
    <profile_background_tile>true</profile_background_tile>
    <notifications>false</notifications>
    <geo_enabled>false</geo_enabled>
    <verified>false</verified>
    <following>true</following>
    <statuses_count>16</statuses_count>
    <lang>en</lang>
    <contributors_enabled>false</contributors_enabled>
    </user>
    <geo/>
    <coordinates/>
    <place/>
    <contributors/>
</status>'
```


The bit that starts with <text> and ends with </text> is the actual body of the tweet. The rest is the entire data structure of the status update (official name for a tweet), which is a lot of data for 140 characters.[18] I'm not going to elaborate on the data fields here, except to comment that, unlike text messaging in a mobile network, more fields can be added. What's more, adding extra fields shouldn't adversely affect any of the many applications that consume tweets, so long as they're well designed. That's how come annotations can be added

[18] Actually, you might be surprised to learn that a humble text message in a mobile network is also a lot bigger than 160 (or 140) characters. As with a tweet, most of the extra data is hidden from the users.

in to the data structure so easily. They can be ignored by existing applications and used by "annotation-aware" applications, which will presumably become more prevalent over time.

The annotations part of the data structure would look like this (in XML):

```
<annotations type="array">
  <annotation>
   <type>website</type>
   <attributes>
    <attribute>
     <name>URL</name>
     <Value>http://wirelesswanders.com</value>
    </attribute>
   </attributes>
   <type>website</type>
   <attributes>
    <attribute>
     <name>emotion</name>
     <value>happy</value>
    </attribute>
    <attribute>
     <name>doing</name>
     <value>work</value>
    </attribute>
   </attributes>
  </annotation>
</annotations>
```

Which we can show more succinctly in JSON format:

```
[{"website":{"url":"http://wirelesswanders.com"}},{"mood":{
"emotion":"happy","doing":"work"}}]
```

The annotations take the form:

```
"annotations": [{"type":{"another_attribute":"value",
"attribute":"value"}},
{"another_type":{"another_attribute":"value",
"attribute":"value"}}]
```

We can have as many types as we want and as many attributes as we want within each type. You're probably wondering what's a type and what's an attribute? Well, there are no rules here – anything goes. The idea is just to provide the structure and then let the developers and hackers create the meaning, hoping that certain trends might emerge for the use of annotations. Twitter suggested a number of recommended types out of the box:

1. webpage
2. review
3. song

4. movie
5. tvshow
6. book
7. product
8. stock
9. offer
10. topic
11. event
12. Place

Here's a table showing an example of how Twitter recommends using the Place type:

Attribute	Description
Title	Name of the venue
URL	Permalink to place
ID	ID from source provider
Provider	Unique string for disambiguation of ID systems – for example, "FourSquare"
Place_Type	One of "POI", "neighbourhood", "city"
Street_address	"795 Folsom Street"
Locality	"San Francisco"
Region	"CA"
Postal_code	94130
Country_code	US
latitude	"37.78212"
longitude	"-122.40096"

There is no doubt that Twannotations will take Twitter innovation to a whole new level. Hackers will find myriad uses for annotations, further cementing Twitter as a potentially default platform for connecting the Internet of Things in real time.

3.3.7 *Summary*

The vision of the Semantic Web, a world in which all the data on the Web is usefully annotated, is some way off yet, but rapidly approaching. Telcos need to recognize and act quickly, before their rightful place in the ecosystem of consumer and business related data is lost to other online services, like Facebook and Salesforce.com (or its platform cousin, force.com). However, this requires a change in thinking. Operators need to think much more aggressively about data because the future belongs to data!

3.4 Future Web: "People OS?"

3.4.1 *Introduction*

In the previous sections, I've talked a lot about the evolution of the Web from linked documents to linked data to linked things. But what about people? Not only have social networks become huge platforms in their own right, in some cases growing into fully bloomed ecosystems, but they have caused many *other* types of site, nothing to do with social networks, to become socially enabled or "socially aware." This has been made possible by linking to people via social APIs and similar mechanisms.

Social networks have exerted their influence on the very structure of the Web – their effect is felt everywhere. For example, it has become commonplace when engaging with an online service, such as purchasing a product, to share the experience with friends via Twitter or Facebook, or some other social site. Our actions all over the Web are increasingly triggering a variety of social interactions, which in turn trigger other actions on the Web (e.g. my friend buys the same DVD as me). There are numerous websites that collect or aggregate social interactions between users and can then output a social graph via an API. This mechanism is enabling other sites, and the Web generally, to become social, as shown in Figure 3.11.

The trend is for social features, such as social sharing, to become increasingly incorporated into sites, fast becoming as common as "file saving" is in a desktop application. As we know, in the desktop world, these common features are usually provided by the operating system. In the case of the Web, many of these features are provided by third party platforms, such as Facebook and Twitter, which are acting like an underlying operating system. Yet again, we are

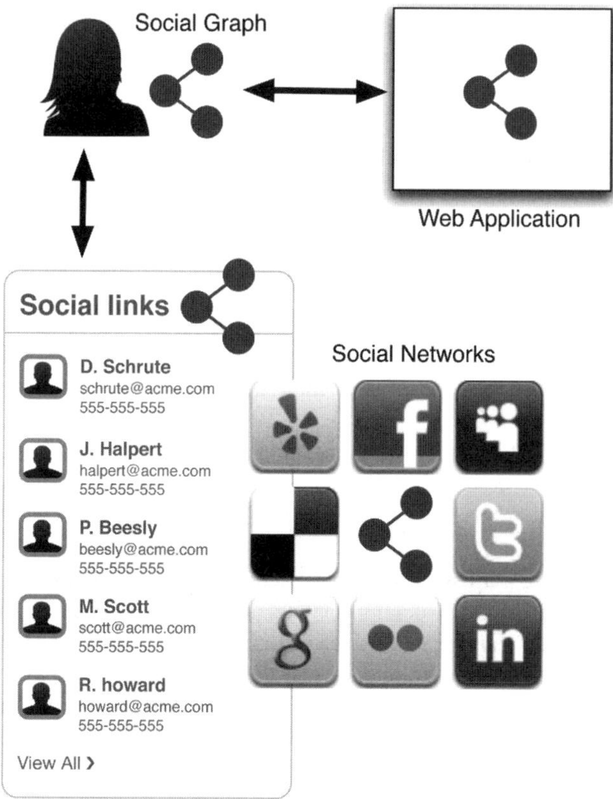

Figure 3.11 Social graphs enable the social web.

forced to consider the idea of the Web Operating System, but this time exploring the question as to how far the Web operating system is actually a People Operating System – "People OS" – which would be a relatively new development in the evolution of operating systems and should be of great interest to telcos.

There is a far more interesting question, which is *where* will the true power of a "People OS" emerge? Surely it is likely to be on the mobile because it is the ultimate in personalized and social experiences. As an example of this trend, observe how the embedded mobile address book is slowly evolving towards a socially enabled address book where the information stored about each of our friends is a combination of local data (the usual name and number couplets that most of us will enter) augmented with remote data (e.g. pictures and status updates) from a social network, such as Facebook. I have long argued that the UI metaphors used for most mobile devices are strangely inappropriate, even on the iPhone with its icons that represent apps and digital stuff, but not people. But this is changing. Some of the new Android smartphones have been released with customized UI layers that blend the online socially networked world with the phone's features, such as Motorola's MotoBlur on the Droid.

Whether the social features of the Web converge as a collection of common features and enablers, underpinned by Web standards, it is perhaps too early to say, but the trend is emerging. Let's now explore the various components of the converged mobile and Web platforms where social computing is coming to the fore.

3.4.2 Social Networks

The foundation of the "People OS" is the social network (SN) in its many forms. Here we shall look at the leading SNs and explore their characteristics from an infrastructure point of view, trying to identify in what ways they enable other services. Any exploration of Web operating systems in the SN domain has to start with the 100-pound gorilla, which is Facebook. These are the features of Facebook, relevant to the People OS notion:

1. **Networking** – In the first instance, Facebook provides the underlying social linking foundation – the ability to create graphs of friends. This is its original social network function and has dramatic network effect. Telcos have the same graph-creation capability, but it remains implicit and not externalized by any APIs or platform models.
2. **Facebook Connect (FBC)** – Provides a mechanism for allowing its users to log in to other sites. When a user logs into another (non-Facebook) site using FBC, that site is able to pull the user's social graph information into that site (implicitly permitted by the user's choice to log in using FBC). In other words, when I log in to a service using FBC, I'm saying "Let me in please, and here's a list of my other friends so that you know who I'm connected with." This list of friends – the social graph – is accessed via an API on the Facebook site. While logged in, users can do more than just connect with friends via associated APIs, they can also post information and updates to their Facebook profile. Developers can use these services to help their users connect and share with their Facebook friends on and off Facebook and increase engagement for their website or application.

A site that enables users to authenticate using their FBC account is typically more likely to get subscribers to join than if they have to set up an entirely new login ID. Some new ventures don't have any built-in authentication, asking users to log in only via FBC or some other external authentication provider (e.g. Twitter or OpenID). Of course, telcos should be concerned that their customers (many of whom will be using Facebook) are placing increased trust in Facebook to act as their primary login ID for a range of sites. A telco might think that they too can be a trusted ID provider. This is not a new idea. The OMA long ago set about trying to establish its Liberty Alliance standard exactly for this purpose. Thus far, it has failed to be implemented. Most telcos have not ventured into the ID management world, even though they have a very sophisticated infrastructure already in place to do this via SIM cards, which is how users "login" to the cellular network.

1. **Facebook Profiles** – Provides a mechanism for a user to maintain his or her own "webpage" – we can very much think of Facebook as a place where consumers, and increasingly businesses, maintain their primary web presence. This is important in foundational terms because for a People OS to be useful, it ought to have a collection of data about the people it supports. Providing people the tools to generate personal content will only add weight to the platform, increasing its social value. Telcos have not managed to create any meaningful Web destinations for their users, especially as a place to create content.
2. **Facebook Apps** – Provides a mechanism for other services to plug into the Facebook platform, which is via direct hosting in the Facebook UI. Many games are provided in this fashion. Once plugged into the platform, these applications can be shared along the user's social graph, enabling social-sharing of the experience. In other words, if I want to build an online casual game that can be shared amongst users, then I can either go launch the game and hope that users invite each other to the experience, or I can plug into Facebook and use the existing social graph infrastructure to turn my game into a social game. Not only is the infrastructure provided, but the users too! Again, there is no reason why telcos couldn't provide a social-sharing mechanism. After all, texting already plays this role, but lacks any platform features.
3. **Facebook Credits** – Provides a means to charge users for the use of services that hook into the platform. This is provided by Facebook Credits. Facebook Credits is a virtual currency that users can purchase and then use to buy gifts and virtual goods in games and applications on Facebook that support the currency. Again, there is a low-friction appeal here – it is easier to plug my game into Facebook and use its currency infrastructure rather than build my own. Yet again, telcos have the ideal ingredients for currency exchange, but not in any platform version.
4. **Social Plugins** – In his book "What Would Google Do?" Jeff Jarvis talks fondly about "Google Juice," which is his own slang for talking about the way Google can spread its services way beyond its own sites, mostly in the form of ads and mashable content liberally sprinkled across the Web. Of course, lots of other sites have cottoned on to the power of spreading their wares beyond their primary sites. Facebook is no exception. It spreads its juice using Social Plugins to add Facebook sharing features into the host site. These are as follows:
   A. *Like Button*: The Like button lets users share pages from another site back to their Facebook profile with one click, basically saying "I like this . . ." about the page containing the Like button.
   B. *Recommendations*: The Recommendations plugin gives users personalized suggestions for pages on the hosting site they might like.

C. *Login Button*: The Login Button shows profile pictures of the user's friends who have already signed up for your site in addition to a login button.
D. *Comments plugin*: The Comments plugin lets users comment on any piece of content on the host site.
E. *Activity Feed*: The Activity Feed plugin shows users what their friends are doing on the host site through likes and comments.
F. *Like Box*: The Like Box enables users to like another user's Facebook page and view its stream directly from the host website.
G. *Facepile*: The Facepile plugin shows profile pictures of the user's friends who have already signed up for the host site.
H. *Live Stream*: The Live Stream plugin lets the host site's users share activity and comments in real-time as they interact during a live event.

All of the above features, plus the extensive Facebook APIs generally, combine to create an environment in which services and applications are focused entirely upon people and their social interactions. I have not even mentioned the actual features of the site, such as fan pages, events calendars and so forth, that add up to a very rich environment for socially-enabled services. If ever there were an existing platform that we could think of as an archetypal operating system for socially-aware people-centric applications, then Facebook is surely it.

If we are to contemplate the hypothesis that the Web will increasingly become a series of rich platforms that enable other "meta" services to run atop, then we must surely think of the People OS as a significant component. In this sense, Facebook is a powerful contender for not only the first real sign of an emergent People OS, but the first real attempt at a Web OS. After all, if the platform can succeed in attracting a billion users, say, then it becomes a compelling place to offer socially-aware services by virtue of its vast network effect. The existence of a rich set of platform services, APIs and hooks, only makes it easier for other services to ride atop in a symbiotic fashion.

Telcos need to be concerned about this development because it is not based on open standards and not easily fended off. In the same way that telcos benefit from their oligopoly advantage of the license to transmit, Facebook, and other emergent Web OSes, benefit from the "unfair" advantage of vast network effect and a land grab (or mindshare) of empire-building proportions.

Facebook had been dismissed early on as just another "fad," not unlike Bebo and other social networks, many of which have waned. We all remember Friendster, Friends Re-United and other social-network originators that looked interesting and on the ascendent, only to fade in the shadow of Facebook. Facebook has succeeded however in holding on to its appeal, growing its user base and fans. We must surely ask ourselves why, if we want to learn anything from its goliath achievements. The answer is often hidden from the uninformed observer who remains in wonder of the apparently fickle and unpredictable nature of the Web, at least by his or her reckoning. They still view *all* Web ventures as somehow the same – a bunch of dorm-bound geeks in t-shirts trying stuff out for fun, some of which happens to reach the mysterious Gladwellian tipping point that belies rational, predictable and repeatable explanation. This is not *real* business – this is VC-backed gambling that went ballistic. This generalist view betrays the fact that some of these "dot com" ventures happen to be powerhouses of technological creativity, innovation and talent. Facebook is one of these, as is Twitter. Their ability to continue to scale, provide great performance whilst all the time adding more and more features, allowing vast amounts of platform services to thrive, is the "secret" of their ongoing success.

Facebook engineers don't merely gesticulate with pens at the whiteboard and wonder if, say, instant messaging would be a useful adjunct, as telcos did for years. They simply go build it – and at vast scale, operational efficiency and speed (ironically using telco technology, called Erlang[19]). It is true service creation at the speed of thought, something that escapes telcos because they simply aren't technology companies. They may have large technology departments and directorates, full of "technical" people, but not "technology creators." The difference is vast and this particular chasm cannot be crossed by telcos short of massive acquisition of the talent via a merger or buyout. And I am not talking just about "technology creators," when I talk of the Web Goliaths, I am talking of "supreme technology creators" at every level of the business: operational and development, never mind research. This is raw and refined talent seated in an environment ideally tuned to get the best from its people. These are "Microsoft-at-peak" companies, not "pet-shop-dot-com" companies.

3.4.3 Social APIs and Platform Thinking (Again)

I have talked about Facebook at some length, spelling out why we might think of its platform as a People OS and a major, if not central, part of the emergent Web OS that we are exploring in this chapter. But Facebook is not the only kid on the block. There are other elements of Web that might sit within the category of People OS, which we are viewing here as any Web-based platform that relegates many common features of social and people-centric services to underlying horizontal software infrastructure. Wherever there are lots of users in a single community, such as Google users, Yahoo users and so on, we have to wonder how they might provide a similar People OS function, if not now, then in the future, and if not entirely, then at least partially – we cannot ignore the notion of a distributed OS that straddles a number of common services, which could happen if we had common and open standards. Within the world of large user communities, we cannot ignore telcos either. Locked away in the silos of OSS and BSS infrastructure lies vast swathes of user data and activity, still very much at the centre of our connected lives.

Wherever a digitally enabled community exists, there's always the potential to turn it inside out and think of the community as a platform. This is a starting point for any People OS – the people. Unfortunately for telcos, they have often missed this opportunity. They have been sitting on platforms for years, spending billions to build and operate them, but unable to get any of that connectivity juice flowing into other ecosystems – there's no telco equivalent of Google juice. This is a well documented problem.[20] Telcos are mostly one-sided businesses, which means that their commercial in-flows come from downstream users – that is, their subscribers. There is little revenue flow from upstream users – that is, In the way that Google makes its principle revenues from advertisers. By contrast, a fashionable element of the Web 2.0 trends on the Web has been to always think of two groups of users from the outset: customers and

[19] Erlang is a language invented by Ericsson for building reliable software, such as telco switches. Facebook are reported to have used it to build their reliable messaging infrastructure.

[20] Writers at http://www.telco2.net/blog/ have been talking about this for a while, although their thinking was informed by various exponents of the two-sided business model, or platform model, as widely discussed in market theory.

developers. I have already described in detail how Twitter built a big operation by offering exceptional support and services to the developer community, creating vast synergies and network effects between the two sides of the platform.

For there to be a platform, upstream service providers need a means to empower their services using the host infrastructure. On the Web, this always means exposing an API. Let us turn then to examples of APIs that provide social functions, not unlike the various Facebook APIs just described.

3.4.4 Open Social API – A Cross-Platform People OS?

Google, not wanting to take a back seat to anything happening on the Web, launched the Open Social API. This was an attempt to make it easier to access social enablers via a common API. Noticing a variety of social networks (i.e. besides Facebook), they suggested that a common API to access them might be useful. This is not too hard to envisage because the types of function enabled by the Facebook Social Plugins, such as recommendations and "likes," must surely apply to any social network. Indeed, many of the features of the entire suite of Facebook APIs are potentially generic to any social network.

The benefit of other ventures besides Google adopting the Open Social API is the obvious network effect, which in this case is skewed heavily in favour of the developer who can build a social application that will work across a variety of social networks exposed via the Open Social API. There has been a gradual increase in platform providers adopting the API, such as:

1. *Consumer providers of Open Social API:* Bebo, Engage, Friendster, hi5, Hyves, imeem, mixi, MySpace, Ning, Orkut, Plaxo, Quepasa, Six Apart, Yahoo and others.
2. *Business providers of Open Social API:* LinkedIn, Tianji, Salesforce, Viadeo, Oracle and XING.

These users have formed a non-profit foundation – Opensocial.org – to manage the definition and evolution of the API, no longer an exclusive Google project. Clearly, Facebook is missing from the list of supporters and members. It is not hard to imagine that a great motivation for some of the members here is to act collectively with their "frenemies" in opposition to Facebook, providing a cross-platform alternative to the Facebook Platform. Unsurprisingly, Google led the charge here, concerned that Facebook is a major contender for the role of the ubiquitous Web OS, thus threatening to gorge a large slice of the advertising pie. Clearly, this is just a defensive move because Google is hardly able to control the advertising revenue from use of the Open Social API, as it is just a specification for an API. However, it is clearly a smart move if the overall network effect of the cross-platform API is such that it increases adoption of Google adwords (in any of the member sites and the social apps running atop) and diverts some advertising spend away from the Facebook Platform.

Advertising mechanics are a major interest on the social Web. Lots of useful data flows from a user's search and surfing habits, which drives the Google secret sauce in targeting users with ads. However, advertising effectiveness can be greatly enhanced by the inclusion of personal data, such as what a user likes and dislikes, which is exactly the type of information explicitly collected by the Facebook Platform – look again at those lists of Facebook Social Plugins. So, if your venture is actually a massive personalized ad machine, like Google, then

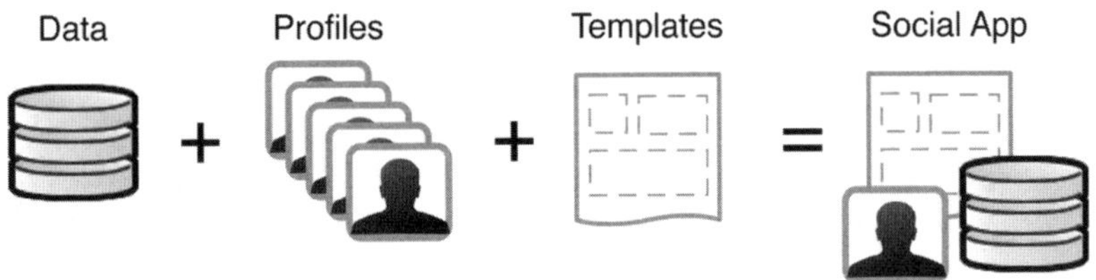

Figure 3.12 Open social – the ingredients.

social networks are vitally important. They, along with the apps that ride atop, present an ideal ecosystem for personalized adverts, which is something that telcos are slowly waking up to.

3.4.5 Open Social API – The Mechanics

OpenSocial enables a number of application architectures. These all stem from an underlying model, which says that a social application typically has the following elements: application data, social data, and a template, as shown in Figure 3.12. These are combined in various ways to eventually present a UI to the user based on data from the user's underlying social network or networks.

In the OpenSocial ecosystem, these components could sit in several places. One approach, similar to the Facebook Social Plugin model, is for the browser-based app to use JavaScript to render API-fetched data into a template, as shown in Figure 3.13.

Social Networks can also expose social data to server-Side applications, which can then take advantage of databases (application data) and server-side frameworks to produce socially-modified outputs without any intervention by the browser, as shown in figure 3.14.

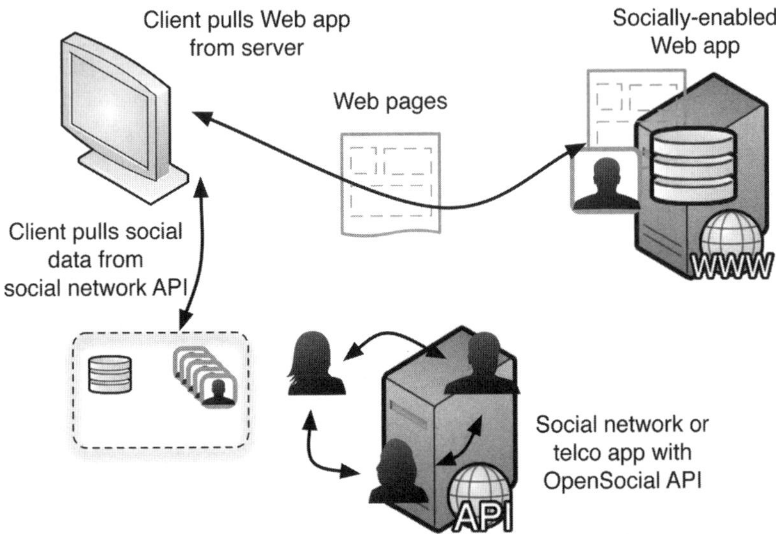

Figure 3.13 Browser pulling social data from OpenSocial API.

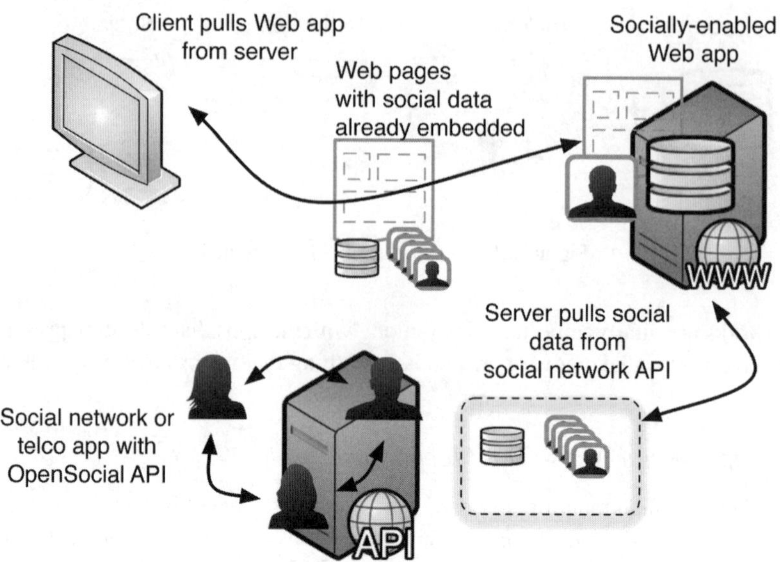

Figure 3.14 Server pulling social data from OpenSocial API.

The other model, as shown in Figure 3.15, which OpenSocial calls "Social Mashups," is an application that runs inside of a social network, just like Facebook Applications. The OpenSocial foundation suggests that social mashups will typically scale extremely well, but might be limited in terms of data storage and/or processing. A mashup is typically created using HTML, JavaScript, CSS and/or Flash – the usual browser programming candidates. OpenSocial mashups always run inside an OpenSocial container, which means any social network that supports the OpenSocial code libraries necessary to make OpenSocial apps work.

However, there is an additional approach called OpenSocial Templates. These are widgets, but the official term is "Gadgets," adopted from the Google Gadgets approach, which are the mini-apps that you might have seen contained inside iGoogle, if you've ever tried it. A gadget is not programmed using Javascript or HTML, but using a gadget-flavour of XML acting as a template language. An example gadget looks like the following:

```
<?xml version="1.0" encoding="UTF-8" ?>
 <Module>
  <ModulePrefs title="Hello World!">
   <Require feature="opensocial-0.8" />
  </ModulePrefs>
  <Content type="html">
  <![CDATA[
     Hello, world!
   ]]>
  </Content>
</Module>
```

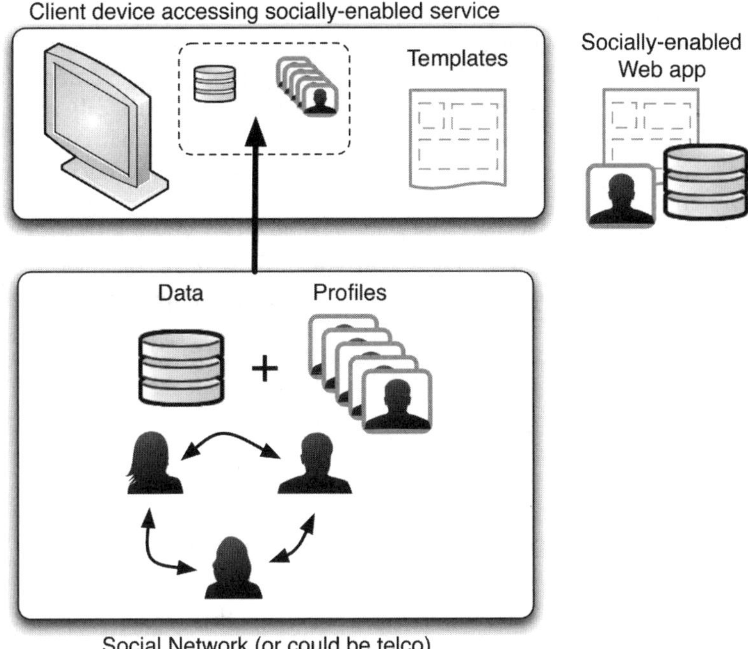

Figure 3.15 Open social mash-up pulling social info from 3rd party site.

This "Hello World" example shows the typical components and skeleton of a gadget:

`<Module>` indicates that this XML file contains a gadget.

`<ModulePrefs>` contains information about the gadget, and its author.

`<Require feature="opensocial-0.8" />` denotes a required feature of the gadget – in this case, the OpenSocial API (v0.8).

`<Content type="html">` indicates that the gadget's content type is HTML. This is the recommended content type for OpenSocial containers, but gadgets for other containers such as iGoogle support other content types.

`<![CDATA[...]]>` contains the bulk of the gadget, including all of the HTML, CSS, and JavaScript (or references to such files). The content of this section should be treated like the content of the `<body>` tag in a HTML page.

If you know HTML at all, then you've probably realized that tags like `<module>` don't exist in HTML, or Javascript. This is part of the gadget vocabulary. You are probably wondering where this snippet of code runs. It is part of a Web page, dished up to the browser. Of course, a browser doesn't understand gadget-speak, but that's where the OpenSocial Container comes in. It is the container that turns the gadget code into displayable HTML and CSS using an OpenSocial code library that could reside either server-side or client-side.

The theory is that a developer can create a gadget that will work across a variety of social networks because the OpenSocial container is always the same. Of course, reality seldom follows theoretical intent. As with all specifications, the execution is determined by the one following the spec. Not all of the gadget features are implemented in all containers. The containers themselves might vary in terms of layout – what displays well in one social network might look a bit squashed in another. Look-and-feel will change too – a font on one site might not work too well on another.

As with all cross-platform development, testing on the major target platforms is essential. A developer can't assume that a gadget will work equally well in all containers. However, the fact remains that the OpenSocial ecosystem is a cross-platform effort supported by a number of interesting players, major and minor.

3.4.6 Emergence of a Person OS at the UI Layer

Take a step back from the typical mobile user interface (UI) and you might notice something – lots of icons. Most of those icons represent things, like apps, messages, settings, browsers and so on. However, much of what we do with mobile devices is interact with people, not things, although this will evolve as we do indeed begin to interact with *real* things, like barcodes and objects in the camera viewfinder. Although there are lots of "design folk" in telcos[21] who talk about UI design and even User Experience (UX) design, they usually mean something more mundane, like branding and its design guidelines, but within existing design metaphors – lots of icons, so long as they look pretty.

What I'm pointing out is that people are remarkably missing from the UI of a device that is essentially people-centric. Apart from the usual contacts-related elements, like the address book, there is often little else to do with people. If 30 per cent, say, of my device usage time is spent interacting with a few close friends and family, then surely those people – "my people" – ought to be more prominent in the actual user experience, not the one that I superimpose on top of the UI. Surely "my people" should propagate throughout the entire applications plane, so that every app, where applicable, has the concept of "my people" built right in. For example, if I open my turn-by-turn navigator, then give me an option to navigate to a person amongst "my people," not just places. If I open an email application, then give me the option to view activity by person, not some relatively arbitrary last-in first-out queue of messages.

When I built the Zingo mobile portal[22] for Lucent Technologies, as far back as 1997, the entire UI was tabbed by [Home | Office | Travel]. If I went inside the "Home" tab, then the entire UI reconfigured itself to display only the people and information related to what I do at home or with people from my home, which included friends and family. When I switched to "Office," the apps switched to displaying only office-related data. In each mode, I had implemented a VIP list. These were the top 5–10 people in each of my home, work and travel

[21] The kind of folk who talk about this in telcos are seldom from a design background. They might have read a few books, but are otherwise re-purposed marketing folk. This shows how much they have failed to grasp the genuine importance of design.

[22] This was the first time that a mobile Web portal had ever been demonstrated. It was built using a Toshiba Libretto and wireless LAN card (per WiFi) to simulate a smartphone experience. Later on, we moved it to early Windows CE devices with wireless modem cards.

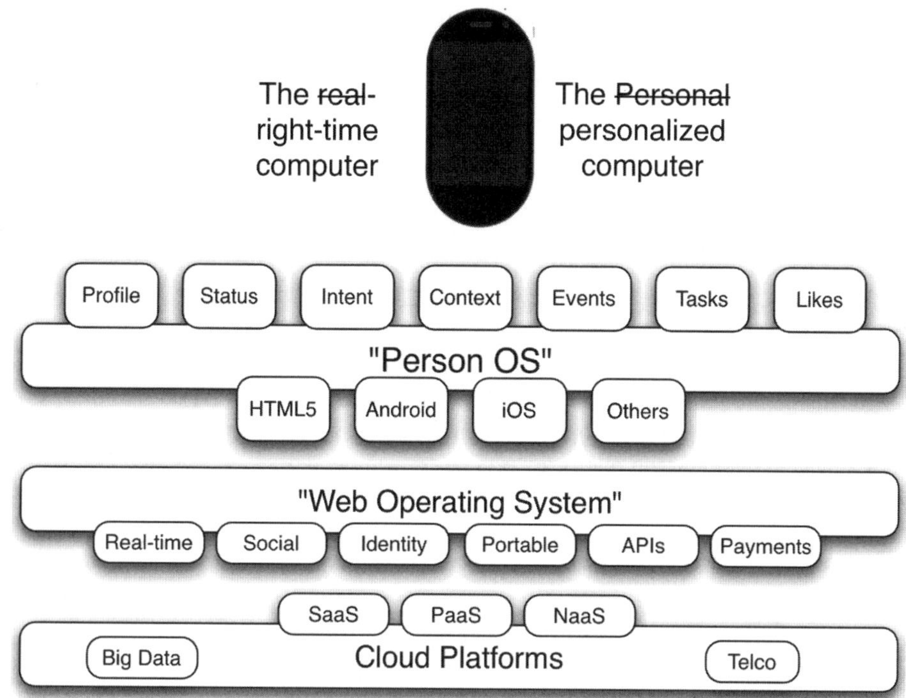

Figure 3.16 Personalized and right-time computer.

circles. Each application, such as the unified messaging app, displayed data related to the VIP list (which could be switched on/off in order to show all my contacts).

Of course, in recent times, driven by social networks, we have seen the emergence of devices that do indeed focus on people, or, more precisely, on social networks. INQ Mobile released the INQ device, which has an integrated address book feature that links with live data from Facebook, Windows Live Messenger and Skype. However, I wouldn't call this a "people-centric" device. It's more of a social-network, or Facebook-integrated device, which aims to give the best integrated Facebook experience on a mobile. Facebook for INQ is built right into the applications stack. Facebook friends show up in the INQ contacts list. Facebook messages, pokes and requests drop in the phone's inbox. I designed a similar applications stack for Extreme Mobile when they were thinking to launch a service in the UK, but pulled back from the venture. I took the idea a step further though, which was to enable "social sharing" in every application, wherever the idea of sharing made sense. Of course, this mechanism has since been widely implemented, such as the iPhone "share to" option (the arrow from the box icon common to many applications).

However, my vision for people-centric UI, as shown in Figure 3.16, is really part of a wider vision for people-centric computing generally, or what I call the transition from the Personal Computer to the Personalized Computer (or People Computer), as shown in the above diagram. The vision is for a wholly people-centric UX and UI to emerge that becomes the dominant way of thinking about computer interfaces, leaving behind (at long last) the "operating system" wars forever, along with their tired metaphors. Users don't care for Windows Mobile 7 versus iOS versus Android, even though these ideas continue to surface in consumer marketing chatter in some circles. What users really care about are people and apps. Everything else is irrelevant. I would put people centre-stage because, as I hope is becoming clear by now, we are moving (have moved) into an era of socially-enabled or socially-aware computing. Most apps will have a social element to them by default.[23]

The question now remains as to how this people-layer will work and where it will reside. As shown in the diagram, I envisage that the primary UX is strongly attached to a number of people-centric elements, like profile, status, intent, context (which would include the user's social-graph, context adjusted) which surface in the UI to a greater or lesser extent in each application space that the user enters. I talk here of application spaces rather than applications because I believe that the container for an application will also shift away from the confines of an executable program on the device to the wider space of a cloud-based platform, which might be Facebook say (should proprietary platforms win the day) or might be some kind of Open Social container (should platform "standards" win the day).

3.4.7 Privacy and Personas

You probably have the same doubts about all of this that I frequently encounter from telco associates – what about security? How will we keep our users safe in this people-centric world? How will we know who is who? Identity is core to the telco apparatus because it defines the end-point for charging and billing. Of course, the issue of charging in a mobile network is really about charging someone who will pay for the use of a particular number used on the network. This brings into focus the historical emphasis of an operator network, which has always been on the telephone numbers, or MSISDNs,[24] as they like to call them.

This emphasis has for too long been about numbers and connections, not people. This explains, at least in my view, why operators have failed so miserably to capitalize on the major asset of their network, which is the people network, or social network itself, rather than the physical network. However, the question of privacy remains and I will deal with it in the next section where I explore the opportunities for "Social Telcos."

3.5 Social Telcos and the Web OS

3.5.1 Where are the Telcos?

The principles of OpenSocial should be of relevance to telcos, who historically have an interest in cross-platform services approaches (or interchangeable standards), as pursued by the efforts of cross-industry bodies like the Open Mobile Alliance (OMA). Of course, when telcos talk of cross-platform, they typically mean between each other. They seldom think in terms of interoperability with Web platforms, whatever that might mean to a telco (as it seldom gets

[23] I have actually extended this model to more than just people, but I only want to focus on people for the current discussion.

[24] MSISDN is a number uniquely identifying a subscription in a mobile network. It is actually the telephone number of the SIM card used in a phone. This abbreviation has several interpretations, the most common one being "Mobile Subscriber Integrated Services Digital Network number".

considered in my experience). As an example, the OMA spent a long time working on a set of specifications for an inter-operator Instant Messaging service, thinking of it as an extended version of texting (indeed, early IM implementations were presented on the mobile as "longer text messaging.")

I was asked by the GSMA to conduct an analysis, comparing the OMA IM standard (called IMPS[25]) with the various open IM standards on the Web, such as Jabber (which went on to become XMPP) and others. This was really an afterthought. Ultimately, nothing ever happened with IMPS because the mobile industry mostly resigned to the fact that the existing (and growing) IM communities were unstoppable and what consumers really wanted. Well, in the absence of a compelling alternative, why wouldn't they? Since then, the IM experience has evolved again, such as the embedded chat facility in Facebook. Again, operators are nowhere to be seen in the debate or thought leadership around this topic, having resigned even further to the dominance of Web ventures. This is very sad when we think that the world's most successful messaging standard by far – a wildly giant phenomenon – has been texting.

Much of the recent effort in exposing network functionality to the Web community has been embodied in the OneAPI initiative, which has a very limited set of APIs, none of which relates to an underlying social enabler in the network (in the direct sense – sending texts between people doesn't really count). The first version (1.0) contains APIs for messaging, which is unsurprising for a telco as the focus is clearly on exposing underlying network functionality that has a clear charging model.

I won't dwell too much on the OneAPI idea here as I intend to discuss it more thoroughly when looking at Network-as-a-Service models (NaaS), later in the book (see Section 9.1 Opportunity? Network as a Service). All I need say here is that, as of today, the telco world hasn't responded meaningfully to the surge of social network activity in any innovative fashion. Telcos remain absent from endeavours like OpenSocial.org, even though they might well occupy a rightful place in the formulation of social API standards.[26]

At this point, I'm guessing that my earlier speculation about the emergence of a "Social Telco" model (see Section 1.3 Six Models for Potential Operator Futures) now seems a bit far-fetched. But all is not lost. Recall that a "Social Telco," were one to emerge, is, at its simplest, a convergence of telco platforms with social network platforms. If any operator is serious about social relevance, then they ought to be devising strategies for social convergence as much as service convergence.[27] What might this convergence look like? That is what I shall explore now, at least in outline. Of course, this is now speculative, because much of this is yet to happen. I shall press on regardless, given that a goal in writing this book is to stimulate "Web thinking" for my telco audience.

3.5.2 *Telco Social Graph and APIs*

Telcos should not be afraid to adopt an open stance towards social data, so long as the requisite opt-in mechanisms are in place to give consumers control. Telcos should expose their extensive social graphs via social-graph APIs, which could be consumed by other service providers and

[25] Instant Messaging and Presence Service.

[26] Of course, telcos might argue that they have their social standards, as in the directory services related to 3GPP IMS, but we have yet to see these adopted in any noteworthy way.

[27] Which means convergence across different service types, like TV, Mobile and Web.

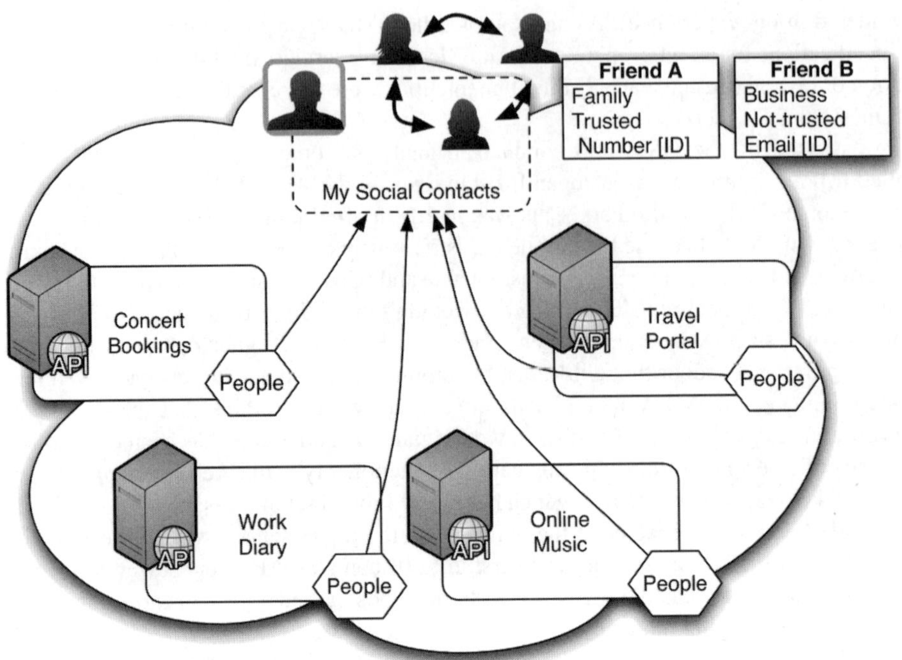

Figure 3.17 Telco as the social hub for socially-enabled sites across the Web.

developers as part of a Network-as-a-Service play (see Section 9.1 Opportunity? Network as a Service). This could be aligned with the OneAPI initiative, standardizing on a single set of APIs for social-graphs and socially-related data, with appropriate security protections mechanisms (like OAuth authorization).

Of course, it would be quite a stretch to suggest that telcos take on Facebook in an attempt to provide an alternative social networking platform, especially in the sense of being a destination for users. I am not suggesting this approach at all, although I believe that providing destination services on the Web should not be overlooked. I am not advocating that telcos become social networks, but that they recognize that they *are* social networks, of a type, and that they should therefore exploit this as far as possible. The opportunity, as I see it, is in becoming a reliable and trusted social-connection hub for other services, as shown in Figure 3.17. The telco could – and probably should – be the one place and venture that a consumer trusts with keeping her social data along with the necessary privacy controls that are not likely to be challenged by shifting commercial interests, as we have seen at times with some of the Web-venture social networks.

Telcos would instinctively want to go off and define their own standards for exposing social data. This would be a mistake, requiring years of lengthy debates in committee meetings peopled by folks who like to travel and talk a lot. Rather, why not adopt an existing standard and work on improving it? OpenSocial would be as good a starting point as any. It's where I would start and then start to add extra features to leverage operator assets.

For telcos, the social API should be a nexus for exposing more data other than the raw connection data of who is connected to whom. With careful thought and investment, the social API could become a valuable source of data that allows other services to deliver much more personalized and tailored experiences. As I have indicated in the diagram, the social nodes in the graph could be enhanced with qualifying data, such as how "trusted" a particular contact is. This would allow other services, with user permission, to query a contacts list and filter out only the most trusted contacts, or the business contacts, or family and close friends etc. Let's say I sign up for a new premium credit card service that offers extended benefits to up to four other family members of friends. I might want to offer that to some of my closest friends, but don't want to type in their data etc. It should be possible, via the telco social API, to ingest suitable qualified social-contact data to make this possible, as shown in Figure 3.18:

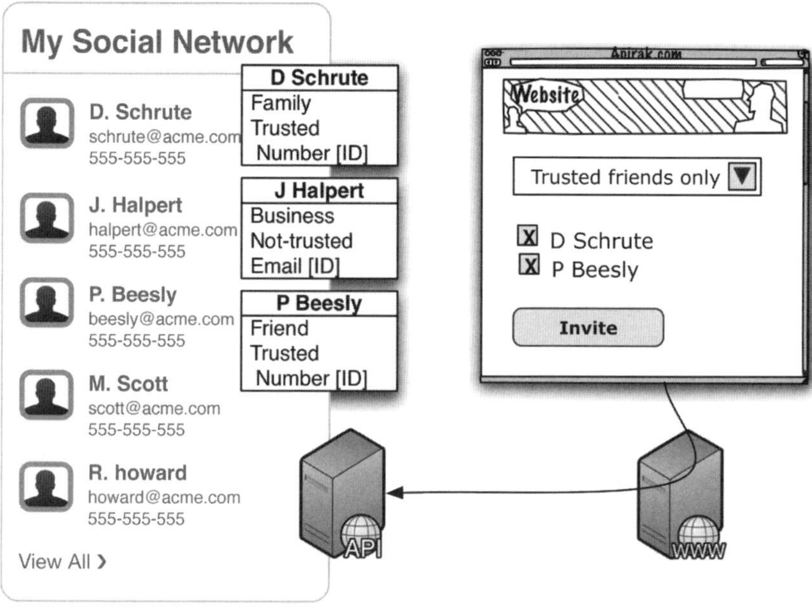

Figure 3.18 Importing trusted friends list via social API from telco.

It seems plausible that a consumer would rather only keep one *authorized and authoritative* source of social data and contacts, especially for dealing with more critical services like government, banking and making any kind of booking. That being the case, where should the data be stored? Of course, social networks are natural contenders because that's what they do today – they store our social data. However, how far are they trusted by consumers? How far can they be trusted? These are not technical questions, as in how secure and reliable are their IT systems. These are probably marketing questions. At the moment, it probably doesn't occur to most telco customers that their telco could fulfil any other role besides selling them texts and minutes.

I believe that there is a significant opportunity for telcos as keepers and managers of our social data. I think that this opportunity is not as a Web brand, like Facebook, but as a data keeper. This is an idea that has yet to enter the consumer consciousness, but I believe that it is rapidly approaching. If consumers can be forced to think of the question: "who should be looking after my data," then it would be similar to the question "who should be looking after my money?" No one is going to say Facebook or Google, I would guess. Consumers want banks to look after their money, notwithstanding all the financial disruptions of late. Similarly, I think that when it comes to data, it is going to be the "bricks and mortar" brands that have a role to play, like telcos, who can demonstrate a history of reliable data protection with stringent user controls.

One of the issues that is going to arise sooner or later is the concept of digital identity and trust. How can I identify myself on the Web to my bank, to my healthcare provider, and to all these linked-data services in the Internet of Things (see Internet of Things), hoping that they're not about to spew out my precious data without some hefty identity and security checks. I don't want any old Joe accessing my home security status, my car, and my vital signs! The issue of identity has to be related to social data and social networks. How do I know that Paul G is related to Chris G? Are they really brothers? Who says so? How does my local doctor's surgery know that this is the right Paul G trying to gain access to his healthcare records? Digital identity is the next big opportunity on the Web. It is a key component of the Web Operating System, as is social network data. It is also a huge opportunity for telcos.

3.5.3 Identity and Security

As I have just explained, in the world of connected services, digital identity is an increasingly important concept. Telcos have the upper hand here for two reasons:

1. Security is ultimately related to trust. It doesn't matter how many bits you use in your digital security certificates if we can't be sure that the certificate really belongs to Paul G, Healthcare Provider X and so on. Brands that evoke trust have an opportunity.
2. Connected services of all kinds are increasingly going to become mobile-centric, whether accessed remotely or physically (e.g. Using NFC – see Section 7.2 Sensor-net: Mobiles as sixth sense devices). Telcos, at least in many regions, still control or heavily influence the key transactional link with the mobile customer.

In my view, telcos should stand behind established and emergent born-on-the-web standards like OpenID and aggressively pursue the opportunities to become trusted identity providers, bringing their brands to the fore in user's minds, creating a strong association between brand, trust and security transactions. It is the rightful place of telcos to do this because they have been running networks and businesses for years that are built on these principles. What they haven't done is thought of how to extend the underlying security, identity and trust apparatus to become a platform. Again, we need more platform thinking!

Whether or not it's OpenID or some other technology, that isn't too important. In fact, it is relatively easy to introduce new standards for authentication on the Web. The starting point for me would be to occupy the role of identity provider in some really critical areas of life, like banking and government services, such as booking hospital appointments and even gaining access to patient records. Adjacent markets would be next, like "well-being" services we are likely to see emerge (see Section 7.2 Sensor-net: Mobiles as 6th sense devices). In all these areas, telcos should insert themselves as *the* provider of identity and authentication services, taking a role that will firmly cement telco brands as "trusted brands."

4

Big Data and Real-Time Web

```
SELECT nosql FROM BIG_DATA WHERE tech IS trendy
```

- Big Data is a collection of ideas, trends and technologies that enable Web ventures to exploit the value in massive data sets that exceed the confines of the conventional storage and processing limits of single computers.
- Big Data is about making value out of unthinkably large amounts of data.
- Facebook's Cassandra and Amazon's Dynamo are examples of "Big Data" platforms that bring unique business advantages to these Web ventures.
- It's not just the technology that matters – "Data Geeks" are important.
- The buzzword in Big Data is "No SQL," but it's a very mixed bag of storage technologies and solutions.
- A common fallacy is that so-called "No SQL" solutions can replace more conventional (e.g. Relational) data stores.

4.1 What is Big Data and Where Did it Come From?

4.1.1 In Search of the New Big Data

Where did it all begin? It's difficult to say, as is what we're talking about exactly when using the term "Big Data." It's not as if handling large amounts of data is new, nor big data sets. In science, finance and government projects, large amounts of data are routinely collected, analyzed and, sometimes, lost.

But those domains offer us a clue as to what's new with "Big Data," rather than "Big IT," and where I'd like to take you in this chapter. Science, finance and government are all the domains of REALLY BIG projects, as in lots of money. Enterprise IT projects costing billions aren't unheard of. Telcos are no strangers to large amounts of data either. A "big data" scenario here would certainly include the hefty billing systems or the CRM. These IT projects can easily run into hundreds of millions of Euros over time in order to reliably handle hundreds of millions of records per day.

Connected Services: A Guide to the Internet Technologies Shaping the Future of Mobile Services and Operators, First Edition. Paul Golding.
© 2011 John Wiley & Sons, Ltd. Published 2011 by John Wiley & Sons, Ltd.

But our domain of interest for this book continues to be the Web, where such scales of investment in IT projects are rare. But we know that very large data sets do exist, such as in the world of search: "1 Trillion pages indexed and counting," or so the strap line used to go. And where there's search, the data-goliath Google instantly comes to mind. That, then, is a good place to start our Big Data story.

For the current discussion, we can think of Google's search engine as a big table of data. Here's what an ultra-simplified search table index might look like for the search term "Mobile":[1]

Keyword	Rank	URL
Mobile	1	amobiletelco.com
Mobile	2	somemobilephonecompany.com
Mobile	3	anothermobiletelco.com
Mobile	4	mobiletelcodeals.com
Mobile	5	mobiletelco.com

Now, given that Google's index went past the 1 trillion mark back in 2008, we can well imagine how big that table is by now – and our example is an extremely simplified snapshot of what the actual data logs might be. Yes – it's going to be one extremely big table. No coincidence that Google has built its own proprietary database technology to store this table and – wait for it – named it "Big Table." More on the inner workings of that particular technology later.

The search index isn't the only data in the Google "digital brain." The search engine also knows what users are searching for. That's a lot of search terms and a lot of data, which can also be ranked into order of popularity, such as how many people search on the word "money," or whatever. The most popular search terms are publicly shared by Google as part of its Zeitgeist press pages. But that's not all.

The search engine also keeps track of which links the searchers click on, helping to tune the ranking over time and to personalize it for each user. All that data gets stored in big tables too. And I haven't even mentioned the ad-words data yet – which ads get dished where and clicked on when. And on it goes – more and more data in bigger and bigger tables. I think you get the picture. This really is BIG data.

4.1.2 The Business of Big Data

Over time, Google has developed great expertise in handling large amounts of data. In a remarkable feat of its engineering team, most of this data processing takes place on what is essentially commodity hardware. Google Engineering is renowned for building its own hardware from low-cost commodity parts, enabling them to keep the costs down (and the temperature up[2]).

[1] The domain names here are supposed to be fictitious and only for illustrative purposes. If they do exist, I'm not offering a real page rank here ☺.

[2] Supposedly, Google are able to run their data centres at higher temperatures than usual, which allows them to save massively on energy (cooling) costs.

So, what business is Google in then? The search business? I'm sure that you might say "yes," although some of you will say "no – they're in the advertising business." Both are correct in their own context. However, there's another point of view. What really enables Google to make money from search and advertising? Surely it's their incredible "brain," at the core of which are these various big tables. Google are in the "big data" business. In our story, this gets us to an interesting place because the name of the game with big data is adding value. In this "big picture" context (please forgive the endless play on the word "big") I prefer to think of big data as being about:

```
Making value out of unthinkably large amounts of data!
```

Let me say that again:

```
BIG DATA is about making value out of unthinkably large amounts
of data!
```

And what's new about Big Data, different to the "Big IT" projects, is that almost anyone can do it! How so? Because in very recent times, a plethora of born-on-the-web technologies and tools have come into the open source domain that enable processing and storage of large amounts of data using commodity priced hardware or cost effective cloud-computing resources, within the reach of even the most meagre of start-up budgets.

At this point, it is worth considering another aspect of the big data challenge, which is the computational processing overhead. This is an ongoing problem with all computing architectures – how to process data sets efficiently, the larger the sets get. Beginning with the processor, if we can fit the data inside the processor, then things run really fast. While onboard processor caches have grown dramatically in recent years, we're still talking really tiny data amounts in relation to the sizes of tables in our big data story.

Beyond the processor itself, we have the physical memory in the computer. This can easily be extended to something quite large, typically in the Gigabytes of data. We can process some pretty large data sets here, potentially tens of millions of records in a table (depending on record size, of course). However, Big Data is a whole order of magnitude beyond this. It deals with data sets that no longer fit comfortably on a disk! With Big Data, we are dealing with data sets that can only be stored across a cluster of machines, as shown in Figure 4.1. Some commentators say that this is the very definition of Big Data.

The ability to process data sets this big brings with it a whole set of technological and business possibilities. This is incredibly exciting and powerful. We have always known that there's lots of power buried in data, just as the old adage says: "information is power." In many ways, if Web 2.0 was about putting data onto the Web, then what we're seeing now is a laser focus on making sense of the data, finding patterns, exploiting connections, extracting and making value.

How did this low-cost data revolution come about? The same way so many other revolutions happened on the Web, which is via the open source movement. Returning to Google's big data story, Google has created a number of technologies to help them process vast amounts of data. In the spirit of sharing, whatever the real motivation, Google published a couple of papers that explained the principles of some of their big data technologies: Map Reduce and the Google File System (GFS). Don't worry, I'll explain these later, but for the current plot, all we need

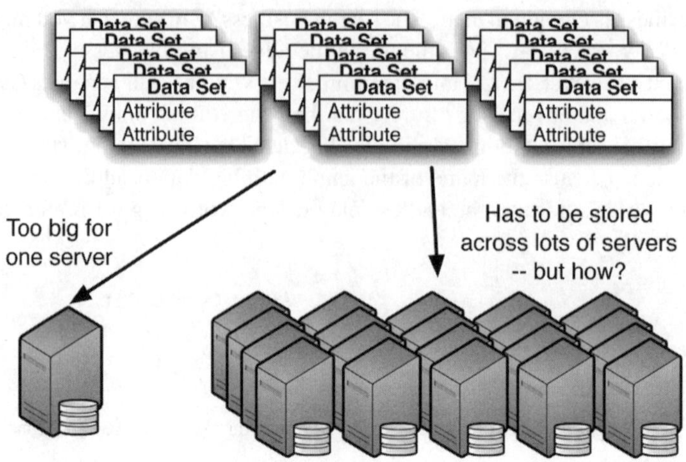

Figure 4.1 Large data sets that will only fit across many servers.

to know is that enough was said in these papers to inspire and inform the creation of an open source Big Data platform called Hadoop, which was adopted by Yahoo.

It's a good job that some clever people contributed to Hadoop and then shared their work in the public domain. Designing systems that can process data across a cluster of machines is not easy. Repeat: it is not easy! Not only does the data not fit on the disk, but the kind of problem solving required to build distributed algorithms doesn't typically fit in the average sized brain. However, once that genius is codified into algorithms that are then coded in software, it's a much easier proposition for a developer to download that software and start playing with their own large data sets.

At this point, it's worth commenting about the nature of many of the Big Data tools that have now been released into the open source domain. They aren't just about storing large amounts of data. This is a relatively easy problem to solve, even if the data exceeds more than one disk drive. Many large enterprise IT projects were forced to solve this problem long ago. After all, a billing system in a telco might produce over a billion call records a day that need to be sifted, sorted and processed. And that's the clue to the difference. If we have a relatively *linear and serial* process of data collection followed by transactional processing, then those patterns are well known and relatively easy to implement.

However, if we now wanted to ask a new question about the data, such as "tell me which of the users have a similar call pattern to this guy," then the conventional data-processing system (e.g. billing) probably can't do it. We might be able to build it out initially, but if I then change what is meant by "call pattern," the system might falter. The key difference then is that these new Big Data platforms can both store and process the data, but in a way that is often flexible, storing the data "as is" in readiness for future processing algorithms as yet unknown. Traditional large storage solutions usually aren't this flexible.

Many of the sorts of problems that we find interesting on the Web, at least for now, are to do with this sort of approach to the data, which, as you might have noticed, is essentially a search problem. And what matters is that they are exhaustive search problems, which means that useful solutions often require the *entire data set* to be scanned and processed in order to find the best fit result. Why are these sorts of problems common on the Web? It's because value in Web applications, especially monetizable value, is usually related to personalization. The more personalized the user experience and the data, the more valuable it is to both the users and the suppliers of services, especially advertisers.

Exhaustive search is difficult because we have to scour the entire data set in multiple dimensions, which means joining lots of big tables together to get a useful results. For example, imagine that we have a movie database in one table, a set of movie reviews in another and then a set of social graphs in yet another, showing the relationships between users. To recommend a good drama movie for you based on your preferences and friends with similar tastes, the recommendation algorithm probably needs to do something like this:

1. Select all the records from the movies table where the genre is drama.
2. Filter these by joining the results with all the records from a search of the reviews of these films.
3. Filter these by joining the results of that search with a search of all friends who have a high "similarity ranking" and who have recently watched drama films (selected from another table possibly of which films they have recently watched).

These selective filtering steps are done by joining one table with another in order to find the overlaps, a bit like our old junior school friend, the Venn diagram, as shown in Figure 4.2:

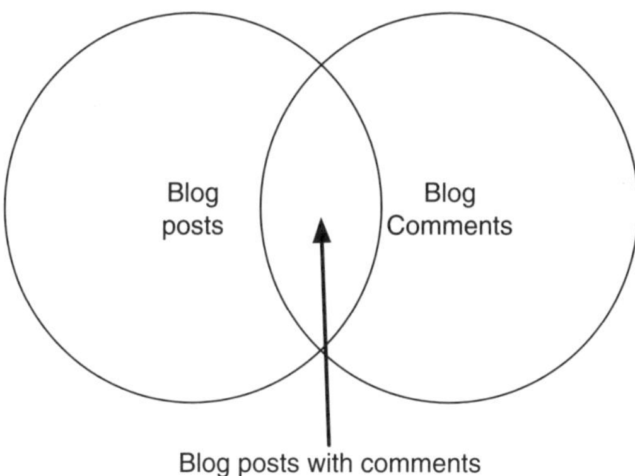

Figure 4.2 Two data sets with a join (overlap).

This joining process has an immediate problem. Each of those tables that I want to join is big, maybe millions of rows long, too big to fit into memory for a single computer to process. Don't forget – they're too big to fit on a disk! It might be possible to break the calculation down into smaller chunks and steps, but this requires taking a process that is essentially one operation (albeit a big one) and turning it into lots of smaller processes that run in serial, one by one, having each time to load and unload the next chunk of data. I'm sure you can see where this is headed – processing bottleneck! That's right, the process is deathly slow. The solution is not to join the data in one place, if possible, and to farm out the processing task across the computer cluster that's collectively holding all the data. Even then, you can imagine that this might be a slow process, relatively speaking. On the Web, response times are everything. We all know what happens when a website is too slow – visitors click off, often never to return.

That's why the innovators in various Web ventures have been forced to create their own solutions. If Google couldn't keep up with its users' demand for search results, it would instantly fail as a business – speed is everything. If Google couldn't deliver relevant ads in real-time, based on the user's context and then priced (to the advertiser) according to the complex auctioning rules, it would fail as a business.

Delivering value on the Web means delivering lots of personalized data very quickly for lots of users. No matter the application, the underlying problem is probably the same, which is some kind of Big Data problem. It is easy to see how Big Data tools and technologies will begin to emerge as part of the underlying Web OS that we explored in the previous chapter. In its overt form, it is search. In its hidden and somewhat abstract form, it is still search – looking for answers across vast sets of data – but it is very much "personalization" too.

The beauty of the explosion in Big Data tools and their following is how accessible they have become to anyone with a need to process large data sets. They are low cost and relatively easy to program for. Moreover, they facilitate search-like queries of the data with very useable response times.[3] This means we can almost begin to "play" with the data and ask of it questions that previously would have been unthinkable or too costly to ask. This is an ideal set of ingredients for a new wave of innovation on the Web, which is exactly what we're seeing.

Before exploring Big Data in more detail in the rest of this chapter, I should add a very important caution to the telco observer. Big Data is *not* just a Web technology. We must avoid falling into that way of thinking. Many of the Big Data tools have come from Web ventures because they have encountered a particular problem – search – that is core to nearly all Web ventures. Search is not new. Every computer science graduate from the 1970s, 1980s and early 1990s – all pre Web 2.0 era – would have cut his or her programming teeth on search algorithms, closely related to sorting algorithms. Finding the lowest cost item in a linked list of shopping items is a sort/search problem. What is new is that we are talking about searching across vast irregular sets of data and in ways that don't easily fit into conventional programming patterns.

4.1.3 *Welcome to the Age of Big Data*

Let's think of Big Data in yet another way, possibly the way that Google thinks about it, at least at the strategic opportunity level. If my venture happens to have a million or so machines[4] that can act as a vast parallel processing cluster to extract value from unthinkably large amounts of data, then the more interesting question becomes: "What else might I use this cluster for?"

It is perhaps no coincidence that Google's Eric Schmidt has a seat at Barack Obama's table. Just think of all that government data, healthcare data and so on! Think of the queries that might be useful. Think of the possibilities. The same goes for so many other domains, including telco where the telco giants process large volumes of customer data on a daily basis. I think that by now you're getting the Big Data **big picture.**

[3] Just to be clear, we often don't expect Big Data stores to respond in real-time. We use them to pre-process vast amounts of data that can then be used to improve real-time responses from associated data sets.

[4] I don't know how many servers Google actually has.

To quote from Ian Ayre's book *Super Crunchers*:

> We are in a historic moment of horse-versus-locomotive competition, where intuitive and experiential expertise is losing out time and time again to number crunching.

So, if you find that you're still driving your business decisions using gut feel and intuition, calling upon your experience, then ask yourself whether or not the data supports your judgement. More than likely, there's data somewhere in your organization that can be sliced and diced to reveal critical data points and correlations that might inform your decisions with actual data rather than a hunch. Moreover, if you're not crunching the data, perhaps your competitors are. They certainly are on the Web.

4.2 Some Key Examples of Big Data

In this section I shall sample a mixture of Big Data technologies and their uses. The two are often intimately linked because many of the Big Data technologies arose in response to a particular domain problem, such as search efficiency or e-commerce availability. I shall outline the essential nature of the problem so that the reason for the solution becomes clear. I won't elaborate on the technical details too much as they get very involved very quickly and the purpose here is to give examples of where Big Data is being deployed and why, not to explain its inner mechanics and mathematics.

4.2.1 Statistics Collection at Facebook

There are quite a few Big Data stories that have emerged from Facebook, including the widely reported invention of Cassandra, one of the rising stars of the so-called "No SQL" data storage solutions. I'll get to NoSQL and Cassandra in a minute, but I'd like to start with a very simple use of Big Data in Facebook that reveals its power as a foundation for business intelligence and adding value. Don't forget, our moniker for Big Data is "making value out of unthinkably large amounts of data."

Facebook, like many of the major Web ventures, is a frenzy of Web activity: clicks, status updates, wall posts, friend-joins, likes, app usage, and so on. Each of these events causes data to be generated and collected. Like many Web ventures, Facebook collects all of this data, whether it has an immediate business value or not: everything is logged. Some of this logging is essential, as it might be needed to undo actions, verify changes and so on. Some of this logging is just enriching the giant data pool, ready for some future analysis and use. So long as data can be collected easily and cheaply, then why not store it? That's more the attitude of Web ventures these days – store unless there's a reason not to.

As reported by Facebook's Itamar Rosenn,[5] Facebook's Data Team used statistical analysis[6] of their data to answer two questions about new users:

1. Which data points in the data sets could be used to predict whether or not a new Facebook user would remain an active user?

[5] At the conference – http://www.predictiveanalyticsworld.com/

[6] Using a tool called R – http://www.r-project.org/

2. For those users that stay, which data points would predict how active those users would still be after three months?

These are fascinating questions. It's also fascinating to think that somewhere the answer is buried in the various big tables amongst all that logging data. In case it isn't obvious, I must point out that had Facebook not logged data in the first place, they would not be able to answer these questions. Perhaps it might have been left to more traditional means, like email surveys of those who left and stayed, or maybe just the intuition of a product manager.

For the first question, Itamar's team used a technique called recursive partitioning to infer that just two data points are significantly predictive of whether a user remains on Facebook:

1. Whether or not the new user had more than one session.
2. Whether or not the user completed the basic profile information.

For the second question, the team used a relatively new technique in statistical analysis called **least angle regression** and found that the level of activity after three months was related to variables that indicated three classes of behaviour:

1. How often a user was contacted by others.
2. The frequency of using third party applications.
3. What Itamar referred to as "receptiveness," or how forthcoming a user was on the site.

What is fascinating about this example is that the team were able to drill into the data because it was available for analysis. This is a characteristic of most Web ventures – the built-in and habitual collection of data in readiness for analysis in order to gain insights into user behaviours. As we shall see in Section 10.3 – Key Web Start-up Memes, the essential activity of an early stage start-up is the search for a viable business model, trying to home in on a successful business model via an unfolding series of experiments with the service, guided by perpetual data analysis, or what has become known as the discipline of analytics.

I started with this example because of its simplicity and obvious value. The team were trying to answer very simple questions that have an obvious business value. We could ask similar questions in so many other ventures:

1. What data points predict when a mobile phone user is about to leave the network?
2. What data points predict when a coffee drinker is likely to order a Danish pastry with the beverage?
3. What data points predict when a buyer of product X is likely to buy an extended warranty?

What's deceptive here is the relative ease with which it is possible to ask these questions once the data has been collected and stored on a platform that enables the appropriate analysis of the data. Once again, I reiterate that these tools are within the reach of start-ups running on the smallest of budgets, notwithstanding the necessary expertise to structure the question and know which analysis technique to use. Big Data platforms and tools are not in any way dripping in magic sauce that automatically conjures up answers to probing questions. Competent programmers and analysts are still required to run these platforms and find meaning in the data. The key point though is that these questions can be asked against very large sets of data that would have previously remained beyond the reach of any type of analysis at almost any cost.

4.2.2 *Real-Time e-Commerce at Amazon with Dynamo*

In 2007, Amazon's CTO, Werner Vogles, published an academic paper[7] describing the invention of a new storage system called Dynamo. In my previous book, I gave numerous examples of large-scale systems, including e-commerce. In those examples, I included a data storage layer, but didn't elaborate on the performance requirements at scale, where a system, like Amazon, needs to serve tens of millions of users at peak times. I described the general principles of scaling via clusters, load-balancers and so forth, but just assumed that it all works perfectly well. The problem is that it doesn't when it gets too large. Herein lies the reason for Amazon's quest to invent a new storage system using the latest thinking from the world of distributed systems. To quote from the paper's abstract:

> Reliability at massive scale is one of the biggest challenges we face at Amazon.com, one of the largest e-commerce operations in the world; even the slightest outage has significant financial consequences and impacts customer trust.

So that's the problem: reliability at scale. And it has real business impact. As we all know, Web users are fickle and the competition is always just an easy click away. Once on the Amazon site, Amazon wants to keep the user there for as long as it takes to generate revenue from the visit. Those unfamiliar with the mechanics of an e-commerce venture will perhaps assume that Amazon's business is essentially an online catalogue of products – browse the catalogue, pick the products, check out and await delivery. Sure, that's a good headline summary of their business, but it doesn't reveal what the core of their business is really about.

The core of Amazon's business is **maximizing the conversion of visits to revenue.**

In order to do this, Amazon's massive Web IT infrastructure uses hundreds of loosely coupled software systems to render the right information at the right time to the visitor's browser. These processes must co-ordinate to bring the information together reliably, providing the user with the best possible site experience. One system might be the product catalogue, another the shopping cart, another the recommendations, another the reviews, another the product history, and so on. The architecture looks something like that shown in Figure 4.3.

One thing to appreciate about a system of this scale is that it will be running across thousands of computers connected via thousands of network links. Inevitably, some of this hardware and the underlying system software is going to fail. With a system of this size, something will always be in failure mode, which means that some of those hundreds of co-ordinated e-commerce processes are going to be affected. This is unavoidable! Amazon sets a high bar in terms of the level of service that it wishes to offer customers, seeking a relatively high 99.9 per cent availability in production.

[7] Giuseppe DeCandia, Deniz Hastorun, Madan Jampani, Gunavardhan Kakulapati, Avinash Lakshman, Alex Pilchin, Swami Sivasubramanian, Peter Vosshall and Werner Vogels, "Dynamo: Amazon's Highly Available Key-Value Store", in the Proceedings of the 21st ACM Symposium on Operating Systems Principles, Stevenson, WA, October 2007

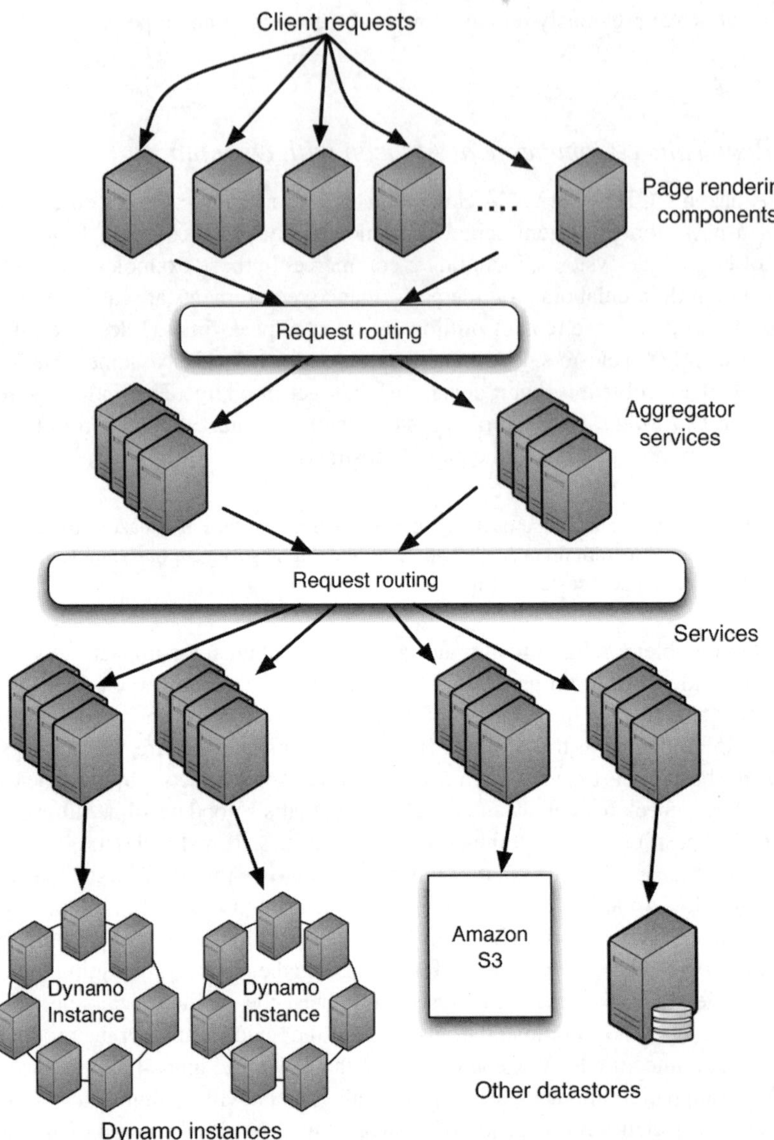

Figure 4.3 Representation of Amazon's distributed architecture.

During execution, each of the e-commerce processes will be storing state information, such as:

1. The most recent item looked at.
2. The last product added to the cart.
3. The most recently viewed page.
4. Changes to cart quantities.
5. Delivery address information, etc.

Many production systems store their state in relational databases, which are designed to enable relationships to form between tables. However, many of the processes in Amazon do not require relational storage models. The data can be stored in a simple table, almost like a spreadsheet. This poses an opportunity. Most relational databases are quite complex beasts because of the need to support complex querying across multiple tables of related data. This complexity comes at a cost, both in terms of software and hardware, plus relatively expensive skilled support labour. These problems are greatly magnified when attempting to scale relational systems. Without elaborating the details, it is easy to appreciate that trying to query tables that span multiple machines is going to have its problems. The opportunity then is to design a much-simplified solution that avoids relational queries with all their overheads. This was the opportunity that Amazon exploited with Dynamo. Without relational data, it is possible to implement a simple key-value store.

Anyone who has designed a Web application will have already noticed that it is often the case that processes don't require relational storage. Nonetheless, the data is still stored in a relational store because these are the dominant stores in use, such as the ubiquitous MySQL and Oracle product lines. Ordinarily, no one is going to consider designing a non-relational alternative because it doesn't matter – the overhead of the relational store is simply not felt. Now magnify the requirement to Terabytes of data that needs to be highly available across multiple data centres, and suddenly the pain of a relational store becomes apparent.

The design objective of Dynamo was high availability across a massive number of processes running across a massive number of commodity machines, some of which will inevitably be in failure mode. In other words, they sought to design a solution that assumed a constant presence of failure.

Let's take a brief look at what such a system might entail and how Amazon solved the problems. Firstly, there's the issue of availability to consider. What does this really mean? There are two sides to the availability story: writing data to the store and reading data from the store. In either case, it is obvious that a single storage node (per e-commerce process) isn't going to work. If it fails, then reading and writing are blocked and the user experience rapidly falters. For example, it would be unacceptable if the user couldn't update her shopping cart. The solution is clustering, providing more than one node to support the read and write actions. If node A goes down, then we can write the data to node B instead, as shown in Figure 4.4.

Looking at the read problem, it is obvious that if node A fails, then node B can only be used to read the data if a recent replica exists on that node. This replication process is one of the features of Dynamo. It happens transparently to the application. When the app writes to the store, the data ends up on a node and is then replicated automatically. The application doesn't need to keep track of which node to write the data to. The same is true for reading data. The app isn't required to find out which node is available with a copy of the data. All of this is managed by Dynamo, which is capable of replicating data across as many nodes as it likes in order to achieve the required level of availability and performance. Nominally, the nodes are considered to sit in a ring (see Figure 4.5).

To achieve performance across a large number of commodity machines, the replication process happens asynchronously, which means that data is copied to other nodes in a kind of best efforts sense until all copies are identical. There is no attempt to guarantee that all copies are consistent at exactly the same time. Such guarantees are expensive to implement. What Dynamo offers is a replication process that is *eventually consistent*. The point here is that Amazon is not a bank or a flight booking system where inconsistent copies of data might

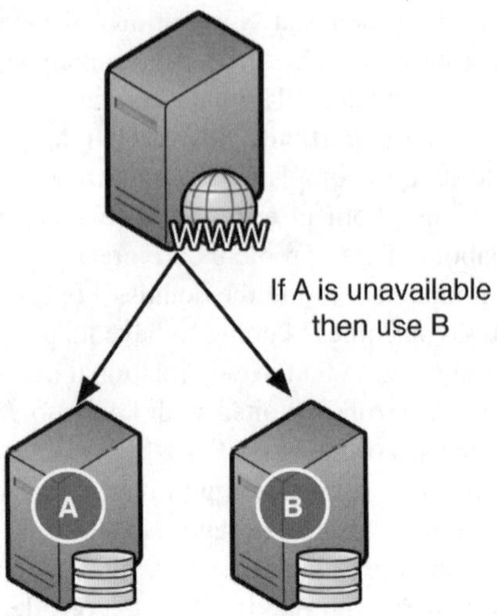

Figure 4.4 Node A or Node B cluster.

cause all kinds of unacceptable hiccups. Worse case, and very infrequently, you might miss an item from your cart, which you can always put back.

The system used to distribute the storage across the nodes is called a Distributed Hash Table. It is a mechanism that proved successful in so-called P2P (Peer-to-peer) architectures where lots of computers could collaborate in the storage and retrieval of data without any one computer being in charge of where data gets stored. All computers share the same algorithm

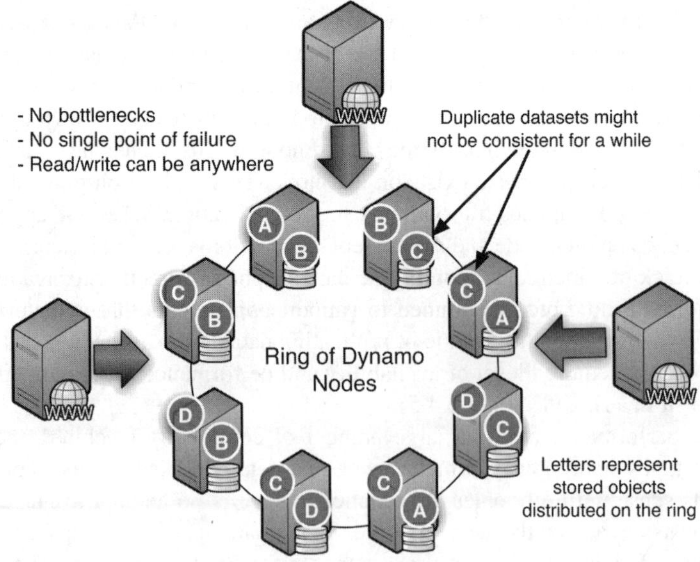

Figure 4.5 Distributed storage in an Amazon Dynamo ring.

for generating a hash table that says given a particular key value, this is where the data ought to be stored and/or available. The advantage of a DHT approach is that nodes can fail, or new nodes be introduced, without having to restart the other nodes and without disrupting the existing pattern of distributing data across nodes.

4.2.3 Amazon's Dynamo Features

4.2.3.1 Data Model

The query model is a simple read and write operation to a data item that is uniquely identified by a key.

Key = Paul Golding; value = Author;

Key = Steve Jobs; value = Biz Guru;

Key = Linus Torvalds; value = Software Guru;

This is a simplification. The "value" is actually a binary object (blob[8]) and can be anything that the application wants to store against the primary key value. In other words, there is no table with column headings to fill out against the key. Dynamo is aimed at applications that need to store relatively small objects (usually less than 1 MB), which works well for the data types frequently found in an e-commerce system.

4.2.3.2 ACID Properties

ACID (Atomicity, Consistency, Isolation, Durability) is a key acronym in the relational database world as it says a lot about the "strength" of the solution for many demanding applications, like finance and other data-critical domains. The ACID set of properties guarantee that database transactions are processed in a predictable fashion. As I said earlier, in a banking application, having multiple copies of data that are inconsistent would cause all kinds of disruptions. A user (and the bank) needs to know how much money is *really* in the account, not what the best guess is based on a number of data replicas that need reconciling. In the ACID world, there is no reconciliation – the data is simply always reliable, always consistent.

According to Amazon's Vogels, experience in Amazon has shown that stores with ACID guarantees tend to have poor availability, a fact supported by the database industry and academia generally. Dynamo is designed to support applications that operate with weaker consistency (the "C" in ACID) in order to achieve high availability.

4.2.3.3 Efficiency

The system needs to function on a *commodity* hardware infrastructure. According to Vogels, Amazon's e-commerce services have tight latency requirements measured at the 99.9th

[8] BLOB stands for Binary Large Object.

percentile of the distribution.[9] Given that access to state information plays a pivotal role in these services' operations, it is the storage system – Dynamo – that must be capable of meeting such stringent service level agreements (SLAs). Therefore, where required, services must be able to configure Dynamo such that they consistently achieve their SLA requirements.

4.2.3.4 Other Assumptions and Summary

Dynamo is used only within Amazon's walled garden, inaccessible to external parties. Therefore, there are no security related requirements such as authentication and authorization, which would significantly add to the complexity, especially across a distributed system. Also, each service will have its own Dynamo ring, so that scalability is limited to the service level, not the entire Amazon aggregate of services. This sets the likely limit to scalability to hundreds of servers, not thousands.

In summary, Amazon's Dynamo is a highly available and scalable data store that has achieved a consistent level of performance (>99.9 per cent) across Amazon.com's e-commerce platform. This example of Big Data is an engineering solution to the problem of delivering a high quality Web experience across a vast number of input processes that combine to render a highly personalized interface, allowing Amazon to focus its business resources on the job of converting visits to revenue rather than the expensive job of maintaining relational stores.

It is worth a mention here that Yahoo have built a similar infrastructure called Sherpa, Google have their BigTable and Facebook built Cassandra, which is inspired by Dynamo (with one of its designers having worked on both projects). With the exception of Cassandra, which was open sourced by Facebook, these technologies remain proprietary solutions used internally by their inventors. However, inspired by publications about these technologies, various engineers have created similar solutions that are open sourced from the outset. I will explore some of these later in this chapter.

4.3 Say Hello to the Data Geeks

A significant part of the Big Data story is extracting value from the data in order to make sense of it. Without meaning, the data is useless. It might as well be noise. So, how do we go from data to value? Step one is to search for information in the data. It's a bit like breaking a code where one goes in search of fragments of familiar patterns that might indicate nuggets of buried information.

By way of analogy, to illustrate the complexity of the task, we can think of this "code-breaking" as finding words in a word search puzzle, except that we're probably dealing with a puzzle that would easily cover the side of a skyscraper. Much of the data is invisible, beyond our reach, so we need tools to bring the data into view. But, there's a second problem. Let's say that we could scan the entire data set in search of hidden words, we then have to mark where those words are and what they say. We have to start grouping the words, trying to piece them together in a way that informs our understanding of the data. These words, groupings

[9] A typical SLA required of services that use Dynamo is that 99.9 per cent of the read and write requests execute within 300 ms.

and patterns are themselves unthinkably large. It's not like we can list them on a single piece of paper in search of patterns while we hold the data in our heads.

And there's yet a further dimension to consider. Once we have all the patterns and can start to make sense of them, what we often want to know is something about the *trends* in the data, either the historical ones or what we might postulate to be the future ones, based on some kind of regression. Much of this agonizing over the data belongs in the domain of statistics.

Hal Varian, Google's Chief Economist, said the following in the January 2009 McKinsey Quarterly:

> The *sexy job* in the next ten years will be statisticians ... The ability to take data – to be able to understand it, to process it, to extract value from it, to visualize it, to communicate it – that's going to be a hugely important skill.

This quote inspired Michael E. Driscoll's talk at the Google IO Ignite event, where he mused about how sex appeal and statistics might go together to create a new breed of superhero – the "Data-Geek."

Driscoll said he believed that the folks to whom Hal Varian is referring are not statisticians in the narrow sense, but rather people who possess skills in three key, yet independent areas:

1. Statistics
2. "Data munging"
3. Data visualization

Let's briefly explore these skills, per Michael's elaboration in a blog post.

> **Skill 1: Statistics:** Perhaps the most important skill and the hardest to learn. As those of you who studied stats will know, this is not the easiest of subjects to grasp, perhaps because of its dryness, although that might abate somewhat in this Big Data era that we now find ourselves in. However, just like everything else, where there's money, there's interest. As kids begin to figure out that there's money – and power – in understanding data, they might be tempted to study stats.
>
> **Skill 2: Data Munging:** According to Driscoll, this is a slang term used within geek circles (you can find them, apparently, via a Twitter search for #rstats), to describe the gruelling process of cleaning, parsing, and proofing data sets in readiness for analysis. Real world data is messy. Many telcos will know this, with myriad systems containing various shards of the user's profile and activities. The data will come in all kinds of formats, none of them likely to be stats-analysis friendly.
>
> Of course, the tools of Big Data, like Hadoop and its map-reduce, are now the tools of data mungers – this is how they process the data, by dumping it all in a store and trying to slice and dice it. Much of this munging requires a good deal of programming or scripting and a level of comfort with the various file systems and underlying operating system operations that manipulate them.
>
> **Skill 3: Visualization:** This third and last skill that Professor Varian refers to is probably the most exciting, although deceptively difficult. Many of us will

> know how to throw together a quick pie-chart in Powerpoint, especially with the latest version that comes packed with various visualization widgets. However, in the world of Big Data, we aren't talking about viewing a 10 by 5 cell table. We're talking about possibly millions of multivariate data points where there is no obvious way to visualize the data.

As Driscoll points out, there are two aspects to visualization. The first is the data-geek's bag of tricks to get a "feel" for the data, to poke around at its edges, dive in its middle and try to make sense of what's going on in the data soup. As I pointed out before, human imagination is still very much a part of the world of Big Data. Whilst Big Data might well be about making value from unthinkably large amounts of data, the value proposition itself clearly does have to be thinkable! But, with unthinkably large amounts of data, one can intuit about the meaning in the data. Tools are needed that play to the data-geeks senses, namely the eyes.

This leads to the second aspect of the visualization quest, which is how to communicate the value of the data. Being able to present the data in meaningful ways, in order to convey the value proposition, is just as important. It might also require a different approach to the kinds of visualization used to play with the data and explore its surfaces. Here, we can expect to see animation and even 3D presentation to play an increasing part in the communication of Big Data patterns.

4.4 "No SQL" and Some of its Flavours

4.4.1 No SQL Means No SQL, But not Much Else

The so-called "No-SQL" class of databases is actually a misnomer. What these data stores really have in common is the absence of any relational query capability, or relational schemas. A schema is a way of structuring data by stating how the data is related. Typically, in a relational world, a schema might look like that shown in Figure 4.6.

Table A is a table of people, which I'm sure looks familiar, or at least plausible. It is indexed using a primary key field – "ID" – which is a unique identifier. No two rows of data in the

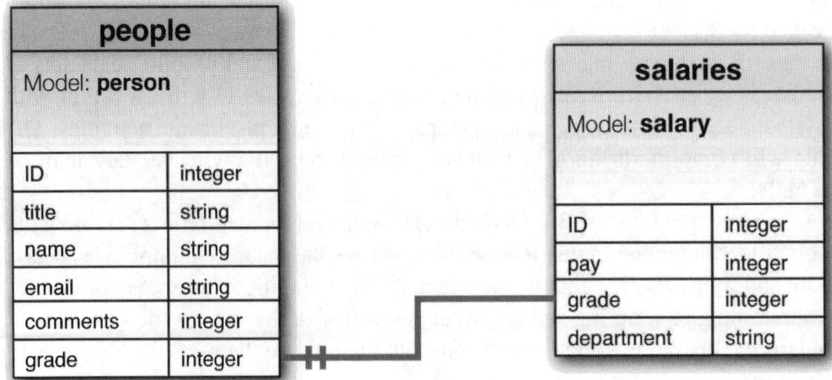

Figure 4.6 Relational data schema to show how tables are related.

table can have the same ID. This is how each entry is uniquely identifiable. For example, there might well be two people with the name "Paul Golding" in the table. However, each "Paul Golding" will have a different and unique integer in the ID field, thereby indicating that these are different records.[10]

Table B is a table of pay grades against pay. Again, an ID field uniquely identifies each pay grade here, which is the primary key for this table. In theory, we don't need this ID field if the pay grade names are all unique, as the grade names could act as the primary key. Notice that table A has a column for recording the pay grade of each person. In order to link pay grades in A to actual pay rates, we have to link the pay grade column in A to the corresponding pay grade row in B, as shown by the arrow. This connection is a deliberate and formal relationship between the tables. The system using these tables relies upon a relational database management system (RDBMS) to enable these tables to be joined in this manner. The data is typically extracted using a query written in a language called Structured Query Language (SQL). An example here would be:

```
SELECT pay FROM salaries WHERE grade > 10
```

A RDBMS is designed to process this SQL statement and read from the tables to extract the required data set. Or, we can insert data using an SQL INSERT command.

Relational schemas can get very complicated very quickly, spanning multiple tables. One can easily imagine this in a telco that has user records, call records, billing records, texting records, CRM records and the like, perhaps with complex relationships throughout. There's nothing wrong with relational schemas and RDBMS solutions to support them. Traditional database solutions like these have been around for a long time and work very well. They are robust and enterprise grade.

The "No SQL" storage solutions typically avoid relational schemas and, by extension, the use of SQL, as this query method doesn't really make sense in a store that can't respond to many of the queries that SQL assumes are possible by virtue of an underlying relational schema. Beyond this characterization of alternative stores, the actual stores themselves don't have a common set of characteristics that make them a unified family of solutions. In fact, the variation in the "No SQL" group is quite large and growing all the time. Moreover, some of the characteristics that were assumed to be common initially, such as the absence of ACID[11] compliance, are no longer true.

As just mentioned, there is a growing number of variants in the new breed of database solutions that have emerged in the Web 2.0 period, many of which are closely related to the overarching theme of Big Data that I explored at the outset of the chapter. Let's now explore some of the variants and mention examples from each.

Before we do, a brief word about the use of the term "No-SQL movement," which has been used often by various commentators, or should I say detractors? It is often used as part of a straw man fallacy to misrepresent what these solutions are all about. They are *not* about replacing SQL or RDBMS solutions, nor are they claiming to be something entirely novel (we all know that various non-relational stores have been around for some time). Also, advocating a particular solution, such as MongoDB, is not to say that everything else is rubbish. Many

[10] In fact, they could be the same person if this is a system where Paul Golding has registered twice.

[11] I discussed ACID when looking at Amazon's Dynamo solution in some examples of Big Data solutions.

systems that use MongoDB will also use MySQL too, or some other store. However, the term is meaningless. There isn't really a movement that represents or advocates the use of No-SQL solutions instead of SQL ones. As said earlier, the term No-SQL is vague anyway. Nonetheless, we can still elaborate on some of the particular No-SQL solutions, or solution types, as follows in the next sections.

4.4.2 Key-Value Stores

When looking at Amazon's Dynamo, we already got a feel for what a key-value store is all about. A piece of data – the value – is stored in the database, which, depending on the store, could be any kind of data object, such as an array, a string, or a binary object (e.g. an image file). This object then needs to be accessible for future retrieval and updating. This is done by giving it a unique name, which is what the key is for.

In accessing a key-value store, the mechanism is usually as simple as issuing a set(key, value) function to store the data, or a get(key) function to retrieve it. It could be that the actual data returned has some enhancements, such as a version number. In some of these stores, the multiple copies are replicated throughout the cluster. However, this process isn't instant or what the computer-scientists call atomic, which means it doesn't happen in one go. Therefore, after data is written into a ring of storage nodes, it might slowly ripple through to the replica nodes. Meanwhile, if an application tries to read the same data, it is now faced with multiple versions of the data, as shown in Figure 4.7.

This is the downside to some of the key-value stores. Whilst the engineering is much simplified, resulting in higher efficiency and lower costs, the result is the possibility of potential

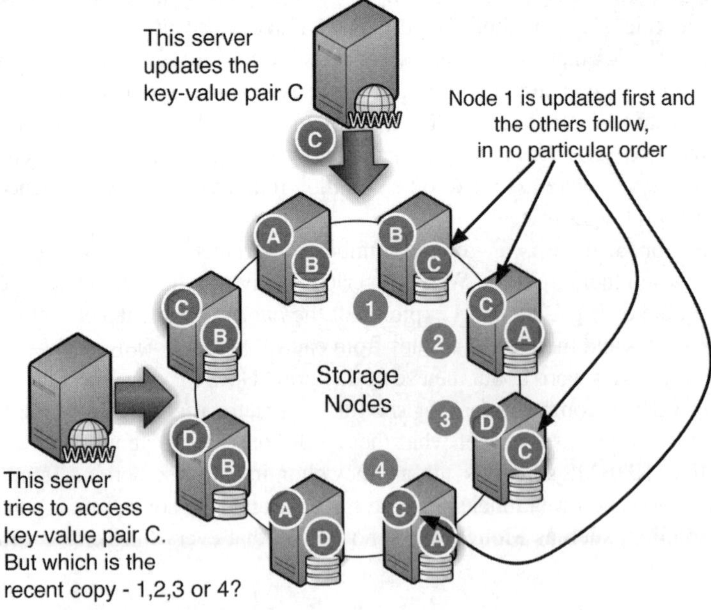

Figure 4.7 Possibly inconsistent data sets across nodes.

conflicts in data during retrieval. However, in many Web applications, it is often relatively easy for the application itself to resolve the conflict. For example, in the case of a shopping cart, the more recent version could be taken as the correct one, but offering the user the chance to confirm that the item is correct, such as by flagging it in a different colour, asking for confirmation. This is in many ways a better experience for the user than having the item disappear altogether (i.e. by assuming that an older version of the cart should take priority).

Of course, the best possible solution is to have such a robust replication method that the problem of version conflicts never arises. But this was one of the trade-offs that I explored in the examination of the Amazon Dynamo solution. A replication model that is eventually consistent dramatically improves availability of the data overall, even at the cost of transient version conflicts. As said, these conflicts are often tolerable in many Web applications. We are not dealing with banking and finance systems that need the full consistency and isolation of data.

Examples of key-value stores are:

1. **Amazon Dynamo** (proprietary, used by Amazon only).
2. **Cassandra** (open sourced by Facebook, as used by them and Twitter and a growing number of Web ventures).
3. **Voldemort** (as invented by and used by LinkedIn).
4. **Tokyo Cabinet** (used by Mixi, Japan's largest social network).

4.4.3 Document Stores

Document stores aren't about storing Microsoft Word docs, or similar. They are similar to key-value stores in a way, except that they group a bunch of key-value pairs into a collection, which gets called a document. This turns out to be useful for a wide set of Web applications, especially social sharing ones where the content being shared is often in the form of a blog, wall-post, comment, or some such similar construct. Here's what a blog post might look like as a document:

```
"Subject": "Those Big Data Stores Again"
"Author": "Paul G"
"PostedDate": "10/1/2010"
"Tags": ["data", "big", "storage"]
"Body": "Document stores are all the rage on the Web."
```

As you can see, we have a series of field names and contents. The contents are fairly unstructured strings for the most part, except for the collection or array of strings in the "Tags" field.

In a standard relational store, we would have to define a schema that named each of the fields and specified the contents data type and length. The problem with this approach is the rigidity of the schema. Later on, I might decide that I need a "comments" field, or even something a bit more complex like a comments table in order to store a whole series of comments per blog post. This would require some major updating of the overall database schema and is often quite difficult to do across a distributed set of storage nodes.

The beauty of No-SQL document stores is that it is possible to update the document structure at any time. In fact, it isn't specified anywhere – there is no schema. There's nothing wrong with inserting a million documents with the above structure, only to insert the next document with an additional "comments" field, populated with an array:

```
"Subject": "Those Big Data Stores Again"
"Author": "Paul G"
"PostedDate": "10/1/2010"
"Tags": ["data", "big", "storage"]
"Body": "Document stores are all the rage on the Web."
"Comments": [ {"Jim", "This blog sucks"};
             {"Geoff", "Nice blog Paul, as ever"} ]
```

Moreover, any of the existing documents could be extended in this fashion – we could update them later with a bunch of comments or an entirely new field. This flexibility works really well for a lot of Web applications and enables an efficient, available and reliable store to scale as needed.

Many of the document stores allow the data to be queried using Javascript, which is widely known amongst the Web programming community. Solutions such as CouchDB allow very complex "views" of the data to be created using Javascript. These Javascript views are called Design Documents and can be stored on the server in order to allow the server to create an index of the view to support rapid retrieval across the entire data set. For example, we could create a view that assembles all of the top comments across a particular set of blogs, so that these comments can be pulled out in a feature on a website, like "Most commented" or "Popular comments" etc. Storing the view and building an index would allow the documents to be ranked in some order (e.g. most popular by number of comments) and then retrieved in that order very quickly.

Examples of document stores are:

1. **CouchDB** – Written in a robust language called Erlang, used by quite a few Facebook apps providers.
2. **MongoDB** – Written in C++ and used by FourSquare, among others.
3. **RavenDB** – Written for .NET/Windows platform.

4.4.4 Graph Stores

In the age of the Social Network, we've seen a lot of interest in a relatively new type of data store – the graph store. This is designed to make storage of social graphs easy, or any graph for that matter (social graphs aren't the only graphs on the block). The main issue with storing social-graph data isn't the records of the people in the graph, it's the ability to traverse the connections between them and to continue following these connections from one graph to another.

To deliver a tweet, for example, Twitter needs to be able to look up someone's followers and then traverse through them quickly. Consider that some Twitter users have over a million followers and many have thousands. Scaling this up to millions of users rapidly becomes a daunting task that pushes more traditional database stores to their limits.

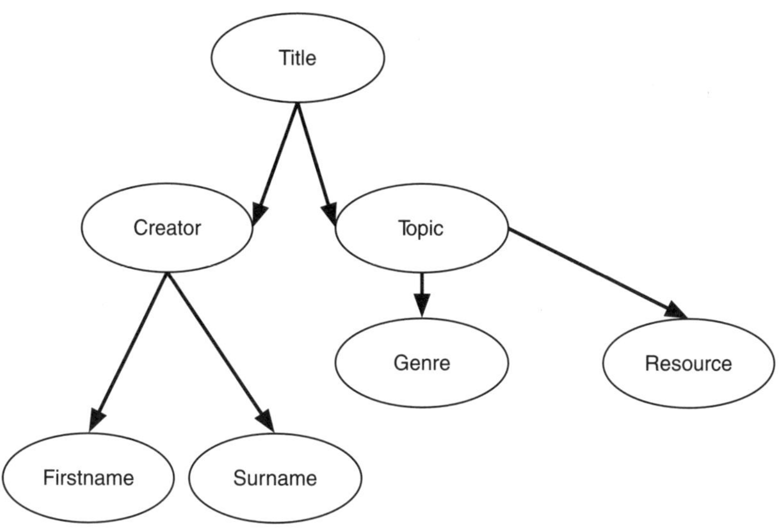

Figure 4.8 Graph structure of RDF-XML schema.

Twitter found that they could only solve this problem by developing their own in-house graph store, called FlockDB, which has now been open-sourced to the developer community.

Example of graph stores:

1. **FlockDB** – Proprietary store developed by Twitter for managing graphs of followers etc.
2. **Neo4j** – An open-source store, programmed in Java, can be used for any graph-based storage problem. It has many features of robust stores, such as ACID compliance.
3. **AllegroGraph** – This is a reference implementation for a new query language called SPARQL, which is a standard from the W3C for querying data stored in the RDF format. Recalling our earlier discussion of RDF, an RDF file is used to describe relationships between data and semantic fields. Here's an example:

```
<div xmlns:dc="http://purl.org/dc/elements/1.1/">
   <h2 property="dc:title">Why tagging is important</h2>
   <h3 property="dc:creator">Paul Golding</h3>
   ...
</div>
```

Notice that the fields can be embedded, or nested, which actually maps to a graph structure, as shown in Figure 4.8. In other words, RDF documents are actually graphs. Not surprisingly then, some of the work in graph stores, like AllegroGraph, is aimed at making the world of "linked data" come to life, helping to bring us a step closer to an infrastructure capable of supporting the ideals of the Semantic Web.

5

Real-Time and Right-Time Web

```
<push time="real">Websockets</push>
<push time="real">XMPP PubSub</push>
<push time="real">PubSubHubbub</push>
<stream timeline="now">Twitter</stream>
```

- The movement of data on the Web has migrated from an on-demand pull mode to a "just in time" push mode. Data increasingly flows across the Web as it becomes available – in real-time.
- Twitter is a real-time service on the Web and has influenced many of the real-time patterns and technologies that have emerged in and around the Twitter ecosystem.
- A number of protocols have emerged to support real-time streams, such as Web Sockets and pubsubhubbub.
- The combination of real-time platforms, "Big Data" and smartphone platforms enables the right data to be delivered at the right time, leading to the "right-time Web."
- Connected services are increasingly "right-time Web" services.
- Telcos need to move quickly to ensure that telco platforms are fully integrated into the right-time Web.

5.1 Real-Time Web and Twitter

5.1.1 Web Becomes Real-Time Thanks to Twitter

Initially,the Web was a relatively docile place. Data moved on the Web when a user clicked on a link. Data was pulled, not pushed. The pace at which surfers wanted to consume and read data, which wasn't all that frenetic, drove the Web's pace. By contrast, mobile data started off as a push system. The original data service on mobile networks was text messaging. Data moved at the whim of the sender: think, tap, deliver!

Connected Services: A Guide to the Internet Technologies Shaping the Future of Mobile Services and Operators, First Edition. Paul Golding.
© 2011 John Wiley & Sons, Ltd. Published 2011 by John Wiley & Sons, Ltd.

Interestingly, the texting system was driven by user content from the start. Users texted each other with their own content, joke-forwarding aside.[1] Later on, the US start-up Unwired Planet developed a system call the UP Browser, which was based on the concept of pulling "decks of cards" of information from a server, a card being a "page" of information, not unlike a Web page. Using the browser, it felt like an interactive version of texting: texts with clickable links, dished up using Web servers. However, rather than evolve into a kind of "interactive texting," which I believed back then should have been the way to go,[2] it became the Wireless Application Protocol (WAP), which was an attempt to shoe-horn the Web browser into a mobile. We all know what a disaster that was in terms of user expectations and actual performance. Adoption was slow, or nearly non-existent.

Move on about ten years, or more, and text messaging eventually begins to take off in the US (where adoption was slow initially) and becomes the germ of an idea called Twitter, a platform for exchanging short messages called status updates. These updates were prompted by the single question: "What am I doing *right now?*" And from that one single question comes a string of possible updates:

"I am writing my book . . ."

"I am on the beach (I wish) . . ."

"I am just about to meet @pgolding for a hack . . ."

"I am bored . . ."

"I can't see the point of this . . ."

That last comment was all too common, or something similar to it, but no need to debate that any more, some 150 million users later (and climbing). Plenty of people told me exactly the same thing about the mobile phone when I was working on the launch of GSM (2G). Oh yes! They said the same about text messaging too, which remains an irony within the operator world, which will otherwise insist on confirmed and concrete user insights before pursuing any new idea.

Two things about Twitter make it interesting and have had a profound effect on the Web and the way many software architects now think about the Web. Firstly, Twitter is based on a "follower" model. When I publish a Tweet, my followers all get a copy of my Tweet (see Figure 5.1).

Secondly, tweets usually hold content that expresses a sentiment, idea or action that is happening **at that moment**. It is a real-time system – an expression of "now." If I tweet out – "I'm having a cup of tea" – then it's probably because I'm actually having a cup of tea at that moment. This differs from blogging, updating Wiki pages and so on. Sure, some blog pages and news pages are updated fairly close to the occurrence of an event, such as some of the journalistic content on actual news sites, but much of the Web is not real-time.

The important point to grasp is that the Tweets are *not* really sent to the users at all! That might be surprising to hear. The Tweets aren't even sent to an inbox or some kind of queue where the followers can pick them up, like an email message. Whenever I tweet, the Tweet is published to a *public* timeline, which we can think of as a stream of everyone's tweets flowing by.

---

[1] Sending jokes and forwarded messages between friends was a major behaviour amongst early texting adopters.

[2] Not instead of mobile web browsing, which was inevitable anyway. If we had "Interactive texting" today, we would have enabled a far richer ecosystem around mobile early on. I said the same about open address books on phones and using FOAF format, all duly ignored by the operator community.

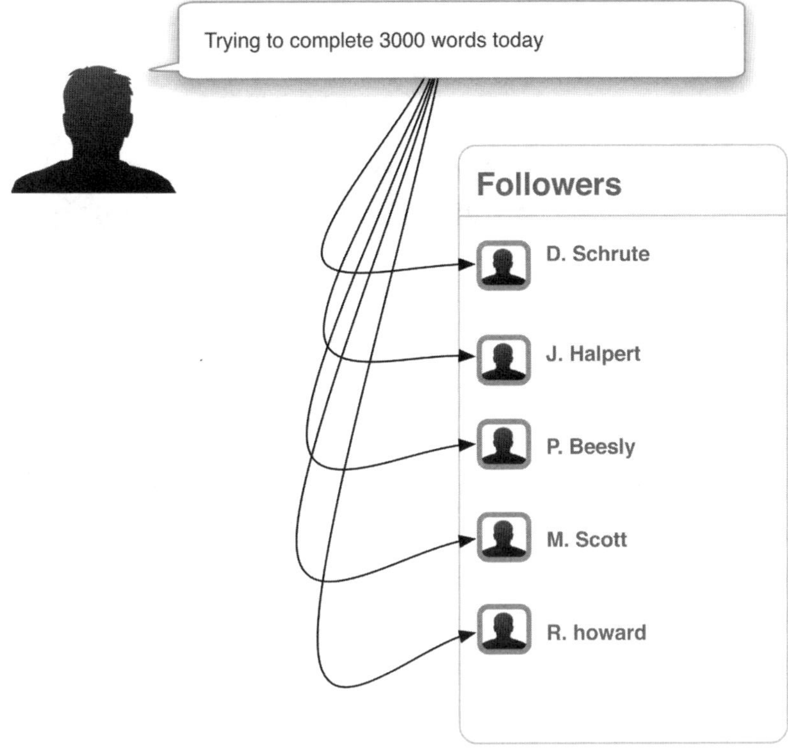

Figure 5.1 Tweeting to my followers.

Just like a fisherman can only catch fish as they swim by at the moment he's standing on the bank, followers can only pick up tweets at the moment they are published to the public timeline, as shown in Figure 5.2.

In this sense, Twitter really is *hard* real-time – either followers capture things as they happen, or not at all.[3] Twitter rapidly evolved to support a vast network of real-time conversations. A follower can post a Tweet back into the timeline a little while later, optionally using the recipient's Twitter username (e.g. @pgolding) to catch his or her attention, as shown in Figure 5.3:

Note that the reply Tweet is still in the public timeline – any one of the sender's followers can see it.[4] Of course, for a "conversation" to take place, the original Tweeter ideally needs to be a follower of the respondent.

The ability to respond to Tweets and post them back out into the public timeline has enabled a kind of "side channel" for thinking out loud, swapping ideas, updating each other of news. Indeed, at many conferences and events, especially tech ones, but increasingly others, Twitter

[3] Actually, all the Tweets are stored on the Twitter platform, enabling a user to go and look at a user's historical stream of Tweets, but this isn't the normal modality of usage.

[4] In fact, any Twitter user can see the Tweets in the overall public timeline for Twitter, where *all* public Tweets are visible.

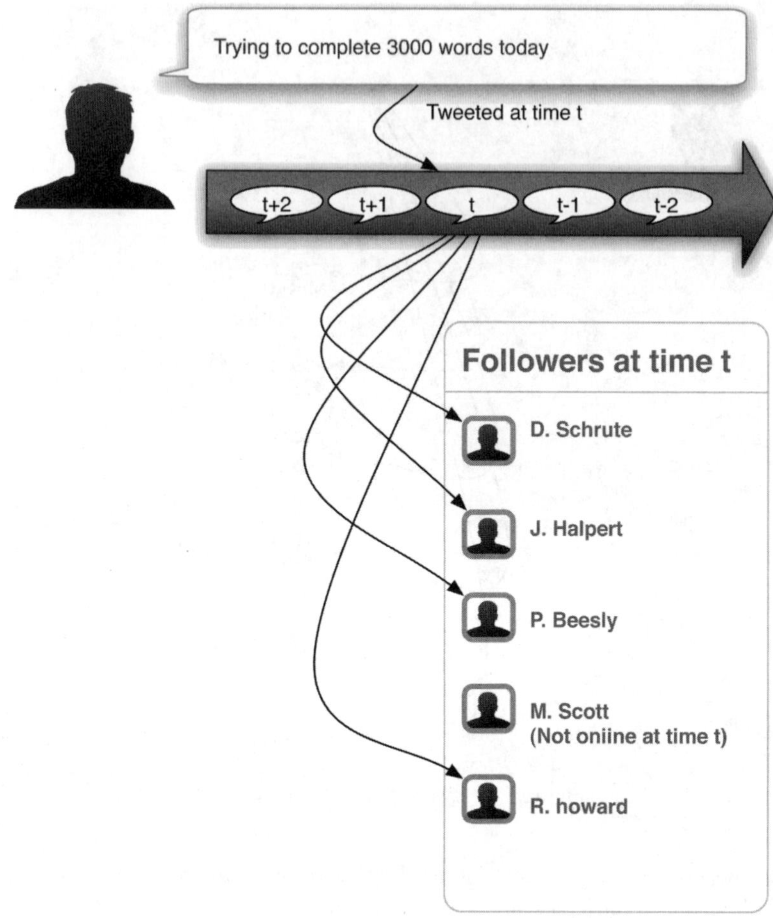

Figure 5.2 Tweeting to the public timeline.

is used as the official "back channel" for conversations. People present whilst the audience tweet, chatting among themselves while involving their followers, most of them physically absent from the conference itself. This can have some fascinating side effects.

I have taken part in many conferences via the Twitter stream. The way it works is that Tweets are marked with an unofficial "hash tag," which is the # character followed by a word, such as #connected or #techcon. Twitter users can set up a real-time search to scour all Tweets in the timeline for the search term, usually a hash tag. In this way, it becomes possible to follow a particular "topic" or "channel." Much of the time, these hash tags appear only for the duration of the conference and then don't get used any further.

Similar to Twitter, other Web applications, such as Facebook with its shared status updates, have gradually shifted usage patterns of the Web towards something more real-time and pervasive. This trend is creeping into the enterprise with tools like Yammer and micro-blogging extensions to the popular Microsoft intranet product Sharepoint. From a technological perspective, the emergence of Ajax-powered user interfaces has enabled the move towards

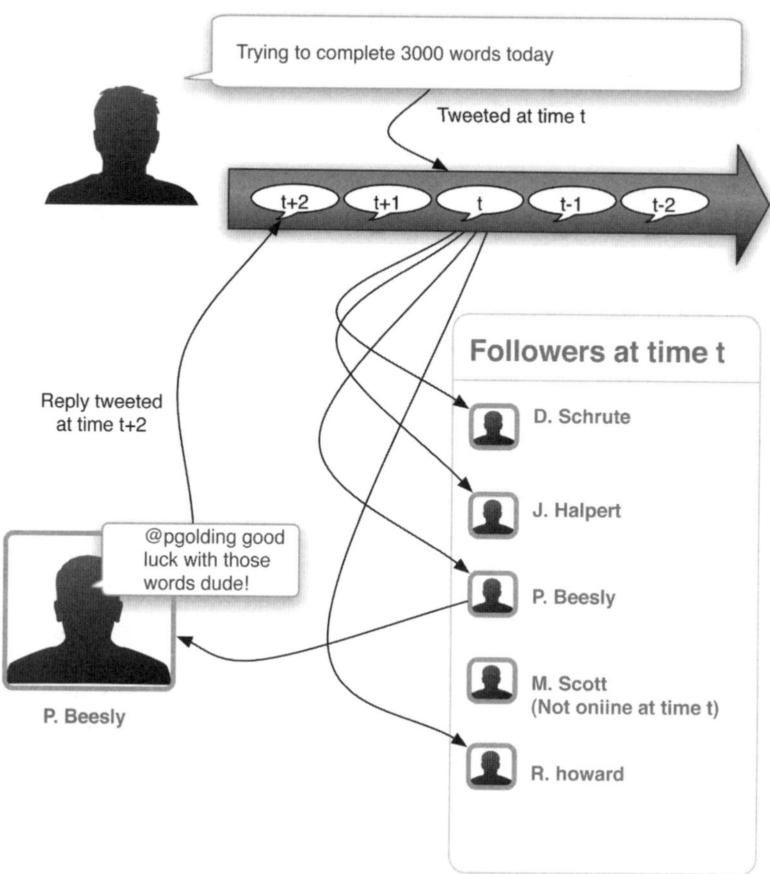

Figure 5.3 Sending replies back to the timeline.

real-time display of information in the browser, which is only set to become easier and more widespread through the spread of HTML5 browsers. Ajax enables a single user-interface element in a Web page to be updated without the browser having to go back to the server for an entire (relatively slow and clunky) page reload (see Section 6.5 Key Platform: The Mobile Web).

Ajax works well and has been widely adopted by many sites. It was first popularized by sites like Google who used it to fetch possible search terms from a server as the user began to type in characters to the search box. However, Ajax was originally designed to use standard HTTP requests back to the server. The HTTP protocol is essentially a pull mechanism – clients have to send a GET request to the server in order to ask for data. This means that events are being driven by the rate at which the client is polling the server. However, in the real-time world, events are happening, like Tweets, at the rate that the source is updating information, like status updates from Twitter users. It would be ideal then if we had a technology that enabled data updates to be pushed out to the browser, assuming that one is available to receive updates. This problem has now been solved with something called Web Sockets.

5.1.2 *Web Infrastructure Goes Real-Time*

5.1.2.1 What About Instant Messaging?

It wouldn't be correct to say that the Web has moved to real-time thanks only to Twitter. It was already real-time with services like Instant Messaging – the clue's in the name! A key difference has been the openness of the technology. Originally, the most successful IM services, like AOL and MSN, were walled-garden services. It wasn't possible to access them via APIs and thereby grow the infrastructure into a wider ecosystem. This hasn't really proved to be a limitation to the relative success of those communities, but it has certainly prevented IM from becoming something greater than the instant messaging that we are all familiar with. For example, we haven't seen buttons all over the Web with "IM This" on them. No. We see lots of buttons with "Tweet this," instead. Did the guys at Twitter sit around and think that we need a "Tweet this," button? No. The idea came from use of the API, open to any developer to innovate with.

It was possible for IM to become something more Twitter-like and it might well have done had innovators been allowed to extend the service. Interestingly, IM was used as one of the early interfaces for Twitter, although scaling the IM support proved to be too cumbersome for them, so they dropped it. But there has been some innovation with IM, thanks originally to the open source project called Jabber, which was an effort to develop a scalable and open IM solution and standard. It eventually went on to form the Extensible Messaging and Presence Protocol (XMPP) standard.[5]

I was asked by the GSM Association to evaluate XMPP and other IM solutions alongside the Instant Messaging and Presence (IMPS) standard developed by an alliance of mobile companies in the Open Mobile Alliance (OMA) forum.[6] At the time I noted the lack of openness as a potential issue for IMPS. You probably won't be surprised to read that IMPS didn't succeed. No operator went on to release it in any successful fashion, mostly because of the way text messaging had become so wildly popular and profitable. Operators didn't want to cannibalize their own texting revenues.

Avoiding cannibalization is fine, but the closed and limited approach is one of the key reasons that the operator technology environment has not succeeded in growing from a transactional and connection-orientated one (crudely known as the "dumb pipe") to a services one. It seems that an openness and APIs are a key ingredient to the creation and sustenance of a healthy services ecosystem. I believe that with the right level of openness for texting and IMPS (or similar, had operators adopted XMPP) things might have been different.

Returning to XMPP, it is an open standard with APIs that potentially anyone can use to build new services on top of XMPP. This means that XMPP can feature as an ingredient inside of a software solution on the Web. This has led a number of innovators to create solutions that use XMPP to carry messages in a way not originally intended, simply because XMPP turns out to be a reliable means of sending messages across the Web. In fact, XMPP is better thought of as a means to send chunks of XML[7] (called XML Stanzas) across the Web. However, it

[5] The Jabber protocols were originally developed within the Jabber developer community in 1999 ("XMPP 0.9") and subsequently formalized by the IETF's XMPP Working Group in 2003 and 2004, resulting in definition of XMPP 1.0.

[6] See my book *Next Generation Wireless Applications* for a comparison of the various IM protocols – pp. 198–205.

[7] See my book *Next Generation Wireless Applications* for an introduction to XML – pp. 227–229.

has to be said that XMPP is more difficult to program for than HTTP. After all, Web servers and frameworks (like Ruby on Rails) are built to use HTTP out of the box. XMPP requires a different type of connection management that isn't supported by default. I shall return to XMPP in a minute when I explore PubSubHubbub, as it is a similar technology pattern to the PubSub feature of XMPP.

5.1.2.2 From Polling to Pushing

Before Twitter arrived, folk were already frantically trying to keep up with changes on the Web – What's new? Who's done what? What's changed? Questions like this have one problem, as do all questions. They can only be answered at the time of being asked. In other words, on the Web, if you want to find out if a page has changed, then you'd have to keep visiting that page to find out – to "ask" the question "What's changed on this page?" This mode of asking to get an immediate answer is called *polling*.

Polling is a technique used widely in software so that one piece of software can get information from another. The GET cycle for fetching Web pages is an example of polling (see Section 2.2 Beneath the Hood of Web 2.0: CRUD, MVC and REST). In this case, the Web browser polls the Web server for some information. It's like asking "Do you have some content at this address?" If the server does, then it replies with the word "Yes" (which in the HTTP world is written "200 OK") and then sends the answer along too.

Polling is inevitable whenever there's a system that only responds to requests ("questions") and has data that is likely to change thereby causing an external system to keep making requests. The Twitter API is another example of a polling interface. If a Twitter client, like the popular TweetDeck desktop interface, wants to display the next set of Tweets in my timeline, then it has to poll the Twitter API to ask for the next set of Tweets.

As our appetite for information and services has increased dramatically on the Web, our expectations for real-time information updates have driven the performance and architecture of various software components towards real-time behaviour until, in some cases, the limits of polling have been reached, forcing developers to create alternatives.

For example, bloggers have popped up everywhere to add vast amounts of information to the knowledge pool on the Web. Keen readers of blogs want to get to know when the next blog post is available, but it simply isn't viable to keep popping back to see if the blog has changed. Developers produced a solution to this problem, which was called RSS.[8] It allowed a single summary of all the blog posts to be aggregated into one file. This file can be accessed using an RSS reader that works on behalf of the user to check for updates on blogs by monitoring updates to the RSS feeds. This turns out to be a convenient way to monitor lots of blogs in real-time, as the RSS reader is polling the blogs for us.[9]

Often, polling is the only way to handle the transfer of information between two software systems. If the "producer" of information doesn't know who might request it, then it can only sit and wait until someone (a "consumer") asks for it. The producer can't deliver the information to consumers it doesn't know exists.

---

[8] RDF Site Summary, the first version of RSS, was created by Guha at Netscape in March 1999.

[9] Note that the RSS reader is what enabled users to get a "push like" experience of blog updates. The RSS format itself was invented for the purposes of allowing blogs to share (syndicate) content with other sites.

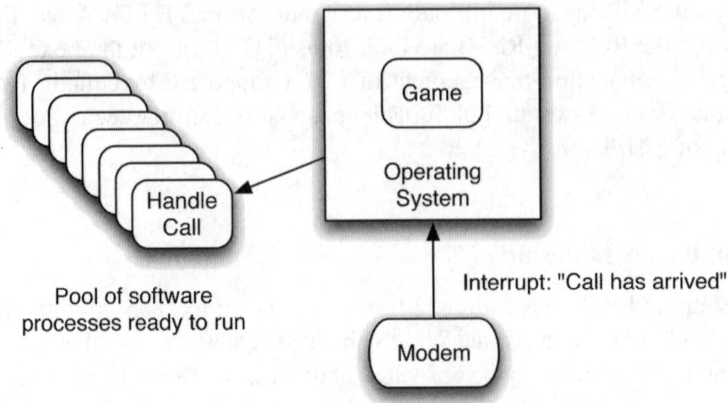

Figure 5.4 Interrupt mechanism on mobile operating system.

However, there is another technique, which is generically called "push." This has always existed deep inside the bowels of systems like mobile phones. For example, a mobile has to be ready to accept an incoming call. It doesn't have to poll for a call. The network will send out a paging notification to the cell where the mobile is currently located. The modem inside the mobile will pick up the paging request and now has to tell the software running on the processor that a call is arriving. The software on the phone might otherwise be busy doing something else, like running a game on the phone. It wouldn't make sense for the game to keep "polling" the modem to say "Do you have a call for me yet?" This would be very inefficient, likely to slow the game down. Besides, most of the time, the answer will be "no – sorry, no calls yet . . ." and polling will just be a waste of time.

What happens in a mobile is that the modem will interrupt the operating system to say "Hey, something's going on here – you better deal with it." In response to that "interrupt" (which is also the technical name for it), the operating system will pause the game and set the "Handle call" software routine running instead, as shown in Figure 5.4.

This method of push has been largely (though not completely) absent from the Web infrastructure, which was built around the pull-based HTTP protocol. Even email, which appears at first to be push (because the mails just appear in the mail client) is a pull system, using the POP3 protocol to pull emails from the server. The mail client (like Outlook or Thunderbird) continually polls the server to ask "Do you have any new emails for me?" The polling tends to be every minute, thus giving the illusion that things are happening in real-time.

The rapidly evolving trend towards real-time flow of information has led to the creation of either entirely new methods of interchange, based on push, or to the adoption of existing techniques, such as XMPP, adapted for other uses (or simply discovered for the first time because of the growing interest in the real-time experience of the Web). In the next sections, I'm going to look at several push techniques:

1. Websockets
2. XMPP PubSub
3. PubSubHubbub

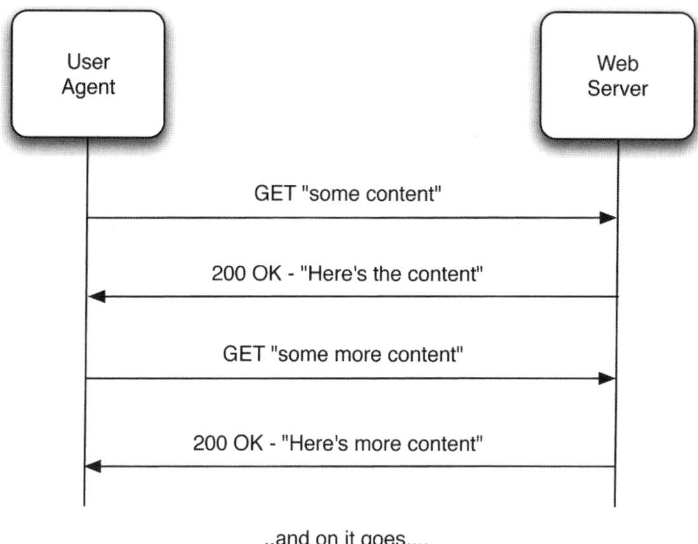

Figure 5.5 GET polling from a browser to a Web server.

5.1.2.3 Pushing Data with Websockets

I love the HTTP protocol. It's enabled the Web to flourish. I discuss it in detail in my book *Next Generation Wireless Applications* (pp. 160–167). We can skip those details and get to the essence, which is that HTTP is a *polling* protocol where the HTTP client (called a User Agent[10]) asks (using a command called GET) and waits for a response. That's it. The two ends become disconnected and have nothing to do with each other. If the user agent wants more data, it has to issue another request, as shown in Figure 5.5.

But what if we want to establish a flow of data between the two end-points, such as a constant stream of Tweets from Twitter, or a stream of updates from an Instant Messaging (IM) session in a chat room. It's inefficient for both ends to keep up this malarky of endless GET-request-response cycles. A better way would be to open up a channel that stays open and allows the data to flow as and when it's ready from the server or client, such as IM messages.

This is exactly how Websockets works. The initial request (called a handshake) from the user agent (e.g. To http://example.com") is to open a permanent Internet Protocol (IP) connection, otherwise known as a socket. If the server can support the socket connection being requested, then it upgrades the connection to a socket and both the end-points switch over to this connection (e.g. To ws://example.com). The data can then flow freely, meaning that it gets pushed automatically by either end-point without the opposite side having to request it, as shown in Figure 5.6.

[10] This is the official term used in the HTTP specification. HTTP can be used by any software object – it doesn't have to be a browser.

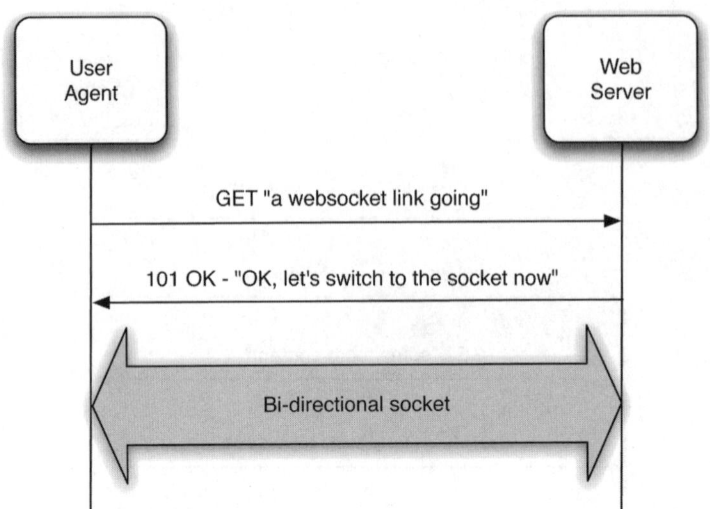

Figure 5.6 Websocket handshake.

The advantage of Websockets is that real-time communications can take place between two HTTP end-points. It also allows the use of port 80[11] for the socket connection, which is the same port used for HTTP itself. This means that the Websockets connection should work through corporate (and home) firewalls, which are already open on port 80. Otherwise, Websockets connections would most likely be stymied by firewalls.

You are probably wondering what takes place over the Websockets connection itself. The answer is anything that the two endpoints want to engage with, so long as the data being sent is text-based (i.e. not binary, like video). The data gets sent in frames with easily detectable beginnings and endings (see Figure 5.7).

The communications takes place via the message frames for as long as the two end-points want to engage in a dialogue. Control frames are used to manage the ongoing connection, such as the "Ping" and "Pong" frames that are just used to check if the connection is still open – one end-point sends a ping for the other to respond with a pong. The control frames are also used to signal the end of a connection, requesting it to be closed. Once closed, the socket is gone and the two end-points go their separate ways.

If the end-points wish to set up a new Websocket connection, they will have to go through the HTTP handshaking again. Of course, a Websockets connection assumes that the endpoints are able to consume frames from each other. However, this is implementation detail. The Websockets protocol merely says how this type of socket-based (push) connection can be established via two HTTP end-points. It doesn't say how the connection should proceed thereafter. It is a very generic method of establishing bi-directional push connectivity.

[11] IP socket connections are assigned to ports, which you can think of as a house address on a particular street. This is so lots of IP connections can exist between two end-points without colliding – for example, HTTP on port 80 and email on port 25.

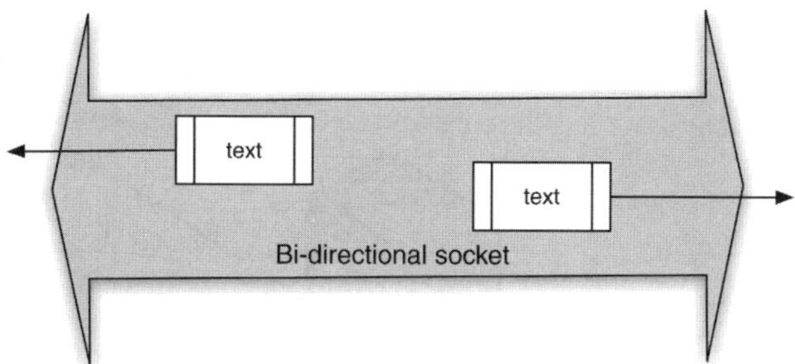

Figure 5.7 Websocket frames.

5.1.2.4 Pushing Data with XMPP PubSub

I just described how a Websockets connection can be set up between two HTTP endpoints, which then upgrade the connection to a Websocket connection allowing ongoing asynchronous communications – messages can be sent in frames, pushed over to the end of the socket to the opposite endpoint without it requesting the data. This is a generic protocol that will address a growing number of real-time push-based services that are emerging on the Web. The Websockets protocol is an ideal companion to HTTP and is emerging in response to the growing evolution of the Web to a real-time environment. However, before Websockets came along, real-time push-based communications was already possible, although with particular application domains in mind, such as IM and chat-rooms.

One such existing mechanism for pushing data is the XMPP-PubSub protocol, an extension of the XMPP family of base protocols for streaming XML data over the Web. In PubSub, a user publishes some XML data to an XMPP server (let's call it a node – you'll see why in a minute), which supports the PubSub extension. Other users are then able to "subscribe" to this node. Whenever the node changes, a notification will be *pushed* to all subscribed users. The origin of this mechanism is multi-chat, or the idea of online chat-rooms hosting a number of participants. For a chat-room to work, it is essential to have a reflective behaviour whereby any message sent to the chat-room is instantly pushed out to all the participants, as shown in Figure 5.8.

The software clients (XMPP endpoints) for each participant do not want to keep polling for messages. That's fine because XMPP will support pushing messages to end-points, just like Websockets. However, how does the node know where to send the messages? That's taken care of by the PubSub protocol. The idea of the protocol is to allow an XMPP server to set up a topic, which might be the name of a chat-room for example. Once set up, our XMPP server becomes a "publishing node" on behalf of that topic, ready to publish any data it receives (which in our case would be chat messages typed into a client, say).

The PubSub protocol allows clients to "Subscribe" to the topic, telling the node that they want to get copies of any data posted (pushed, or published) to the node. This is the essence of PubSub – end-points can publish data to a node (or topic) and end-points can subscribe the node (or topic). End-points don't have to be both publishers and subscribers. Clearly, in a chat-room scenario, all end-points would naturally want to be publishers and subscribers. Not much fun in a chat-room if you can't dialogue! However, for more generic usage as a

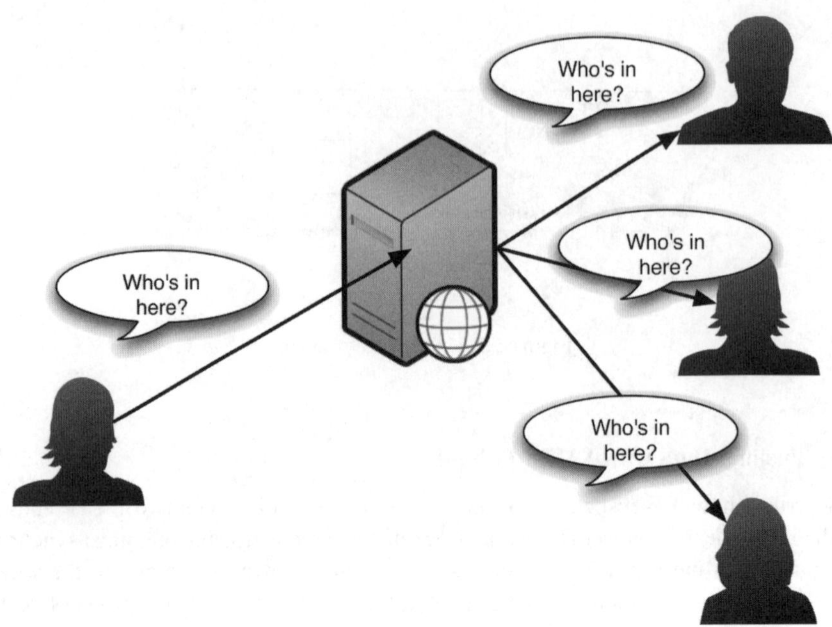

Figure 5.8 Message reflection in a chat room.

way to push data updates to end-points, it makes sense in a number of scenarios that most of the end-points would be subscribers. Any one XMPP server could host a number of PubSub nodes. This makes sense if we think of the chat-room example again, where we'd probably want to host a number of different chat-rooms on our server.

5.1.2.5 Pushing Data with PubSubHubbub

PubSubHubbub is similar in principle to PubSub, except that it was intended to add a push mechanism to RSS (or Atom[12]) feeds. End-points that understand the PubSubHubbub protocol can get real-time notifications via HTTP when a topic (RSS feed URL) they're interested in is updated. Let's break this down a bit.

Let's imagine that we want to monitor updates to a blog, which is hosted on a blog server. When the blogger updates her blog with a new blog article, the blog software automatically updates the RSS feed to include the new blog post. The RSS feed for this blog declares its hub server in its Atom or RSS XML file via a special tag in the file called a link tag, which looks like this:

```
<link rel="hub" some_link_to_the_blog_hub>
```

[12] Atom is a similar principle to RSS feeds.

The hub might be run by the publisher of the feed, or might be a shared hub that anybody can use. In Figure 5.9, we're showing it as a separate server.

Server C fetches the RSS feed but is unable to consume PubSub notifications. Therefore, as shown, it continues to poll the RSS feed the old-fashioned way. However, servers A and B can both handle PubSub. Upon seeing that the RSS feed has a Hub address (the embedded link tag), they both register (subscribe) to the Hub. They do so by each giving a unique "callback" address, which is a URL that the Hub can use to reach either server in order to push the notification that a new topic has been published via the RSS feed.

When the blog next updates the Topic URL, the publisher (blog) software pings the Hub saying "hey – I've just updated my RSS feed dude." The hub efficiently fetches the published feed and multicasts the new content out to all registered subscribers using their previously registered callback addresses. This is done via HTTP (the callback addresses are URLs) thus avoiding the need to implement any additional protocols, such as XMPP.

Although the XMPP PubSub and PubSubHubbub protocols look similar, it is clear that originally they have been designed with different purposes in mind – multi-chat notifications versus RSS notifications. However, there is no reason why the developer needs to keep to these application domains when deploying either of the techniques. For example, PubSub could be used to push any type of information over the Web where there is the need to notify disparate and distributed systems with some kind of information update. For example, it could be used to push real-time weather updates or financial information. The end-points don't have to be interested in syndication of content either. They can consume it however they want, depending on the application.

5.1.3 The Real-Time Nature of Mobile

The drive of the Web towards real-time has caused a much greater interest in mobile as *the* device for interacting with real-time services. Texting, in particular, is already an ideal means to convey real-time data, as exploited by numerous notification services already (and long before Twitter arrived on the scene). As noted above, Twitter set out to exploit the real-time nature of mobiles, including the fact that we carry them with us at all times – a fundamental and obvious prerequisite for engaging with real-time services.

There is a wonderful arc of innovation to be marvelled here. Twitter started out as an essentially mobile service, but didn't get much traction initially. However, by exposing its functions through an open set of APIs, innovators started to play and ended up creating desktop clients, like Tweetdeck, rather unexpectedly. This enabled users to really grasp the potential and unique attributes of Twitter through allowing a Twitter client to stay perpetually running on their desktop, presumably while they sat there using it to do other things. In this mode of usage, it became what I call an "ambient" experience – always on and sitting most of the time in "the background," but still within our consciousness, enabling us to dip in and out as required. Through desktop discovery of its subtleties and potential, Twitter became more popular.

Moving back to the mobile, Twitter clients then followed their desktop contemporaries and have become a much better way of consuming the service on a mobile device. This in turn has provoked further, and very significant, innovation, particularly the inclusion of pictures

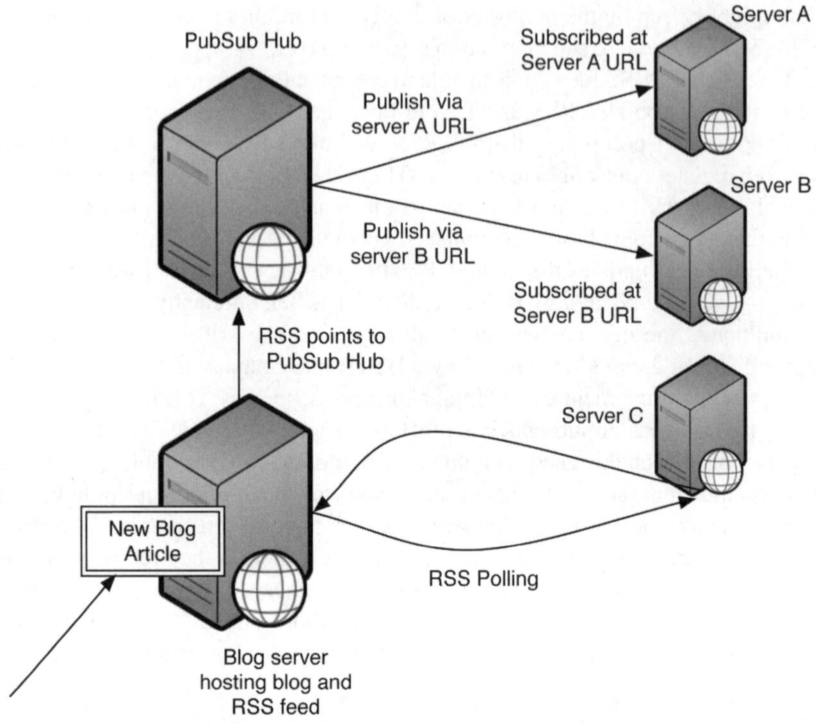

Figure 5.9 A typical PubSubHubbub configuration.

and location, exploiting the camera and GPS capabilities of mobiles. For many users, these features – unintended by Twitter – have become an integral part of the experience. However, perhaps more significantly, the location feature was absorbed back into the mainstream Twitter service, which now officially supports geo-location of Tweets. In fact, unknown to most Twitter users, a Tweet is much bigger than its 140 characters. Beneath the service, a Tweet is a far large message structure, as shown in Figure 5.10.

As you can see from the map of the Tweet, the actual 140 characters of text is a tiny part of the overall data structure. The geo-tag feature has been fully absorbed into the data structure. Twitter now encourages (and offers) a variety of location-enabled Twitter services. On top of this, the popularity of hash tagging and the vast number of uses adopted by Twitter users has led to the idea of adding annotations to Tweets, called "Twannotations," not shown in this map. This enables meta-tagging of Tweets so that, for example, we can tell that a Tweet has come from a music application, like Spotify, and thereby aid in the creation of a sub-system ecosystem of music-related Tweets.

A further development of Twitter has been its adoption by many other Web services as a kind of real-time "hailing" channel, with so many sites offering the facility to "Tweet this . . ." and push out site-related nuggets of information into the so-called Twittersphere. This has become particularly common with blogs, for example, where a new blog post pushes a Tweet out to announce its availability. For many blogs, Twitter has become an important mechanism for "advertising" articles. I certainly use this myself for wirelesswanders.com. Interestingly,

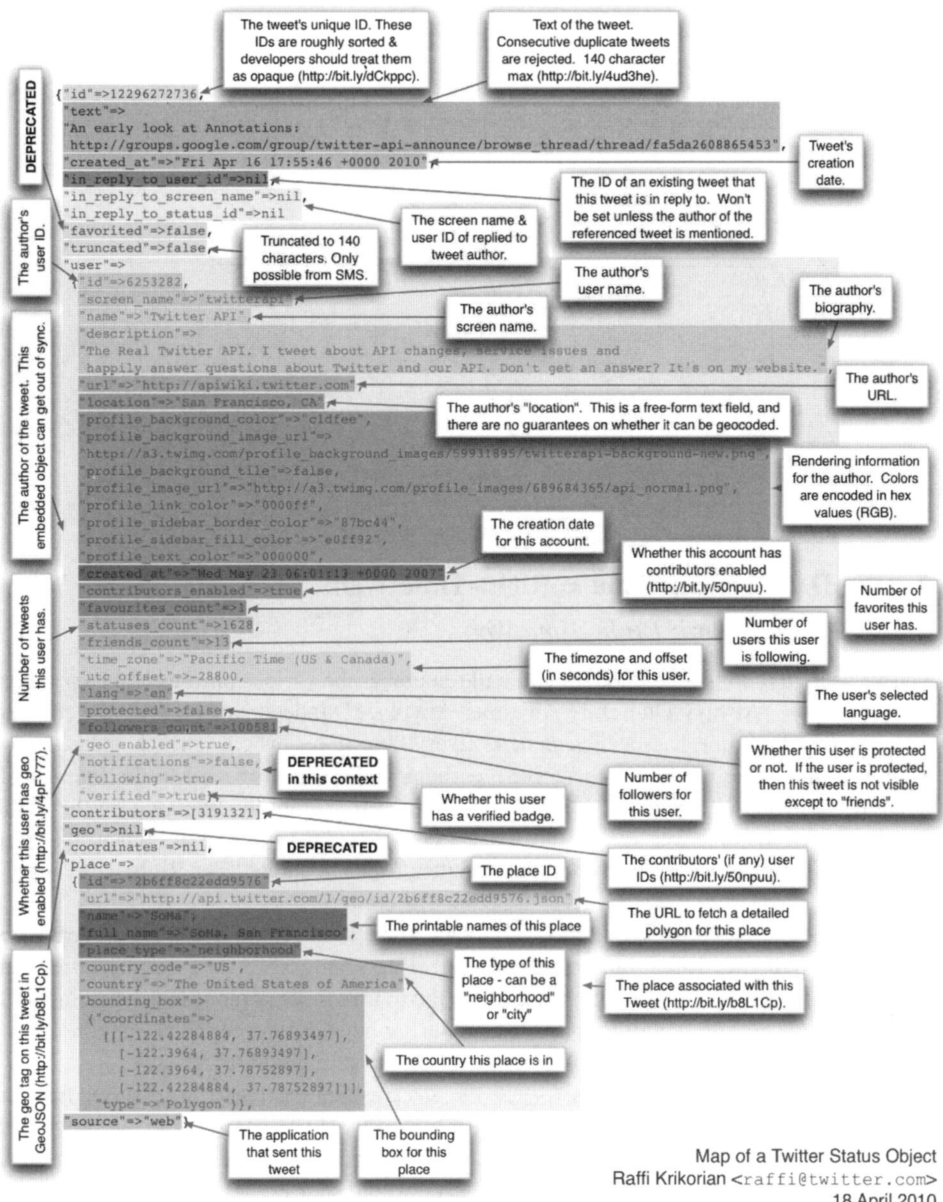

Figure 5.10 Map of a Tweet. Courtesy Raffi Krikorian from Twitter.

comments on my blog have mostly been replaced with chatter and commentary now taking place via Twitter.

In conclusion, we have come full circle from mobile to desktop and back to mobile in the evolving story of Twitter, which has extended our use of the Web into a very real-time experience, aided greatly by the real-time nature of the mobile.

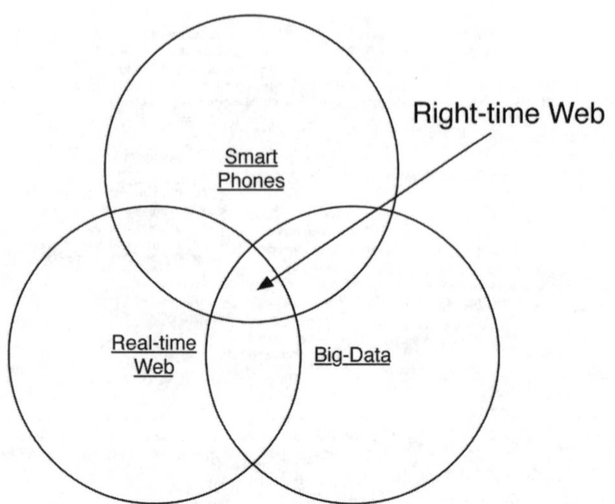

Figure 5.11 Right-Time Web.

5.2 Big Data + Real-Time = Right-Time Web

5.2.1 New Buzzword: Right-Time Web

If you've been in mobile long enough, you will remember the various phrases and buzzwords like, "Ubiquity," "Always On" and "Anytime, Anyplace." Indeed, in the UK, the merger of Orange and T-Mobile has been renamed "Everything, Everywhere," which indicates an evolution of how ubiquity is now perceived, perhaps with a tip of the hat to the Internet of Things. It is interesting how this idea of "right-time Web" is rapidly emerging on the Web. When I attended the first ever Twitter developer's conference (Chirp) in San Francisco (April 2010), Dick Costello, the COO (now CEO) of Twitter, used the phrase "The Right-Time Web," which he got from David Pakman, a VC attending the conference.[13]

The concept is about articulating the need to add some kind of filtering, or intelligence, to the real-time Web (i.e. Twitter and other streams). Real-time streams are getting fatter and faster. It's possible to sit at the desktop and watch the murmurings of thousands whizzing past our screens. We can also get all that activity on our mobiles. However, it's not that useful if we don't need or want the information contained in those streams all of the time.

What we probably want to do is to fish the streams for the right information for our circumstances, or context. For example, I probably don't want to hear how great the barista is in the new downtown deli if I'm out of town. But, when I am downtown and happen to be looking for a great place to eat, drink and surf, then it would be useful to know that a great new barista is in town. Similarly, when I am in the market to buy a digital camera, I would love to know who in my social and information networks has recently researched similar items and what they did about it.

It seems that we now have all the ingredients in place to allow the real-time Web to become the right-time Web: smart phones (with location), real-time Web and Big Data as shown in Figure 5.11.

---

[13] Who claims that he got it from a colleague, Brian Ascher, at his venture firm.

The opportunities at this intersection are huge! The availability of the right information at the right time – and in the right place – is enticing. This is more than just an idea. We now have the infrastructure and tools in place to make this happen. It is an idea that operators have been talking about for a long time: "Context-aware Services." I wrote liberally about it in the first edition of my book *Next Generation Wireless Applications* when I discussed location-based services. I updated the thesis several times, including a lengthy article commissioned by Vodafone's Receiver online journal, which I titled "Riding the Timeline with Widgets."

What I hope is becoming clear to you, and will become clearer throughout this book, is that targeted streaming of real-time info is no longer just an idea. It is a reality. Let's explore the key ingredients and trends that are making the right-time Web possible. I'm only going to summarize the components of the right-time Web here because each component is described in detail elsewhere in the book.

5.2.2 Key Components of Right-Time Web

5.2.2.1 SmartPhones

We use the term "smartphone" all the time, as if we know what it means. We mostly do, of course, but in the context of right-time Web, the question is what makes a smartphone smart? The following diagram gives us a clue (see Figure 5.12).

1. **Location** – The ability to locate the phone accurately is a key part of right-time Web. There is still a lot to do here because not every phone on the market has accurate location capabilities, such as GPS. However, the market trends are towards greater smartphone

Figure 5.12 What makes a smartphone smart?

penetration. What we're missing thus far is a better mechanism for dealing with proximity, which has two dimensions. First is nearness to objects, plus the ability to interact with them. Second is nearness to any point of interest (POI) on a map, both outdoors and indoors. The proximity issue is partially being addressed by the "checking in" behaviour seen with services like FourSquare, which I explore in 7.1 Real or Virtual Worlds worlds? However, much of the problem is being dealt with by sensors.

2. **Sensors** – The ability to interact with the physical world is increasingly what makes a smartphone smart. The camera is an obvious example and we are beginning to see the emergence of powerful image-processing capabilities on some devices, enabling object detection and recognition, folded into augmented reality experiences and services (see Chapter 7. Augmented Web). However, sound, temperature and movement will become more important, as will the ability to extend the phone's sensor-sphere using peripherals, such as personal health monitoring sensors worn on the inside of vests, as discussed in 7.2 Sensor-net: Mobiles as sixth sense devices.
3. **Persistence** – We all know about the importance of persistent connections, but what really counts these days is the ability for applications themselves, rather than connections, to be always on, running in the background and ready to come alive at the right time, bubbling up information to the user's attention. Various elements of modern mobile operating systems are making this possible, which I will explore in Chapter 6. Modern Device Platforms.
4. **Connectivity** – Of course, without connectivity, especially to the universal dial tone of "http://" smart phones would be anything but smart. Physical connection is important. Networks are getter faster and fatter, although they are often unable to keep up with the pace of data demand for voracious data users clustered together in dense downtown areas. The iPhone has driven a lot of network traffic wherever it's been launched, often exposing weaknesses in the underlying infrastructure's ability to cope with data-hungry applications like video, or simply with the large amounts of "app connection chatter". The outstripping demand for data has led to a back-peddling from the position of "unlimited data" towards a more realistic capped-data offering. This is likely to continue and there will be an increasing need for application designers to become more aware of data consumption when designing applications, perhaps leading to a system of data-friendly ratings, like the eco-friendly ratings of electrical equipment.
5. **Rich Applications** – The power of modern device platforms, like iOS, Android and HTML5, is opening up a whole new category of possibilities. Combined with the powerful processors needed to run these platforms, we see possibilities to incorporate heavy number-crunching not just within the cloud, but within the applications themselves – image processing is an obvious example. Without the ability to make sense of the data in real-time, the right-time Web will not be as responsive as it needs to be. We have seen time and again that users are extremely sensitive about delays, which is why local apps often win over browser-based apps, although with HTML5 the distinction between native and browser based begins to blur. I shall look at this in some depth in 6.5 The Mobile Web.

5.2.2.2 Asynchronicity and Streaming

With the ability to push data around the Web, it becomes increasingly easier to decouple processes in time. It's no longer necessary for one system to co-ordinate with another in time (i.e. synchronize) in order to exchange or grab information. Whenever the information becomes available, the source end-point can push the data on its own way towards where it needs to go. I've reviewed this in previous sections when looking at various real-time mechanisms on the Web like XMPP-PubSub and Websockets. This asynchronicity ripples all the way out to the user because smartphones are capable of handling an increasingly wider set of push channels, from the original (and still well used) text message channel to specialized "side bands" like the iOS push service.

The Web is becoming increasingly data driven. Changes in the data, whether triggered by other systems, people or events, are what cause the data to move around the Web. This is all made possible because of the powerful software infrastructure that makes up the emergent Web Operating System (Web OS), which we have already explored in some depth (see Section 3.1 Why is the Concept of a Web OS Important?).

5.2.2.3 The Web OS and Right-Time Web

The underlying infrastructure of the Web has evolved from something very transactional in operation and architecture to something more stream-based – PubSub, Twitter, Facebook, XMPP, and many systems (or protocols), all support streams of data flowing across the Web. This mechanism is a key underlying component of the right-time Web, as shown in Figure 5.13.

What are the key technological themes and foundations that make this stream-driven architecture work? They are as follows:

1. **Cloud computing –** The availability of large amounts of processing power in the cloud is a major ingredient to the innovation that has led to so many software services being available to power the Web OS. Services like Amazon Web Services are responsible for making an entire new breed of start-ups and software engineering endeavours possible. Powerful, always available and elastic computing resources are at the heart of the modern right-time Web revolution. Put simply, real-time streams need an engine that's always collecting and pumping out data. Cloud-computing gives us the engine power in abundance.
2. **Open APIs –** I really ought to say "Open Innovation," which is more about the ethos of so many Web 2.0 projects and services. What works, and what powers the right-time Web, is the low-friction APIs that enable services to easily share data. Without APIs, there would be no Twitter and no successful stream-based services on the Web. We can contrast the text-messaging world, which is closed,[14] with the IP world, completely open, and witness how literally thousands of new messaging services and paradigms have emerged on the Web versus mobile. With mobile we are still stuck with the one mode of messaging, which is texting, virtually unchanged since its introduction in the 1990s.[15]
3. **Open and Portable formats –** This often gets overlooked, but I think that it is going to become increasingly important, reaching a tipping point soon. If we didn't have base data

[14] I know that there are APIs to send and receive messages, but the actual infrastructure and modes of messaging are closed to innovators. If I wanted to extend or build on texting, I can't, although I have been attempting to change this by the introduction of the #Blue service that I led at O2 in the UK (see Examples of NaaS Connected Services).

[15] There have been some relatively minor changes to the protocol, such as "Extended SMS," but the user experience is largely unchanged due to lack of innovation.

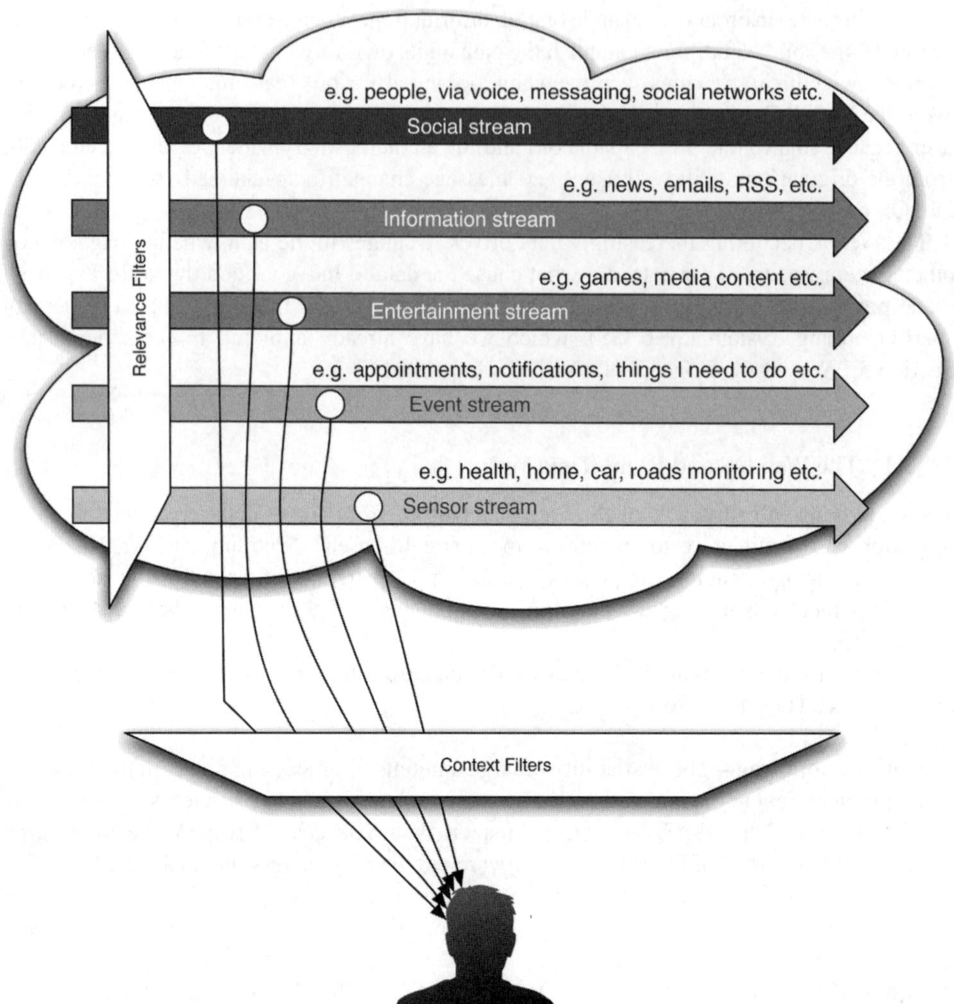

Figure 5.13 Stream-based nature of Web OS.

formats like XML and JSON, we might well be stuck in a swamp of indecipherable data formats that would be difficult to move from one system to another. In other words, with XML and JSON-derived messages, it's as if the whole Web decided to speak English, just like the business world (by and large). And, just as with business, the availability of a common intermediate language has had significant effect on productivity. Layered on top of this, we have micro-formats, RDF and other ways to bring structure and meaning to these JSON-ified communications across the Web.

4. **Social Web –** The transformation of the Web, from its 1.0 "info-centric" nature to the more encompassing 2.0 world of "people-centric" and "service-centric" architectures, paved the way to the right-time Web. As Figure 5.13 shows, when we talk about "Right-Time," we automatically ask "right time for whom?" Clearly, we mean people, not things. Ultimately,

it is people who want the right information at the right time. Machines help the information to get there, but it is people who use the Web! The emergence of deep social structures on the Web has had profound impact on the way developers build services. Oddly enough, users are now becoming a lot more important in our thinking when designing Web software systems and applications. What I mean is that by tapping into the user's context and connections, which is becoming easier, developers can offer even more compelling services than they otherwise might. The social Web can transform and invigorate even a relatively mundane online service, like purchase ordering, adding new meaning to the supplier-customer relationship.

5. **Telco 2.0 elements –** If we're talking about people and social connectivity, then we have to include the massively successful phenomenon that is mobile, especially voice and messaging. After all, whilst data is exciting and very much the future, it is still a relatively small number of the overall mobile community who use data services in any substantial way. The majority of mobile users are still making calls and exchanging texts. Of course, this is slowly changing, but voice and reliable person-to-person messaging will be a strong requirement for all of us for the foreseeable future. The death of voice is much exaggerated (by a small band of geeks who simply don't want to answer their phones anyway[16]). However, in terms of the right-time Web, then it will become a lot more powerful when we can mash the voice services world directly with the Web. This is happening slowly with companies like Telefonica buying Jajah and opening up new types of service. It is happening with platform plays like Tropo and Twilio, which I discuss in Section 9.1 Opportunity? Network as a Service.
6. **Big Data –** Big Data's role in Right-time Web should hopefully be clear. As Figure 5.13 shows, we not only need to maintain and support all these vast streams of data, which are ultimately flowing from one data store to another, but we need filters! There's simply too much information to handle. Users are overwhelmed already. What we need is the ability to process the streams and extract the *right* information. This is a multi-variate problem as it applies across all the streams at once. Finding patterns across streams and extracting meaning is most definitely a Big Data problem! Doing it in real-time, which is non-trivial, but possible, is what gives us the right-time Web.

With these six elements, plus others we could explore, the Web OS has the right architecture to support a "right-time" modality – an array of streams between source end-points and destination end-points that are increasingly converging on the mobile, particularly the smartphone with its various smarts that make it an ideal filter and consumption device for these streams, able to feedback context information that modulate the streams both at source and the destination.

[16] The death of voice is usually mentioned within the context of teenagers who have apparently abandoned talking. This is simply not true and we should be cautious about jumping to conclusions. For example, we were told that TV is dead for young people too busy on Facebook to watch TV, whereas in 2010, viewing by young people has gone up in the UK and US, not down.

6

Modern Device Platforms

```
app = "Game"

platform = case app
  when "Game" then "Native"
  when "Blog" then "Web"
  when "Comms" then "Either"
  when "Weird" then "Widget"
  when "Everything" then "Text"
  else "Web"
end

puts "The " + app + " should be built using " +  platform
```

- In thinking about device platforms, it is sometimes more useful to think of connected devices rather than mobile devices.
- "Being mobile," is increasingly about being able to stay connected to a number of key data streams at all times.
- The digital streams that flow from Web platforms are at the heart of connected services, but streams from the physical world (i.e. Sensors) are becoming more important and frequently involve mobile devices.
- Although not the only device platforms, iOS, Android and Mobile Web are currently central to the device platforms story, each important in their own way.
- Android and iOS, explored here, go way beyond our historical understanding of mobile operating systems. They are platforms for mobile computing, or for connected services computing, which is something much bigger than telephony and messaging.
- The Mobile Web is a major platform that continues to increase in its power to deliver rich connected services, although always one step behind native OS platforms.
- HTML5 and its associated technologies and standards will deliver a substantial step increase in the power of the mobile Web.

Connected Services: A Guide to the Internet Technologies Shaping the Future of Mobile Services and Operators, First Edition. Paul Golding.
© 2011 John Wiley & Sons, Ltd. Published 2011 by John Wiley & Sons, Ltd.

6.1 Mobile Devices or Connected Devices?

6.1.1 What is a Mobile Platform?

Before we get into a discussion of modern device platforms, we ought to review exactly what a mobile platform is, if indeed that is the best way to think of devices. In my last book, *Next Generation Wireless Applications* (see Chapter 11), I described in some detail the principle components of any mobile operating system, by which I meant the underlying software layers that take care of abstracting the hardware from any software services that run atop.

For example, to send data across the wireless network requires the use of various wireless modem sub-systems in the underlying chipset. These modems require all kinds of configuration and management before sending data. The data itself will be exchanged using low-level packets that are eventually mapped to IP packets and one of the higher IP protocols, like TCP/IP and then onto HTTP. However, an app developer mostly doesn't care or want to know about these low-level operations.

An app developer writing an app to connect with Facebook or an operator's API is only interested in dealing with HTTP, probably using a clean RESTful service (see Section 2.2 Beneath the Hood of Web 2.0: CRUD, MVC and REST) The developer doesn't want to know how the low-level modem works, nor, most likely, how TCP/IP works, or even the actual TCP/IP packet assembly of a HTTP call. They just want to do something like this in Ruby, say:

```
url = URI.parse('http://graph.telco.com/0751522128')
 res = Net::HTTP.start(url.host, url.port) {|http|
    http.get('/index.html')
 }
 #Now do something with res.body
```

Those few lines of code are enough to go grab the contents sourced by http://graph.telco.com/0751522128 and start doing something with the data in the app. As I suggested in an earlier section (see Section 3.1 Internet of Things), the data returned might be a chunk of JSON script, like so:

```
{
"id": "559089368",
"name": "Paul Golding",
"first_name": "Paul",
"last_name": "Golding",
"link": "http://www.telco.tel/paul.a.golding"
}
```

This is a chunk of data about subscriber 559089368, which happens to be me in this example. Following the "link" field might bring me to a "home page" for Paul Golding on this telco network, where you might find my phone number and other data of interest, all available in JSON.

Getting back to the plot, the way this code works is via a stack of software layers, like so (see Figure 6.1).

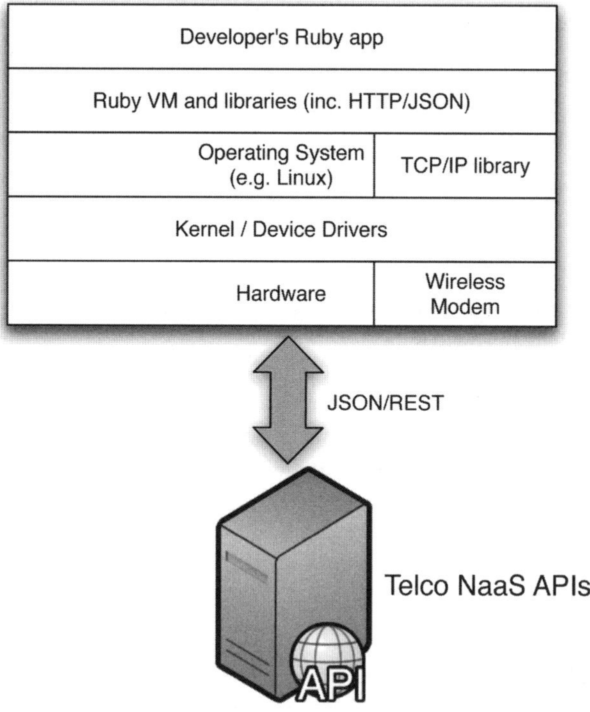

Figure 6.1 Typical software stack for mobile platform.

The developer really doesn't need to know how the modem works or the device driver and operating system that abstracts the device and its capabilities, presenting it in a more convenient interface to the overlying software environment, which in this example is Ruby. Within the Ruby implementation itself is the virtual machine (VM) that translates the Ruby code into code that will run on the underlying platform (e.g. C code). Accompanying the VM is a collection of Ruby libraries[1] to implement common functions that the software developer is likely to need, including a library of software to handle HTTP, which is actually part of a wider library to handle network tasks (called "Net").

In this case, we might be tempted to think of the Ruby[2] environment as "the platform," similar to how Java Micro Edition (J2ME) might also be considered a platform. Other developers might think of Flash as a platform – it enables applications to run on many devices, but it just happens to be contained in a browser rather than run on the operating system directly. However, there is no single or authoritative definition of an application platform on a device.

Recently, there has been a shift in thinking about this issue, moving towards a categorization of platforms that is more commercially motivated, as in which "environment" can I get best market penetration when writing an application for mobile devices. For example, developers

[1] I described the whole mechanism of VMs and code libraries when I wrote in detail about Java in my book *Next Generation Wireless Applications*, 2nd Edition, pp. 309–409.

[2] Note that Ruby hasn't been ported to most mobile operating systems yet, although it could in principle.

think of the "iPhone platform," meaning both the operating system and the iTunes marketplace. Also, we should be aware that the edges of operating systems, language runtimes and associated mechanisms aren't so clear cut as they used to be, such is the case with Android, which is a Java environment layered on top of a Linux kernel. Let us take our cue from the VisionMobile developer survey, where it listed the following common mobile "platforms" (in no particular order):

1. iOS (Apple)
2. Android (Google)
3. S60/Symbian (Nokia/Symbian foundation)
4. Windows Mobile (Microsoft)
5. BREW (Qualcomm)
6. Linux
7. Blackberry (RIM)
8. Web – CSS/HTML/Javascript (Cross-vendor)
9. WebOS (Palm)
10. Flash or Flash Lite (Adobe)
11. Java ME (Oracle/Sun)
12. Java SE (Oracle/Sun)

These are all mobile device "platforms" of one sort or another, but not necessarily mutually exclusive. Note that the "Web platform" (meaning a mobile Web browser) is to be found on most of the other platforms, which we could think of as a subset of each platform, or a superset of all platforms. But let's drill further by asking developers, to see what they think.

6.1.2 Developer Mindset About Mobile Platforms

As far as developers are concerned, they want to create software applications of varying kinds, to solve various problems. When mobiles first came out, it became possible to write "mobile applications" of some description, but the options were very limited by the underlying technology. For example, when Java Micro Edition was first launched, along with its MIDP 1.0 profile, it had a very limited subset of the Java (Standard Edition) language underpinned by a very limited access to the underlying capabilities of the device. For instance, it was not possible to access the address book, believe it or not. Devices and possibilities were extremely limited, so most developers didn't bother with mobile apps. Moreover, there was no way for most consumers to access those apps, unlike desktop apps that could be bought in retail outlets or downloaded over the Web.

Most developers back then were still writing desktop or enterprise applications and had begun to shift their attention towards writing Web apps, not mobile apps. What they were often doing though was to take existing software applications and ideas and then to bring them to the browser. In other words, they were "Web-enabling" their software ideas. As the Web platform evolved to include much more powerful connections, hardware and software libraries, developers began to exploit the various capabilities of the Web ecosystem itself (such as APIs), causing a shift towards creating products and services that were "made for the Web."

Mobile device platforms have advanced so much that it is now possible to create very rich experiences on mobile devices. Moreover, it is possible to get these applications in the hands of users because of the emergence of app stores and a greater trend towards openness, getting away from the "walled garden" world that the operators had previously constructed. There seems to be two types of mobile software activity:

1. Creating "made for mobile" apps that are only meant to work on mobile devices, like Shazam.
2. Creating "mobile extension" apps that extend the reach of an existing Web service, like Facebook.

Almost always, the motivation is tied to a certain opportunity that is often associated with a particular platform. For example, developers who wanted to try and make money in the iPhone app store "gold rush" turned their hand towards writing iOS apps, which was the only way to access the rush. This meant learning Objective-C and using the Apple XCode developer tools, possibly having to buy a Mac first. It would be hard to imagine that many such developers were attracted to the Objective-C language first, or OS X (as a "platform") and then went in search of devices to satisfy that need. In other words, the motivation to use the "platform" was mostly a commercial consideration, which is one of the dimensions of choosing a platform, as shown in Figure 6.2.

However, existing Objective-C developers would be attracted to the iPhone platform for other reasons perhaps. Sure, they might be tempted by the overall gold rush, but they might also have existing Mac desktop apps that could be easily extended to the iOS platform (iPhone and iPad) in order to exploit an existing product. It is not uncommon now to see a number of Mac developers extend their product ranges to include iPhone, not as standalone products, but as companion products to existing desktop applications.

What the VisionMobile survey revealed was that the choice of mobile platform is increasingly influenced by commercial considerations, particularly the availability of a market that is willing and able to discover and buy apps, as is the case with iOS and Android. Even though Symbian devices were (are) far greater in number, the perception (and mostly reality) is that money can be made more easily on these two newer platforms. Of course, the underlying capability plays into the mix. Both platforms are rich in features and possibilities, with vast

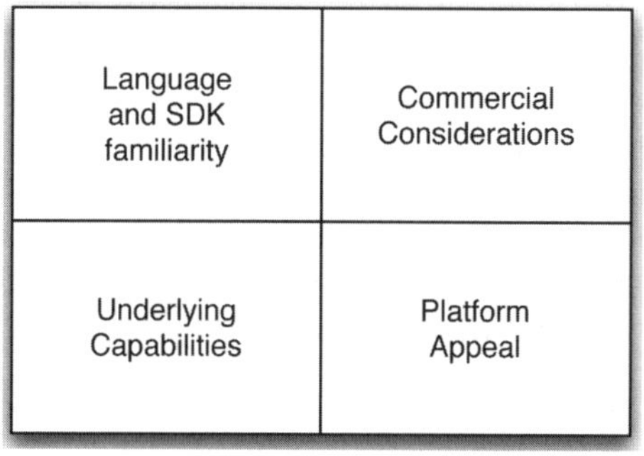

Language and SDK familiarity	Commercial Considerations
Underlying Capabilities	Platform Appeal

Figure 6.2 Dimensions to choosing a mobile platform.

numbers of APIs and exciting handsets to exploit all that these APIs can offer. If the number of possibilities weren't so dazzling, then these platforms would have far less fans than they do.

6.1.3 Mobile Device or Connected Device?

We might want to distinguish between portability and mobility, without getting into the heavy definitions that I explored in the first edition of *Next Generation Wireless Applications*, which was written in a pre-iPhone era and in a time when many folk (especially in the operator worlds) still doubted that there was ever going to be a mobile apps future at all (in the absence of a so-called "Killer app"). Barbara Ballard, mobile design experience guru, has always been clear that the "Carry Principle" is mostly what matters.

Devices that are always carried, or can always be carried, tend to have certain characteristics, often limiting ones (e.g. smaller screen size) in comparison with their non-carried counterparts, which users are willing to put up with in exchange for the utility of always being available. However, in practice, such concerns don't matter too much to developers. So long as they can find a motivation to use the platform, then they will tend to think about other motivators, such as the chance to make money, as discussed in the previous section.

However, one might ask if this "carry" question really matters. I increasingly think not. What really matters is connectivity. Clearly, that's what makes mobiles mobile in real and practical terms – the mobile industry is about providing voice connectivity, then texting and now, increasingly, data. On the other hand, if a device can access the Web, then it opens up a certain number of possibilities, regardless of portability and mobility. But I think that it goes further than this. Connectivity in the age of Web 2.0 and ubiquitous mobile telephony/ messaging is:

1. **Real-time access to voice** – Can I speak with my connections right now?
2. **Real-time access to messaging** (of various kinds) – Can I notify and be notified?
3. **Real-time access to my content** – Email and anything "personalized" on the Web.
4. **Real-time access to social networks** – Can I access my "social streams"?

From 1 to 4, we can see a progression, which is from the mechanics of connectivity – connecting device a with device b to establish a voice circuit – towards the uses of connectivity, which is the desire to connect with content and people that matter.

In the very near future, there will probably be a fifth dimension to connectivity, which is real-time access to things (car, home, office printer, etc). In terms of higher-level abstractions, I find it useful to think of the world of connectivity as being hooked into streams, mostly in "The Cloud," as shown in Figure 6.3.

"Being mobile," then, is being able to stay connected to these data streams at all times. Some streams are entirely Web-based, others will need intimate "Personal Network" connectivity, such as to health-monitoring gadgets sewn on the inside of t-shirts or attached to the skin (e.g. "Smart plasters").

I don't want to push the stream model too far, but there is often value in rethinking of paradigms in order to understand the future, which is partly what this book is about. I am

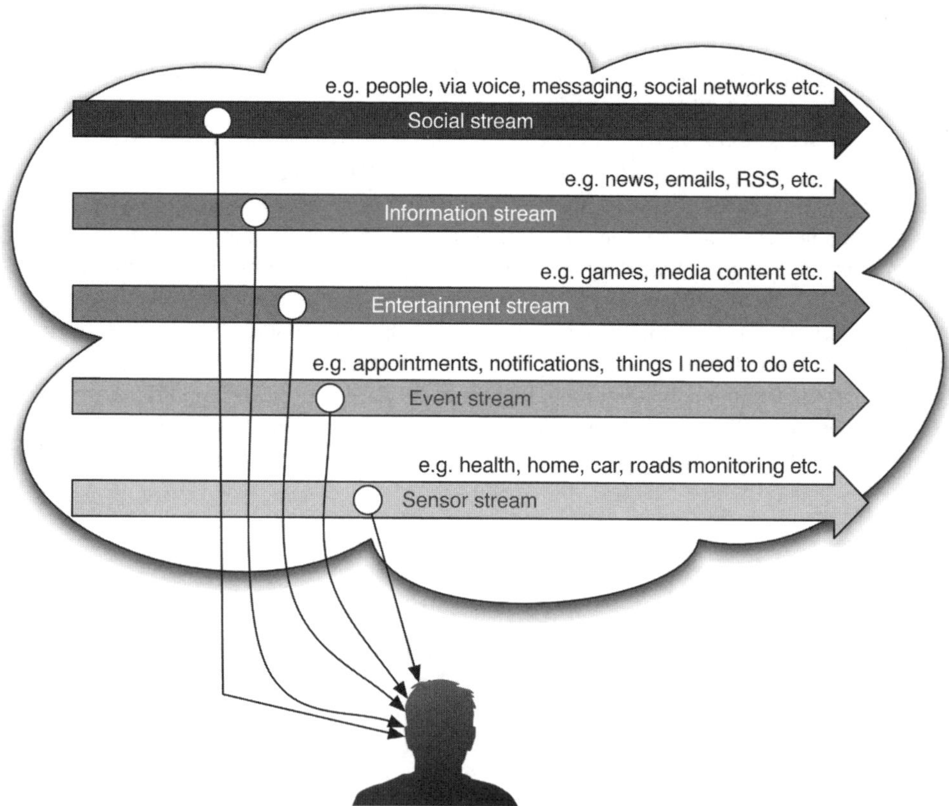

Figure 6.3 Staying connected to streams.

attempting in this book to describe a joined-up understanding of the Web and mobile worlds, which is this overarching concept of connectivity, a relatively new and evolving modality of the applications and mobile world combined. For example, will we really be interested in text messaging in the future, or simply in any means to notify each other in real time. I tend to think the latter, especially as more and more methods of notification become available, such as Tweets, Facebook updates, presence updates etc. More than likely, paradigms like "inboxes" will fade away, to be replaced only by apps of various kinds that consume (and produce) real-time streams. We won't care about text messages from our doctor's surgery. We will care about having real-time and relevant data in our "health app," accessible from anywhere on any device.

What the above diagram really shows is a "telco switch" for the 21st century. In the IP-connected world, there are myriad sources and destinations for data. Voice is just another source. The amazing thing about IP-bound data is that, so long as we can access it, we can manipulate it however we like, including aggregation of various data sources into co-gent streams that enable us to filter and manage how our senses connect with the world around us.

Web apps still live mostly in a world of HTTP, whilst telephony apps live (mostly) in a world of the telco-centric SS7[3] (and increasingly SIP[4]). These two worlds are slowly converging via the IP-enablement of telephony platforms (See Section 9.1 Opportunity? Network as a Service and new possibilities in "Cloud Telephony"). However, with the myriad higher-level data formats, APIs, software frameworks and modern Web-based protocols (e.g. Web sockets), it is easy to see the emergence of a new paradigm, which is this Internet of Things where all we are really concerned about is the flow of information from anyone, anything and anywhere, that is machine readable (see Section 3.2 Internet of Things). In that world, the notion of stream-connected apps is appealing, all running, of course, on the "Web Operating System" (see Chapter 3. The Web Operating System – The Future (Mobile) Services Platform.)

If you were paying attention throughout the book, then you would have noticed that a key attribute of connected services is the real-time element. In a way, this supports Ballard's Carry Principle, as the ability to do anything in real-time *all of the time* clearly requires that a device be on our person at all times. However, it doesn't always have to be the same device. We have seen how, for whatever reasons, real-time has become an increasingly important characteristic of services generally. For example, the stated aim of the Kindle, when being designed by Amazon, was:

> Any book in any language within 60 seconds.

I **love** that goal! Clear, ambitious and inspiring. Thinking about the Kindle experience, which I absolutely love, it delivers the real-time experience in every way. The reading interface, whether on a Kindle device or via a Kindle app (e.g. on iPhone) is easy enough to carry, thus enabling real-time access to content. Moreover, by enabling lots of books to be stored in the application's memory, it allows real-time access to *all* content, not just the particular book that I happen to be reading. Finally, with its wireless connection back to the Amazon store, it enables real-time access to new books – and with the international version, anywhere in the world! The Kindle reinvents the reading experience by adding a real-time "streaming" element – "book streaming," if you will.

The real-time element is, for me, the most powerful component of a connected device. The ability to interact with one of those streams in *real-time* is a key attribute of a connected device. Therefore, when thinking about "Connected Devices," we are interested in all devices that enable a real-time experience across all of our streams: social, information, entertainment, event and sensor.

6.2 Introduction to Mobile Device Platforms

6.2.1 Platforms of Interest

The dominant mobile device platforms, at least as far as developers are concerned (see discussion in previous section), are iOS and Android, so I will explain these in enough detail

[3] This stands for Signalling System 7 and is the name of a major protocol to allow telecoms circuits to be set up between switches.

[4] Stands for Session Initiation Protocol and is the modern equivalent of SS7, constructed using the lingua-franca of the Internet – IP, or Internet Protocol.

to provide you with an understanding of the essentials of these platforms and what makes each platform unique. I will look at other platforms in passing and draw comparisons where applicable. I don't want to promote the view that Android and iOS are the only shows in town, but they are certainly shaping the way that the smartphone industry is evolving.

In my dealings with various folk in telcos and along the way, these two platforms are the ones that get talked about the most and where there's an appetite for understanding how they work. As with most of my attempts to explain particular products or technologies, I will try to make clear the underlying principles, paradigms and patterns, so that you can tell why things are the way they are, thus arming you with the right insights to understand this landscape as it continues to evolve.

I'm not going to focus too much on the technical details of these platforms, which would be more appropriate for a book about device platforms, probably aimed at developers and technologists. I will explain enough to provide a sufficient foundation for telling the much greater story here, which is the wider platform story. This is where I'm headed – I'm going to explain the essential characteristics of each *platform*, which includes the wider ecosystem elements, such as the associated app stores. I will look at the pros and cons of each platform. As these are device platforms aimed at developers, I will inevitably explore them from the developer perspective, so that you can understand how developers view these platforms and appreciate what matters to them. As should be clear by now, attracting and building developer mindshare is absolutely essential for a connected services platform of any kind to be successful.

6.2.2 Brief Explanation of an Operating System and SDK

Here, a brief introduction to operating systems will be useful. Please stick with it, as I will explain some key principles that will get you far in any consideration of device platforms (mobile or not). If I build a computing device, of any kind, and then hand it to a developer, he or she probably won't be able to use it straight away. The first thing that the developer will want to know is how to write a program for it. There are a number of components to this process. Firstly, a developer needs to know which programming language to use. This will depend on which tools I have made to support the various high-level languages available, such as C++, Java and Ruby. None of these languages will run natively on the microprocessor at the heart of my machine. I have to make a decision to port a language to the device.

Microprocessors only talk a very low-level language called assembly language,[5] which isn't a very friendly language at all for developers and most developers these days would struggle to use it anyway. To go from a high-level language, like C++, to assembly language, it is my job, as platform provider, to provide a set of tools that will enable this translation to take place, which is called compiling or interpreting, depending on the approach used to translate.

However, it's not enough just to give a developer a set of tools for supporting a high-level language. What they also need are interfaces, which are called Application Programming

[5] Actually, they only talk machine-level code, which is all 1s and 0s, but there's simply no need to use machine code anymore. Machine code is like talking in phonetics, whereas assembly language is like using real words. However, high-level languages are more like writing in sentences and paragraphs, which are powerful enough to express intent and meaning that developers can get their heads around.

Interfaces (APIs).[6] For example, a developer needs an interface that will enable C++ functions to output data to the screen, or some other user interface. These software interfaces (APIs) will also be needed to access device inputs, such as touch gestures or key strokes. APIs will be needed for cameras and other peripherals on the machine. APIs will also be needed to establish network connections and start talking HTTP.

One option is to tell the developer how the underlying hardware has been designed so that code can be written to send the right low-level commands to the screen, camera and so on. Believe me, this is much more complicated than it sounds. A camera, or other device on the machine, will have a very complex hardware interface with all kinds of settings and configuration options to get it working. However, the better option – and the *only* option if I'm serious about developers ever using my machine – is to write some software myself that takes care of operating the camera and then allows the high-level software to issue simplified commands, like "TakePicture," returning a complete picture file (e.g. JPEG) that can be easily retrieved using the same API, or perhaps a file-handling API.

I hope you can see where this is going. All of this software to control the underlying machine and provide a much neater and high-level-language-friendly interface (APIs) is what we call the operating system. But there's an even trickier thing to take care of first, which is how to run all this software – including any of the developer's apps – at the same time on top of the hardware, co-ordinating their activities. This bit can be deceptively tricky to understand for the uninitiated. Actually, in my experience, a lot of developers still can't really grasp what's going on at the low level of a machine. Having had the chance to design machines for real (I used to design the chips that go in them and the low-level firmware[7]), I feel privileged to understand how the guts of computers and mobile actually work. The point to keep in mind is this:

There is only one processor in a computer![8]

Not only that, but the processor can only run (execute) one process at a time, or one piece of software – a single software function. That's right – the same processor that has to update the UI, also has to check emails (if email is supported by the device), get inputs from the touch screen, camera, microphone, GPS, take calls, and so on. The list of processes that need to run at any one time could be quite large, easily running into tens of processes. What actually happens is that the processor divides its time between these processes, running each one for a short burst of time and then switching to the next process.

Phew! That sounds like a lot of work. I know what you're thinking – won't the user notice that the email stops for a bit whilst the processor goes off to process a phone call, or start playing a music track? Well, yes, if it happened that slowly. Fortunately, processors are blindingly fast and can run millions of instructions per second. Each one of these processes only needs to run a few hundred instructions (say) to get its next job done, like checking emails, taking a picture or receiving a text message. The swapping of tasks is so quick, that the user never notices it.

[6] This is where the term API originates from and most APIs are this kind. These days we have become used to APIs on the Web, such as the Google Maps API that allows a Web program to access the mapping functions and data within Google's servers that run the Google Maps service.

[7] So called "firmware" because it is somewhere between hardware and software.

[8] I don't want to confuse matters by talking about multi-core processors, but the point still stands.

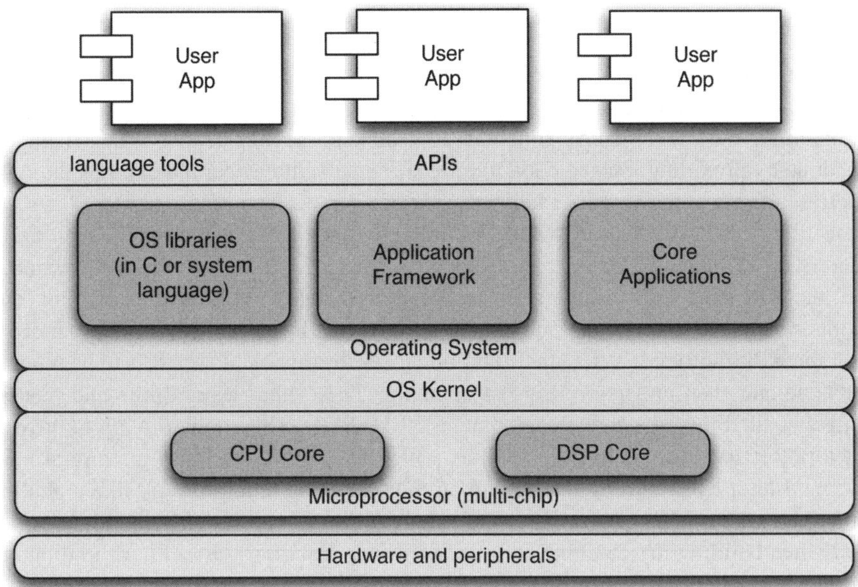

Figure 6.4 Basic operating system architecture.

The swapping of tasks on the processor is the job of another piece of software that I'm going to have to write if I want any kind of developer apps to run on my machine. This low-level management function of scheduling processes – called multi-tasking – is the job of a unique and fascinating piece of code called *the kernel*.[9] It also has some other key functions, such as taking care of loading apps that our developers write.

The kernel and all these lumps of API code, usually called APIs themselves (just by convention) are what collectively we call an operating system (OS), as shown in Figure 6.4.

Once I have an operating system, I can publish the APIs to a developer so that she can use them to write apps. For example, if she wants to write an app to take pictures using the camera and send them off to Twitter, then she will probably need access to a camera API and a HTTP API to allow packets of data to be sent across the network (via a wireless modem probably) to the Twitter API on their site. Those high-level APIs, like the camera and HTTP API will need to be accessible via the high-level language so that the developer can make reference to the APIs in her code. Let's make that a bit clearer. If I write an API to support the camera, then it might have a function called:

```
TakePicture(resolution, zoom, format, flash_mode)
```

This made-up function causes the camera to take a picture with a certain *resolution* and *zoom* using certain flash settings (controlled by *flash_mode*) and then return an image file

[9] I once wrote my own kernel in a language called Modula-2, a language designed and developed between 1977 and 1980 by the legendary Niklaus Wirth. It was one of the most interesting, but painful, software tasks I've ever undertaken.

with a certain *format*, which might be JPEG or PNG format, for example. However, the C++ programming language (or Java, Ruby or any other) doesn't have such a function called *TakePicture*. Therefore, if the developer wrote this into her code and then tried to compile it using the standard language libraries, it would fail. What I need to do is to make the camera API (containing the "TakePicture" function) accessible to the high-level language being used. This is done by providing a set of code libraries (i.e. the actual code that exposes the APIs[10]) and their interfaces to the language being supported.

These libraries then get used by the compiler so that when it's compiling an app and encounters the command *TakePicture*, it knows what to do with the command, which is to "paste" the API code into this part of the program so that it works when running on my operating system. So, when I say publish the APIs to the developer, what I really mean is to give her the API libraries and the other bits and pieces that enable these APIs to be referenced by her code and then understood by the compiler. These libraries and bits and pieces are collectively called the Software Development Kit, or SDK. This too is my job to provide, if I want to attract developers to my machine, which is otherwise a lousy attempt at being a platform. At least by providing an OS and an SDK, it's beginning to look like a platform – something that developers can run their wares on with relatively little friction.

Now, when building my machine and turning it into a platform, I will have to think about what operating system I want to offer my developers. There are a few options here:

1. Build my own operating system.
2. Commercially license an existing operating system from a vendor.
3. Take an existing free (and therefore usually open source) operating system and port as it is.
4. Take an existing OS and modify it to create a new one underpinned by the original.

All of these options have their pros and cons. However, such a discussion is well beyond the scope of this book and the subject of much debate, controversy, success, pain and misery for countless numbers of projects. Of course, these decisions have been made regarding the platforms that we are about to explore, so we can touch on the issues and history. It will help in some ways to understand what's going on and to appreciate some of the details that follow in the overview of platforms like iOS and Android. So, let's crack on with those stories now.

6.3 The iOS Platform

Let's start with iOS, if only because it came before Google Android. However, as we shall soon see, it didn't really come first, and they have similar underpinnings. What really came first was the underlying operating system of Mac OS X, which itself is based on Unix. This often confuses folks, including myself. We've all heard of this thing called Unix, and it's close cousin Linux, and we're generally aware that it's an operating system. If that's so, then what is OS X, exactly, and what, then, is iOS (formerly iPhone OS)? Aren't they operating systems too? Well, yes they are, but not all operating systems are equal, as we're about to find out.

[10] This idea can get a bit tricky because the word "interface" suggests something rather limited. However, the API generally also includes all of the software that actually implements the function exposed via the API, which would include all the software to operate the camera via a camera-control API.

Before exploring the deeper story here, let's be clear that there isn't any authoritative definition of an operating system, so we ought not to get too hung up on the definitions. The principles are more important, as reviewed in the previous section. For completeness, and as these are major systems in our story, we will review the operating systems leading up to the iOS story. However, when we get to the main plot of the iOS story, and Android story, we will focus on what these platforms can do for us, exploring their essential attributes and differences, rather than confine ourselves to some kind of computer scientist's view of "operating systems."

6.3.1 Mac OS X and Unix – The Foundation for iOS

So the myth goes, which sounds entirely plausible,[11] when Apple decided to make the iPhone, its engineering team had to decide which OS to use, similar to my question in the previous section. They had a couple of obvious choices: redeploy the OS used by the incredibly successful iPod family of devices, or use OS X, as used on Macs. We hear there were various debates, but we know the decision was to use OS X and adapt it to the smartphone environment, calling it iPhone OS.

Now that we've seen the success of the app store and understood that an iPhone is essentially a pocket computer with bags of programming flexibility, using OS X was obviously a very smart move. It allowed Apple to start with a base that was already well established for supporting a complex and rich set of APIs and peripherals within an environment of multi-tasking[12] complexity.

OS X was itself built on top of an existing OS base, which was Unix. The core of OS X is an extension of the Mach kernel, which was built using various components from FreeBSD and NetBSD[13] implementation of Unix. This kernel was used to build NeXTSTEP, which was an object-oriented operating system developed by Steve Jobs' company NeXT after he left Apple in 1985. When Steve returned to Apple, it was to introduce a brand new OS to replace OS 9, which was a proprietary Apple OS. OS X was based upon NeXTSTEP.[14]

So, if Mac OS X is really an extension of Unix originally, via BSD and NeXTSTEP, what's the difference? Well, along the way, things have been added. Most notably, NeXTSTEP, and then OS X, added support for the object-orientated language Objective-C and then added an object-orientated API for creating applications, called Cocoa. We can think of Cocoa as what gives a Mac app that "gorgeous"[15] looking UI. Remember that an API is just a big chunk of code that takes care of a useful function that most applications are likely to need when running on the machine. In this case, Cocoa is more like a framework, or a collection of APIs that give the Mac flavour to Unix. For those of you familiar with custom cars, you can think of Unix as the underlying chassis with all its connecting apparatus. OS X is like the new engine and body that makes it a particular style of car different from other styles built on the same chassis.

---

[11] Forgive me for not reviewing the exact historical details, but they don't really concern us here.

[12] We will explore multi-tasking soon, but some commentators mistakingly think that iOS didn't support multi-tasking until version 4. This is incorrect. As I've already pointed out, any OS needs to support multi-tasking – it is one of the main functions of the OS kernel.

[13] These are versions of the Berkeley Software Distribution (BSD, sometimes called Berkeley Unix) which is a UNIX operating system derivative developed and distributed by the University of California, Berkeley.

[14] The company was bought by Apple during Steve's return to the company as interim CEO, eventually permanent CEO.

[15] I couldn't resist using Steve Job's favourite adjective for describing the look and feel of the Mac experience.

Alternatively, if this is getting a bit confusing, and you don't like custom cars, then another analogy might be useful. We can think of Unix as a recipe for an OS. It says what the raw ingredients are for handling multi-tasking, creating files and so on. It is up to the engineer how to write the software that delivers those ingredients. What really matters are those interfaces, technically called APIs, though we have said that these days we tend to think of APIs as the interface and the code behind them. However, so long as the engineer gets the interface right, then a program written for Unix should run on his implementation of Unix. It's like the basic recipe for an apple pie. Chefs might use a slightly different pastry technique, a bit more sugar on the top and maybe even add some spices. However, it's still an apple pie, not a peach pie. So, we don't really mean that OS X is Unix under the hood. It is a flavour of Unix, but a lot more spices and embellishments thrown in. In particular, we have Cocoa to allow "Unix" apps to take advantage of all those graphical interface elements that make a Mac apps unique.

Anyone can ship a machine with a Unix-derivative OS. The designers will then probably add additional components to make the machine more interesting and to differentiate from the competition. The Mac OS X experience is a lot more than just a bunch of APIs. For example, it comes with a number of applications that look like they're integrated into the OS X experience, such as the Spotlight application, which enables a user to search for content on the machine, like files, documents, tunes and so on. Although this is an application that runs on top of OS X, it is an integral part of the experience and can also be accessed by developers via a Spotlight API. Therefore, we can start to think of Spotlight as part of the OS – or platform. Again, it's likely that the ability to search will be required by a lot of applications written for OS X, so its inclusion frees developers from the burden of developing a search facility – it becomes an underlying enabler, which is just what an OS does.

6.3.2 The Mechanics of iOS

Now that we've reviewed the OS X story, it should be fairly easy to see how we go from OS X to iOS, which is just a version of OS X for mobile devices, like the iPhone, the iPad and now the Apple TV.

A smartphone needs to run apps. We all know that now, but it has always been the case, even before the app store frenzy. The baseline iPhone experience includes apps, like email, a telephony dialler, contacts, a media player, a messaging client and a number of integral applications. These are still apps, just not downloadable apps. They all need to access the underlying machine, which they do via a core that is OS X, with a few things thrown out, and then called, unsurprisingly, Core OS with its set of APIs lumped into a framework called the Core Services framework, as shown below. If you like, think of Core OS as being the kernel part that does all the multi-tasking magic and keeping all the apps running smoothly, and think of the Core Services as being all the APIs that a developer can use to do lots of low-level stuff that a smartphone (or smart device) app might need to do, such as manipulating text, accessing the address book, accessing the location-finding enablers (e.g. GPS) and so on.

Let me briefly touch upon what we mean by multi-tasking in iOS. Some commentators have pointed out the iOS didn't support multi-tasking until iOS version 4. This isn't quite true. As I pointed out earlier, multi-tasking is an inherent feature of the kernel of any operating system. There are various techniques for how tasks get scheduled and managed, but if an OS can't multi-task, then it's useless. For example, let's say that an iPhone user is checking her email and then a call arrives, which gets picked up by the wireless (e.g. GSM) modem within the device. It is imperative that the kernel can park the email task for a brief moment in order to allow the "call handling" task to be run instead, alerting the user (visually and audibly) and setting up the display to allow the user to accept or reject the call.

This method of being able to respond to events of greater importance in real-time is called pre-emptive multi-tasking and is a major and necessary characteristic of a smartphone OS, which includes iOS. So, multi-tasking has always been part of the underlying iOS, just as it is a part of the underlying OS X cousin. Another aspect of this multi-tasking capability is the ability for processes to run in *background mode*, which means that an app can continue to carry out certain tasks even when it is not active in the UI.

A good example of multi-tasking is the use of the text messaging application on an iPhone. It can still receive text messages even whilst the user is playing a game, checking email, or on a call. This is because the processor is switching the app on in the background, allowing it to run for a while and then switching it back off again. Or, more accurately, the kernel is aware that if a text message is received via the wireless modem, then it will allow the modem to interrupt the kernel ("Hi – I've got something to do here please") allowing the kernel to invoke the app for a while so that it can take care of the cause of the interrupt (e.g. receive a text message).

However, when running developer-built apps on iOS, as opposed to system apps (like texting) originally it was not possible to support multi-tasking across such apps. This presents a problem, the canonical example being Instant Messaging (IM). If a developer wrote an IM app, such as one to chat via GTalk or MSN Messenger service, then it would only work whilst running and visible in the UI. As soon as the user shut down the IM app, which she would have to do to check an email or text for example, then the IM app was closed and no longer getting any processing time via the kernel. Meanwhile, IM messages might still be arriving, but they would not have any effect on the IM application – the messages would simply get lost and ignored.[16] However, with multi-tasking support for developer apps, it is possible for the IM app to continue running in the background in order to receive messages whilst the user does something else, like play a game. When the user switches back to the app, all the messages are intact. This kind of multi-tasking support was added to iOS 4. A great example of this is a GPS-navigator app that can still update the map position in the background while the user takes a call, which is something I've done many times while using my iPhone as an in-car navigator.

Returning to the major components of iOS, after the core services we then need a bunch of APIs to access the phone-specific features, like telephony and messaging, which is done via a collection of APIs called the Media Framework. And then, ever so importantly, we need a version of the Cocoa APIs that give us the iPhone/iPad look-and-feel, but that supports the multi-touch gesture interface. These APIs are lumped into the Cocoa Touch framework, as shown in Figure 6.5.

Let's briefly summarize what each of these iOS frameworks do for the developer.

[16] In fact, the messages would never make it to the device because the IM app has been shut and so de-registered from the IM service, effectively taking the user offline.

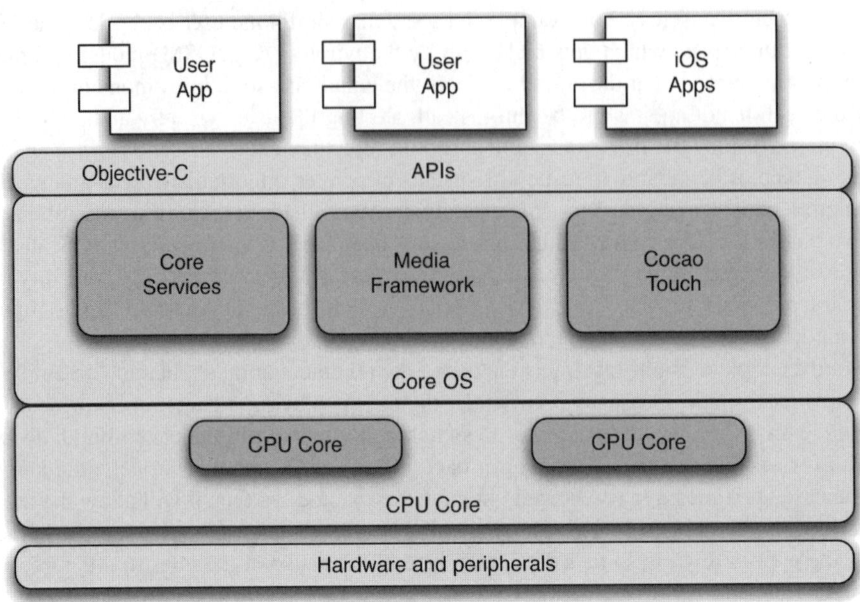

Figure 6.5 Basic structure of iOS.

6.3.2.1 iOS – Overview of Core Services Framework

1. **Core Foundation** – This is a fantastically rich set of APIs to free the developer from all that low-level programming "grungy stuff" that most developers would rather someone else coded for them, such as low-level data management, port and socket communications (i.e. low-level Internet connectivity plumbing), threading, string manipulation, managing of collections of objects etc.
2. **Address book framework** – As it says, allows the developer to access the address book to add, remove and edit entries etc.
3. **Core Location framework** – A set of APIs to allow location information to be explored, especially the positional (lat/long) co-ordinates of the device.
4. **CFNetwork framework** – Objective-orientated abstractions for network programming, such as the ability to open HTTP connections and then fetch data using the CRUD paradigm that we discussed in 2.2 Beneath the Hood of Web 2.0: CRUD, MVC and REST.
5. **Security** – Provides a number of security-related APIs, such as the ability to generate and manage digital certificates and encryption keys, plus perform various cryptographic services important for a connected device.
6. **SQLite** – This is a lightweight version of an entire database management system running on the device. Good for storing data from any application in a structured, relational and secure fashion.
7. **XML support** – It's difficult in the Web era to avoid data being passed between applications using the machine and human readable format called XML. However, manipulating XML in software can take a lot of programming effort, so this set of APIs does it for the developer and is therefore a very handy code library.

6.3.2.2 iOS – Overview of Media Framework

1. **2D and 3D drawing, audio and video handling** – All of the various media-handling software libraries that a developer would expect in order to allow his or her app to consume and produce media files on the device. Clearly, smartphones and connected devices are very often media-centric, so these APIs are essential.
2. **OpenGL ES** – This is a particular set of APIs to enable rendering of 3D graphics using the OpenGL standard, which is widely used in various graphics engines for gaming. This is part of the genius of the iOS – from day one, it was possible to support very rich 3D games, which has made the iPhone and iPod Touch a leading mobile gaming platform.
3. **Quartz** – This API is another type of graphics engine that supports vector-based graphics, which are lines, circles, arcs etc. This is particularly important for a lot of document-based applications that need to support line drawings, such as PDF documents. The PDF viewer built into iOS (and OS X) uses Quartz.
4. **Core Animation** – This is an oh-so-cool API to allow developers to produce applications that support all those various animations built into the user interface, such as when an application collapses into nothing when it's cancelled, or the way that photo-handling apps can support that finger-swiping animation of the images. Animation is heavily integrated into many parts of the iOS user-interface experience and this API allows developers to maintain consistency with that experience without the headache of writing (difficult) code to perform the animations.
5. **Core Audio** – This API takes care of handling audio files plus the various audible alerts built into the iOS experience.
6. **Media Player framework** – This is the real guts of the iOS ability to handle all manner of media formats with great speed and precision, including popular video formats (H.264, MPEG4) and popular audio formats (AAC, Apple Lossless, A-Law, ADPCM, linear PCM, u-Law).

6.3.2.3 iOS – Overview of Cocoa Touch

1. **Foundation framework** – Object-orientated support for collections, file management, network operations, data parsing, web handling and general "programming" stuff that developers would rather not touch. Think of this as a toolkit of stuff, like all the tools (saws, levels, benches, measures) and ancillary equipment (pre-wired sockets, various fixings) that enables you to focus with getting on with the job.
2. **UIKit framework** – This is the real guts of the Cocoa Touch framework. Support for all those glorious UI structures: APIs for windows, views, controls and management of these GUI objects.
3. **PIM[17]/Photo library, camera, accelerometer and access to device-specific information.

More or less, we would expect to find all the above APIs, or similar, on any modern smartphone or connected device OS. This is indeed the case with Android, Symbian, Blackberry and others, but with different API/framework names. The iOS platform doesn't really extend the set of possibilities in terms of APIs, except for its pioneering inclusion of a gesture-based

[17] Personal Information Management – that is, the address book and email etc.

multi-touch interface, which has since been copied elsewhere. The basic mechanics of the OS are similar to the other platforms. There is nothing particularly innovative or outstanding about the underlying OS mechanics that gives developers a special advantage when coding for this platform. The attraction to the platform was initially driven by the shear uniqueness of the device and its user experience, which is an outstanding example of great product design.

6.3.3 iOS – What Makes the Platform Tick

The iOS platform is really a collection of technologies that ensure the optimal smartphone experience that is inclusive of an applications ecosystem:

1. **The device OS** – Using a unified set of APIs and technologies, supported by a dense programming language (Objective-C) it is possible to develop rich gesture-based applications that run on a range of Apple devices.
2. **The SDK and Xcode** – Using an integrated set of developer tools and code libraries, it is possible to create applications efficiently for the Apple devices.
3. **The app store** – Using the iTunes market, it is possible to bring iOS-compatible applications directly to the users of the devices and get paid for it (via revenue share) with very little friction.

The low-friction of the platform is worth highlighting. It is low-friction for both consumers and developers.

1. **Consumer low-friction** – It is very easy to discover new applications via the built-in "app store" button that takes the user to a world of downloadable apps within seconds. Finding, buying and then installing an app is all possible within the holy-grail 3-click limit, beyond which consumer tolerance and enthusiasm tails off rapidly, almost disproportionately so. The importance of frictionless operation for consumers is so widely underestimated.[18] Using an iTunes account, which is itself a low-friction experience, the users can pay for their applications too. It's all very easy.
2. **Developer low-friction** – Despite all the various woes we have heard about the app store submission and acceptance process, which can sometimes have its trials, overall the experience of submitting an app and having it appear in the iTunes marketplace is relatively painless. The fact that the iOS developer tools and platform are linked to an app store at all is itself a huge benefit and removes a lot of the friction in getting apps to market. Without it, developers would have to find their own infrastructure for apps distribution, sales and updates. This has always been a major pain for mobile apps developers.

Moreover, and most importantly perhaps, the developers can get paid! And, they can do so without having any of the payments infrastructure or customer relationship (e.g. billing relationship) that is hard for a lot of (smaller) developers. That said, there are recognized

[18] It is amazing how many product managers, who know about this, will still find excuses for why their users will jump through a few hoops to enjoy the product being offered. Believe me! Consumers don't jump through hoops for anything.

downsides to operating through the store, such as the inability to get meaningful statistics from the store and the inability to communicate with the users.

6.3.4 How Open is iOS?

A question often debated about the iOS platform and iTunes ecosystem is the degree of openness. As discussed in Section 2.4 Open by Default: Open source, Open APIs and Open innovation, the word "open" is widely used and abused, so we need to tread carefully here. Let's look at the various elements of the ecosystem and explore the degree of openness that might apply, or not:

1. **iOS codebase (closed)** – The actual OS, even if it "borrows" from open source components, cannot be extended by any party other than Apple, so it is closed.
2. **iOS native apps (closed)** – The rules are clear that apps running on iOS must be produced using the technologies offered by Apple in the SDK, which means either in Objective-C (or via the Web Browser in HTML 5 – see next point). No other language run-time is allowed. For example, a developer is not allowed to write a Ruby interpreter for iOS in order to support applications written in Ruby. So, in terms of support for other run-times, iOS is closed. Additionally, apps must be digitally signed, which means that they have a unique code that is the electronic "signature" of the app developer. The signing process is inherent to Apple's acceptance into the app store.
3. **iOS Web Apps (open)** – The Web browser can be used to access any website. Apple does not impose any restrictions on this, so the ability to access Web apps is entirely open. Additionally, Apple doesn't use any proprietary features to support Web apps, such as special APIs for the Javascript running inside a Web page. Apple claims to only support the HTML5 set of standards, which is an open standard in the sense of being openly published and without any license restrictions to implement. This is what Apple means when it says that it supports open standards. In this sense, we can see why such claims meet with derision by those commentators who point out that the Apple platform, in every other sense, is essentially closed (per the other points in this list). Those commentators would say that Apple is using the label or badge of "Open", but actually support a closed platform.
4. **iOS Apps distribution (closed)** – The only official way to get an app onto an iOS device is via the iTunes app store. No other route to market is allowed or supported, even though it is technically possible. The digital signing process is linked to app store submission and is used to control application authorization on an iOS device. Unsigned apps won't run on a standard device. In this sense, the distribution for apps – and which apps are allowed on an iPhone is closed. Apple controls it from soup to nuts. Unsigned, and therefore unsanctioned, apps will run, however, on a "Jail Broken"[19] device, but this contravenes Apples terms and conditions for using iOS (as a consumer[20]).

[19] Jailbreaking is a process that allows iPad, iPhone and iPod Touch users to gain the lowest level access to the device in order to unlock the operating system and remove various limitations imposed upon them by Apple, including only supporting signed applications and also being able to access undocumented APIs.

[20] In other words, if a user Jail Breaks her phone, then she has broken a legal agreement with Apple, which is part of the terms and conditions of using the device's technology, which is licensed from Apple, not owned by the consumer.

6.4 The Android Platform

6.4.1 Introduction

Android began life as a mobile operating system created by a start-up in Palo Alto (California) called Android Inc. After being acquired by Google (2005), Google and various other companies set up the Open Handset Alliance (OHA) to develop and release Android as an open-source operating system, available to any handset manufacturer who wanted to license the OS. Early members of the OHA included Google, HTC, Intel, Motorola, Nvidia, Qualcomm, Samsung and T-Mobile.

Google published Android in 2008 as free open source software, making the entire source code available under the Apache License. As Google liked to say about Android, "all apps are created equally." This meant that the licensing terms didn't favour any one application over another, including the system applications bundled with the operating system. A developer is free to replace any of the system applications with his or her own application, so long as the APIs are consistent. This means, for example, that the built-in dialler application could be replaced by a developer-specific one, perhaps to include "Dial Facebook Buddy" as a built-in primitive operation, just like dialling any other favourite from the user's favourite list.

In theory then, from the perspective of customization, Android is one of the most open operating systems on the market. This is very distinct from iOS where all the system applications are protected and can't be replaced by user applications, only supplemented. For example, the precious and important dialler application in iOS can't be replaced or controlled by a user alternative. Moreover, in the case of some core applications in iOS, there are either no APIs or very limited API support. For example, it is difficult to *replace* the media player experience on the iPhone with an alternative, should a developer wish to do so. Apple argues that what matters most with iOS devices is the end-user experience, which has to be compelling and consistent. For this reason, they do not want to give this level of control and flexibility to developers.

The early versions of Android were fairly limited and crude in comparison to iPhone OS (which became iOS). However, the release rate of new versions of Android has been quite aggressive. Each release tends to be a fairly major upgrade with a lot of new features and a lot of strides taken towards a more compelling user experience. Each release thus far has been named after some kind of food item, such as: Cupcake, Donut, Eclair, Froyo[21] and Gingerbread. There is also now a version tailored for tablets, called Honeycomb.

Whilst iOS applications have to be written in Objective-C, Android requires developers to use the more familiar and widespread Java. Note that this is not the Micro Edition of Java (J2ME) that has become widespread on feature phones. J2ME support can be made available via third-party applications, like J2ME MIDP Runner. However, JME is really a poor and distant cousin of Java, unable to support the rich requirements of a smartphone-programming environment. It is interesting that there is no Java Virtual Machine on Android. All Java code gets recompiled into something called Dalvik executable and then runs on a Dalvik virtual machine. This is a specialized virtual machine written specifically for Android and highly optimized for limited resource machines, such as smartphones.[22] Even though smartphones

[21] Which I understand to be short for Frozen Yoghurt.

[22] Note that the Java Micro Edition requires the use of an optimized virtual machine called the Kilo Virtual Machine, or KVM.

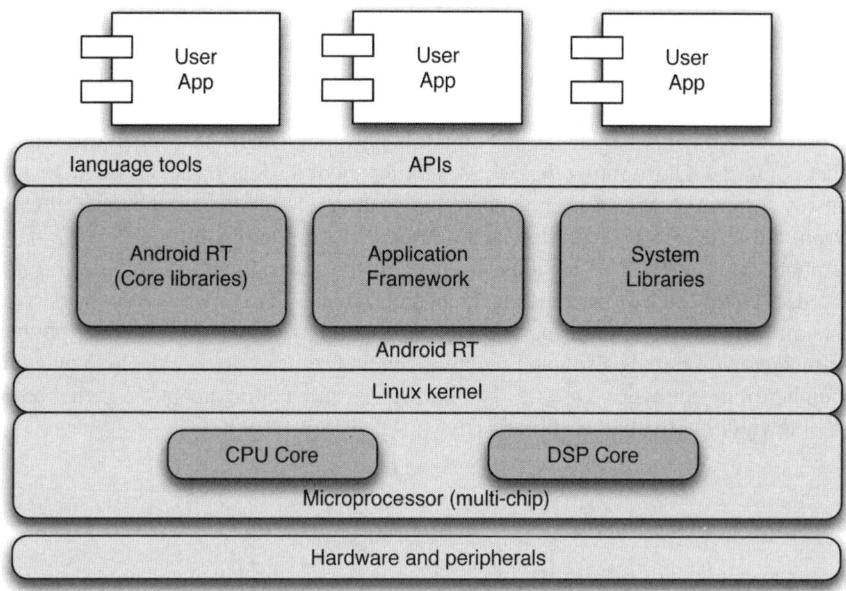

Figure 6.6 Simplified view of Android architecture.

are incredibly powerful, akin to laptops in power and architecture, it is still necessary to consider optimizations for things like battery life.

6.4.2 *Architecture*

In the previous sections, I have outlined the fundamental architecture of any OS and then for iOS. As you might expect, Android is not that much different, as shown in Figure 6.6.

6.4.3 *Linux Kernel*

As we explored in the previous section, all computing machines need a low-level kernel that can take care of managing basic machine operations, like running a program or driving a hardware peripheral. As the Linux kernel is already well established and versatile for this task on embedded devices, it makes sense to reuse it where appropriate. Smartphone use is such a case. We have seen that iOS runs atop of a Unix kernel (similar to Linux). Android does the same. It relies on Linux for core system services such as security, memory management, process management, networking operations (e.g. Wireless, ethernet etc) and its device-driver model, which is a means to enable any hardware peripheral to be mapped into the kernel and made accessible to higher-layer programs running on top of the kernel.

The kernel largely does the same job as it would on a desktop machine, supporting low level and fairly generic hardware-abstraction functions. It doesn't really have any customization, or architectural peculiarity, tailored for smartphones. You might be wondering why this is, perhaps thinking that smartphones must surely be quite different beasts to the large desktop and server machines that run Linux, like the machines beneath the LAMP stack underpinning the Web (which we talked about earlier in the book – see Section 2.3 LAMP and Beyond: web frameworks and middleware).

The point to always remember is that smartphones aren't really that much different from any other computer. Don't forget that we are talking about devices these days with incredible amounts of computing power and memory, and with very powerful peripherals. In essence, apart from the absence of a hard disk and a larger screen, there is little difference between a smartphone and a laptop or netbook in terms of the raw computing architecture and its associated software and hardware management requirements. That's why Linux works equally well on embedded devices, like smartphones, as desktop machines. In fact, smartphones are really tiny laptops. Linux is also quite modular, lending itself to flexible deployment across a wide number of device types. Don't forget that we are only talking about the kernel here, not the fully blown OS with all of its libraries.

6.4.4 Android Runtime

Android applications are not written in C, which is the system language supported natively by Linux. While it is possible to write libraries in C for Android, the application framework, including most of its libraries (APIs), have been written in Java. Therefore, Android applications are required to be written in Java in order to use the various Java libraries and APIs shipped with Android.

Without going into the details of how the Java language works (which I described in detail in my previous book), Java does not run natively on the Linux kernel. None of the Linux kernel functions are written in Java and the kernel can't load applications onto the processor that is written in Java. The design of the Java language calls for the use of a virtual machine (like a processor that runs in software, emulating the underlying hardware processor) so that the Java code thinks that it's talking to the kernel.[23]

In Android's case, the virtual machine is called a Dalvik Virtual Machine (DVM), which first converts the Java to an intermediate format called Dalvik. We don't need to explore the technical merits of this approach, but should be aware that each Android software process needs to run on its own DVM, which then gets scheduled onto the processor by the kernel. As you can imagine, this whole process needs careful management to ensure that the right processes are run at the right time, taking into account things like priority of process, such as giving more precedence to answering an incoming call than to playing a media file.

The Android Runtime takes care of this process management function, relying on the underlying Linux kernel to do the heavy stuff, like controlling access to memory, swapping software tasks on the processor, and so on. Linux does all of these functions well, which is why it was chosen. There is no reason to replicate them in the Android runtime. It just ensures that the interface between Java (via Dalvik) and the kernel is as efficient as possible, taking into account issues like battery life and so on.

---

[23] Recall that software programs on a machine do not talk to the underlying hardware directly, the always talk to low-level kernel routines (pieces of code) that in turn operate the underlying hardware. In this way, the behaviour of all code running on the machine can be managed by the kernel, making it predictable and safe.

6.4.5 Android Application Framework

The application framework is a rich set of software libraries and APIs needed to make an Android application. Just like when we discussed the LAMP stack (see Section 2.3 LAMP and beyond: web frameworks and middleware) I pointed out that the PHP language is just a means to write code, just as English is a means to write a book. The same is true of Java. At its core, it's just a language.

However, if you want to do something meaningful with Java within the Android context, like create an application window and put some buttons on it, or play a media file, then you need blocks of code to create those windows, add buttons, play media files, and so on. You will also need code to allow a user to load the application. You will need code to access any of the other application features of an Android phone, like write to the home screen, access the contacts database and so on. All of these features are made accessible via a rich set of APIs in the Android Application Framework. It is the same framework used by Google to create the native applications that come as part of the baseline Android implementation, such as the dialler, the message composer and other default system applications.

The framework was written with the principle of "all applications are equal" in mind. Any application can publish its capabilities into the framework, making those capabilities accessible to other applications using the framework. This makes the framework highly supportive of a component-based architecture that makes components swappable. For example, I can create a new dialler application and publish its capabilities into the framework, allowing it to replace the native (default) dialler application. This architecture also offers a good deal of flexibility to handset manufacturers to offer a more customized (and therefore differentiated) user experience by supplanting ("overwriting") some of the application framework with their own components, which might be a different look-and-feel, like the glossy and glassy ("aqua") visuals used by some manufacturers. It also enables the default user experience to be supplanted by something more powerful, like the highly integrated social-network interface of the Motorola Droid devices.

6.4.6 Android System Libraries

Android also includes a set of libraries (that are written in C/C++, runnable directly on the Linux kernel) that are accessible from within the Java environment and provide access to powerful smartphone capabilities:

1. **System C library** – A BSD[24]-derived implementation of the standard C system library (libc) that ships with Linux, but in this case highly optimized for embedded Linux-based devices.
2. **Media Libraries** – Based on PacketVideo's OpenCORE library of software components to process media files, these libraries support playback and recording of most popular audio and video formats, as well as static image files, including MPEG4, H.264, MP3, AAC, AMR, JPG, and PNG. However, given that Android is open and extensible, other media types could be added, where needed.

[24] Berkeley Software Distribution (BSD, sometimes called Berkeley Unix) is a UNIX operating system derivative.

3. **Surface Manager** – This library manages access to the video display subsystem and seamlessly mixes 2D and 3D graphic layers from multiple applications, allowing developers to construct rich graphical interfaces.
4. **LibWebCore** – This is a modern Web browser engine which powers both the Android browser and can be used to build an embeddable Web view into any application. For example, it is used by the smartphone intermediate framework called Phonegap, which allows applications to be written in HTML, CSS and JavaScript and then ported to iOS, Android or Blackberry. It does so by using the built-in Web browser to display the application interface.
5. **SGL** – SGL stands for "Skia[25] Graphics Library" and is the graphics subsystem used by Android. SGL is the low-level graphics library implemented in native code that handles rendering. It works in tandem with other higher-level layers of the Application Framework (notably WindowManager and SurfaceManager) to implement the overall Android graphics pipeline.
6. **3D libraries** – An implementation based on OpenGL ES 1.0 APIs; the libraries use either hardware 3D acceleration (where available) or the included, highly optimized 3D software rasterizer.
7. **FreeType** – This is a powerful bitmap and vector font rendering engine, able to generate fonts dynamically across all aspects of the user interface.
8. **SQLite** – This is a powerful and lightweight relational database engine available to all applications. Note that, just like the Linux kernel, most of these libraries are from existing open source projects, like SGL, OpenGL, FreeType, SQLite and so on. Equivalents of all of the above libraries can be found, in their own form (and not necessarily open source) in iOS and Symbian.

6.4.7 Android – What Makes the Platform Tick

Android has had appeal to various interested parties: developers, users, device manufacturers and telcos.

Developers seemed to pay the most attention to Android first, as demonstrated by the thousands of apps that appeared for Android before devices were widely available in the market. Without doubt, the level of openness has been a major appeal of Android for developers, giving them greater flexibility when creating apps, greater freedom to experiment. Don't forget that there is a wide spectrum of developers with all kinds of interests and biases. Many of the developers who simply love to play with software, excited by possibilities, have been attracted to Android. As a platform for trying out mobile ideas, it is supremely powerful. It is no coincidence that some of the earliest examples of Augmented Reality apps (like Wikitude) appeared on Android. This was because of the open platform, enabling low-level access to the camera functions and the built-in sensors long before it was possible on iOS. The innovative spirit surrounding much of what Google does can be found with Android and continues to make the platform tick.

Again, now that we know how much developers as a whole are interested in market economics, the allure of the applications economy continues with the Android story. Just as Apple

[25] Skia was a company acquired by Google. It produced a 2D graphics rendering library.

has its iTunes store, Google has its Marketplace, though not as impressive initially because it doesn't come with the millions of associated credit-cards that make payments so frictionless in the iTunes store. This might also explain why so many of the Android apps are still free, even when compared with some of their iOS equivalents.

For users, the interest in Android has been access to powerful smartphones, including an increasingly vibrant app store, at lower prices than the iPhone. This is because so many device manufacturers have been lured by the open source platform, enabling them to produce fantastic user experiences at lower costs. The wide numbers of Android devices is also appealing to users, giving them greater choice, including the ability to migrate to another device in the future without losing all of their apps (because they will still run on Android). Handset manufacturers have also been attracted to Android because of the support of chip companies like Qualcomm and TI, who have come up with pre-integrated Android reference designs that enable Android handsets to be designed and manufactured very quickly and efficiently. The lower cost of design and integration has allowed the manufacturers to focus their efforts on differentiation through alternative UX design. Motorola is a great example, who used the TI OMAP reference design and were consequently able to add lots of additional socially-enabled features to their Android handsets (like the Droid), creating great differentiation and excitement in the market (for telcos and users).

Telcos have been very interested in Android from the start, despite their fear and wariness of Google otherwise, because of the chance to differentiate in the marketplace. The allure of low-cost customization of an open-source platform has attracted a lot of telco attention. Telcos with a software strategy have shown plenty of interest in Android, with many of the Android handsets coming to market as a telco and handset-manufacturer collaboration.

6.4.8 How Open is Android?

1. **Android Software (Open)** – We have already discussed that Android is an open-source platform in terms of allowing liberal access to the various system apps, even allowing developers to replace them, such as a new dialler. This makes Android highly attractive to handset manufacturers and various software suppliers who wish to offer a more tailored device experience. For example, companies like Microsoft can add capabilities to the Android dialler that would enable it to integrate more seamlessly with their Office Communications Suite (OSC) solution, like adding presence to the dialler so that fellow desktop users can see whether or not the handset user is available to take a call.
2. **Android License (Closed)** – However, from a licensing perspective, Android is not open. The open source project is entirely governed by Google. Only Google can add code to the Android code set. As pointed out by Andreas Constantinou on his company's blog (VisionMobile), Google has an "elaborate set of control points" that allow it to control the exact software and hardware make-up on every model and also allow it to bundle its own software into the user experience. To be clear, these control points make the platform closed as far as the porting to devices part of the value chain is concerned. It does not affect the openness of the experience for developers, notwithstanding any additional control points that telcos can introduce (see below). The long-term implications for this are still unclear, but a completely open platform in terms of device porting would be a problem because it would lead to fragmentation. For example, if a manufacturer broke any of the core APIs

in replacing part of the OS, then applications in the marketplace might not work. Google controls these issues through their compatibility guidelines and restrictions.

3. **Telco customization (closing influence)** – From a user experience (UX) perspective, Android's openness can become "closed" by telcos wishing to control the UX in some way, such as the inclusion of their own look-and-feel and their own apps, whether the user wants them or not (so don't confuse open with the user getting to define the experience he or she wants). Telcos can also elect to remove certain capabilities of the OS, such as the ability to download and install applications from anywhere on the Web. Some telcos have limited downloads only to the marketplace and their own app stores, which might run alongside the marketplace. The "control" that some telcos have imposed on Android is not without criticism. For example, updating has become a problem. When new versions of Android are released, some users are prevented from upgrading because the telco won't support the new release right away.
4. **Apps distribution (open by default)** – Whilst apps are distributed via the Google Marketplace, it is also possible to download apps from anywhere on the Web, unless this facility has been locked down by a telco (see above). Moreover, the rules for getting apps into the Marketplace are a lot less restrictive than for iTunes.

6.5 The Mobile Web Platform

6.5.1 Introduction

In this section, I won't be explaining the underlying principles of the mobile Web, such as how the protocols (e.g. HTTP) work, nor how browsers render HTML, CSS and JavaScript. This level of detail is beyond the scope of this book and has been covered in some depth in my previous books. However, I believe that for those working closely with mobile Web products and services, a good grounding in the principles (especially HTTP) is useful, which is why I have often advocated taking the time to review the basic principles of HTTP and the Web paradigm.

Here I want to cover some of the key trends and technological markers, in outline, that are driving the current evolution of the mobile Web, especially in these interesting times of smartphone proliferation and the flurry of interest in native applications magnified by app stores. There is still a lot of chatter about when to use Web apps versus native apps, with many telcos still placing most of their emphasis and resources on mobile Web portals, with relatively limited forays into native apps. This "native versus Web" debate therefore seems like a good place to start.

6.5.2 Native versus Web "Debate"

As far back as the first edition of my book *Next Generation Wireless Applications* (2004), folks in the mobile developer community were debating, mostly amongst themselves and almost doctrinally, about the issue of "Native versus Web." Even Steve Jobs, before Apple unleashed native iPhone apps and the iTunes app store, proclaimed that the future of mobile was the Web (in the absence of an SDK to develop native iPhone apps). He explained how he expected that most services could and should be accessed via a Web browser, which on iOS at least, would be feature-packed. He urged developers to hone their Web design and programming skills and to embrace the mobile Web future, harkened by Apple's commitment to HTML5. It went nicely with the emergent vision of the time, which was about "everything in the cloud," accessible from any device at any time. Folk were talking about the "mobile cloud," and so on. I will look at cloud computing later (see Chapter 8. Cloud Computing, Saas, PaaS and NaaS).

The same "vision" was extolled by Google, even as recently as the Mobile World Congress 2010 keynote speech, with a fairly similar formula for how our data and services would sit in the cloud, accessible via the mobile browser which, sooner or later, would become the dominant client for cloud services. In fact, Google claimed that for every new product and service in their portfolio, the new mantra was "mobile first." This was explained as meaning that the mobile experience was of paramount importance, with desktop following somewhere behind. Who knows, maybe since the launch of Google TV, the new mantra is "TV first," or possibly second. As it turns out, at least from a design perspective, there might well be merit in thinking about mobile first. I will return to this topic soon.

Native and Web-based are *not* either-or solutions to the same problem. There was no single problem to begin with that caused these two technologies to emerge as competing solutions. The Web existed at the same time that data connections on mobiles became possible. The Web has billions of pages and applications running across its sea of servers, and, more importantly, plays an increasing role in our evolving digital lives. Therefore, including a Web browser on a mobile device makes complete sense – it's primarily there to give access to the Web, not to provide an alternative pathway for implementing the various functions available to native applications.

However, mobile phones, by necessity and through their evolution, happen to be computing devices that can run software. They have screens and buttons, just like computers. Therefore, as we have explored in the previous sections about iOS and Android, it makes sense to allow third party applications to run on these devices by giving them access to APIs and an SDK to develop software with those APIs. This has been possible with mobile devices for some time. It was possible to develop mobile applications long before the iPhone and iOS. It was possible long before Java Micro Edition. It is not a new idea, nor a competitor to the mobile Web in any deliberate sense. In fact, the Web browser itself is just another native application running on the mobile.

As our brief historical perspective has shown, native and Web applications have different origins and purposes. The Web browser exists primarily to give access to the Web. The native APIs exist primarily to give access to the native capabilities of the device. This accounts for the key differences in these two pathways. It is almost always the case that if an application needs to access the native capabilities of the device, this will be relatively easy via native APIs and relatively difficult, if not impossible, via the Web browser.

As you might expect, given different starting points, these two paths have evolved in their own ways, tying them to certain patterns of design that can't be shaken at a whim in order to converge on a single solution. For example, the Web necessarily has to involve a remote connection to the Web server because that's actually where the application resides, not on the handset. This connection to the server requires a wireless connection with all its vagaries, including relatively slow and inconsistent performance. However, as we shall see, the programming environment of the mobile Web is rapidly converging towards a native-like environment. Indeed, there are some frameworks now that allow developers to build apps using Web-design techniques, patterns and languages, and then "compile" these applications to native form (although not quite).

6.5.3 Is Native versus Web the Right Question?

Clearly not! It's an odd starting point, like saying to someone "Train or plane?" before asking where they want to go. If you're setting out to engage a user via a mobile device, then you presumably have a goal in mind. Given that goal and your available resources, you will have to make a decision as to which path makes sense for *your* application, or more aptly, *your use case* – the goals you want your users to achieve within the context that they might want to achieve them.

Your application will start with an idea that you will convert into a series of user stories that describe what the user will be able to achieve using your application. You will also have a strategy for what you want to achieve by realizing this goal, such as adding value to some kind of business or project, possibly expressed via a business model.

Once you have brought all these objectives together, you can begin to make strategic and tactical decisions about how to build a mobile application that gets you what you want. Having consulted on numerous occasions with companies of all sizes and with individuals of all ambitions, I frequently find that the implementation pathway that makes sense, or the one I recommend, is not what the client had in mind. This sometimes includes not building a mobile application at all! For example, I have had to inform clients that the realities of going mobile will fall far short of their project ambitions.

With this in mind, I am going to review the state of the possible with mobile Web applications and examine the emergent trends and patterns, so that you will know what is possible and which types of goal are realistic, now and in the very near future. Before that, let's first review some of the key differences between native and Web applications with today's technologies, so that we can put this topic to rest for a while. The following table compares native versus Web for a number of key project parameters.

Parameter	Native	Web
Responsiveness – how quickly the application loads and how quickly tasks are carried out when invoked by the user.	Very fast – the application sits on the device. Data and user interface elements can be fetched and displayed very quickly. It is true to say that many developers favour native applications for this reason alone. Responsiveness remains a sensitive issue for users.	Relatively slow – the browser has to return to the server to get the data, which includes the UI. AJAX can speed things up, but that technique isn't supported by most mobiles (because they don't run JavaScript in the browser).

Parameter	Native	Web
Richness of the graphical user interface in 2D	Richest possible experience – a native application can usually exploit the full potential of any graphics capabilities present on the device and exposed via the graphics APIs.	Variable – on low-end mobiles, the graphics capabilities in the browser can be very limited. On smartphones, the graphics in browsers is limited to the rendering capabilities of the browser combined with the expressiveness of the HTML/CSS languages. On some devices, the graphics elements exposed via the browser can be quite rich, but not as rich as native graphics API capabilities.
Richness of the graphical user interface in 3D	Richest possible experience, especially where the device supports a powerful 3D rendering library, supported by a dedicated Graphics Processing Unit (GPU). On smartphones, the 3D capabilities exposed via the APIs are usually quite expressive, allowing relatively sophisticated 3D gaming experiences.	Very poor – usually, 3D graphics are not supported in browser, except for pseudo-effects to give a faux 3D appearance, such as pie charts and shadow effects. However, true 3D rendering with its complicated polygon processing is usually unavailable via the browser.
Richness of the interactivity	Richest possible interaction with wide support for input from voice, camera, buttons and, increasingly, a range of sensors (motion, tilt, GPS) on offer in modern smartphones.	Variable, often with only limited interaction via a small set of interactive elements in the web page, with poor access to other input channels like the camera and sensors.
Ease of programming	Relatively difficult – generally speaking, it is much harder to produce a native application than a Web application. Not only are the programming languages and APIs more challenging than HTML/CSS, but the whole cycle of development, testing and deployment is more complex.	Relatively easy – it is much simpler to get a website going, although implementing a large site with complex features can be challenging. Generally speaking, Web programming is accessible to a wider range of developers than native programming.

(*Continued*)

Parameter	Native	Web
Ease of deployment	Has many hurdles – it isn't straightforward to go from a finished application to being in a position where users can download and use it. Deployment requires all kinds of packaging and potentially lengthy weaving through app store submission processes.	Instant! This is where the mobile Web wins hands down every time. Barring the actual deployment of a web app to a live server, which can have its own challenges, once you're happy to go, it's as easy as giving your users the URL to access the app. It couldn't be simpler.
Discoverability	Depends! If your app is in an app store, then it can take quite a bit of user determination to find it there. Otherwise, it's not like the user can easily bump into your app on their phone, as there is no equivalent of surfing in terms of apps.	Also depends! After all, there are millions of websites. There is nothing inherent in the Web that says a user is more likely to discover a mobile Web service, although it has much lower friction in terms of discoverability through trying things out – accessing a Web app is frictionless (apart from any onerous sign-in process, but that's poor Web design).
Monetization	Good prospects! At least mobile apps can be sold through app stores where there is a precedent for paying for something that gets downloaded and installed on the phone. Interestingly, apps are used by some website providers as a means to monetize their site by offering more features, or even the same features, via a native app that is chargeable by being in the app store.	Challenging! There is no inherent mechanism for monetizing mobile Web apps, like the pay-to-play model of app stores. It is up to the site owner to think of alternative monetization strategies and make them work, like subscription or premium features. Business models for Web apps require more effort than taking part in the "app-store economy."
Deployment on a single platform	Easy, if it really is a single platform. If all you have to worry about is one build of your code, then things couldn't be simpler. However, this is seldom the case. Even on one handset, different versions of the code might be necessary to support different underlying versions of the platform OS.	Easy, and with the added advantage that a single platform here probably means a single Web browser platform that might well appear across numerous devices. However, beware! The issue with mobile devices is the variation in screen size and other form-factor issues that can screw things up and blur what we really mean by single platform.

Parameter	Native	Web
Deployment cross-platform	Very difficult! There's no escaping the fact that building native apps for more than one platform can become very difficult and very resource intensive in a short space of time. Even with technologies that were supposed to soften the blow of designing cross-platform, like J2ME, fragmentation has become a thorny problem with lots of devices reporting to have similar capabilities and APIs, but behaving in different ways.	Much easier! If you get the right design with the right baseline of browser platforms, avoiding complex design features, then it's much easier to make a Web application work across a wider range of platforms that it is with a native app. Many Web apps will run unchanged across iOS, Android, Symbian and other platforms. It does depend on the interface complexity though.
Richness of the APIs	Fantastic! You simply can't get a richer set of APIs than by going native. Platforms like iOS and Androids have thousands of API features that can be loaded into applications, ranging from utilitarian to exceptionally complex, like the CoreMotion APIs in iOS to access the gyroscope and other motion-sensing capabilities. Also, graphics APIs simply can't get any richer than going native. This is why games developers go native nearly all the time.	Very limited! The APIs available via the document object model (DOM) in the browser are embarrassingly limited compared with native API exposure. Until recently, it simply wasn't possible to access any of the native features of the phones via browser-based apps, such as the address book, location, camera and so on. This has changed a lot recently and HTML5 is promising to offer a lot more APIs as part of the baseline spec. Also, some platforms are beginning to expose a lot more APIs in the browser.
Battery life	Can be a challenge! Because of the much richer capabilities of native apps, such as the use of 3D graphics, it is often the case that a native app will utilize more processor-intensive APIs that consume battery energy very quickly. Games are a good example. However, it's really a case of get what you pay for because it's not as though a browser can do the same function with better battery life.	No real advantage. Whilst a browser might avoid the complex processing tasks associated with accessing rich APIs (like 3D graphics), it will demand more, potentially, from the wireless connection, which can consume battery energy fairly quickly. Browse for a while and you'll notice the battery drain!

(Continued)

Parameter	Native	Web
Data consumption	Beware! It is tempting to think that a native application won't be accessing the Web and so will consume less data. Yes, a Web app *has* to consume data to even get started, but many native apps these days will connect with a backend app on the Web and can easily consume as much data, if not more.	Can't avoid it! Any Web app has to consume data just to function, but it needn't be a heavy hit like streaming video. Well designed Web apps, designed for mobiles, can be quite economical with their use of data. That said, many Web designers simply don't know to, or don't bother with, any kind of optimization to reduce data consumption. We've grown up greedy with unlimited data plans, even though they're fast disappearing from the telcos.
Market penetration	Limited! With the flurry of interest in platforms like iOS, we might think that these popular platforms rule the mobile world. They don't. Penetration across the entire global user base is still very small. We've all heard how games companies are often forced to developer potentially hundreds of variations of a game to get coverage across a wide set of devices, even running (ostensibly) the "same" platform (i.e. J2ME).	Vast! Most mobile phones released today, in any market, have a Web browser that can at least display web pages in one form or another, perhaps not as richly as the best smartphones. Whether or not users have data tariffs to access the Web is the key question.

6.5.4 Major Trends in Mobile Web

6.5.4.1 Affordable Tariffs

Compared to the early Web browsers on phones (and I mean Web, not WAP), things have come a long way. Most of the progress, it has to be said, has been with the increased affordability of accessing the mobile Web. With cheaper and more transparent data tariffs, more users have flocked to the Web. This is an important trend to recognize because, as a developer, one always wants the total available market that is accessible via one platform or another. As I've highlighted in the previous table, the Web is the most ubiquitous applications platform for mobiles.

At the time or writing, mobile operators are back-tracking on their previous stance of offering "unlimited" data bundles for a flat fee. Due to the rabid success of iPhones and other smartphones on the network, and the widespread adoption of mobile dongles for laptops, flat rate "all you can eat" propositions have become unsustainable as a business model for telcos. However, this is unlikely to have a negative impact on mobile Web adoption and usage, especially as most users on a network will stay within the limited bundle sizes that are replacing the unlimited bundles.

6.5.4.2 Faster Connectivity

Being able to establish a decent wireless connection to the Internet has been a major boom to the mobile Web. The first iPhone would not have been successful without the inclusion of EDGE technology (so-called 2.5G) to offer decent connection speeds, as limited as they actually were in comparison to today's 3G and 3.5G (HSPA) connection speeds. Of course, we have all been victims of our own success, with the iPhone causing dramatic increases in traffic on cellular networks, leading to congestion and a decline in wireless performance. Even so, the combination of faster broadband networks and the wide availability of WiFi have combined to have a massive positive impact on the evolution of mobile Web.

Faster connectivity has led to greater consumption of the available bandwidth. Many developers have started to ignore the principles of bandwidth-conversation that used to apply in the earlier days of mobile Web when shifting bits back and forth was cumbersome and expensive. However, with "all you can eat" tariffs disappearing, plus the ongoing network congestion issues in some metropolitan areas, we will have to see a return to better utilization of the available bandwidth, with Web designers taking this into account when building sites targeted at mobiles. I will return to this topic shortly.

6.5.4.3 More Usable Browsers

Mobile browsers have got better in two ways. Firstly, they increasingly support a richer feature set of the wider Web standards defined by the W3C, having mostly run very cut-down subsets for some time. Browsers on smartphones can more or less support the viewing of websites designed to be accessed by desktop browsers. As we shall see, this is because desktop and smartphone browsers are beginning to use the same rendering software at their core (e.g. Webkit). Secondly, due to the faster processor speeds and bigger memories on most phones, mobile browsers can render pages quicker and can support rapid re-rendering features like pan and zoom.

Both of these trends combined leads to a position where some phones can support access to the full Web without that much restriction, apart from the extra effort to read the sites on a small screen, including constant pan and zoom. This process has been made a lot easier with the introduction of touch screens, especially the multi-touch technology introduced by Apple.

Outside of the smartphone world, we have seen a dramatic improvement in mobile Web access through the use of proxy solutions like Opera Mini, which is one of the most popular mobile browsers in the world. Opera Mini is not a conventional Web browser. It doesn't understand or render pages written in HTML, which is the lingua franca of the Web. Instead,

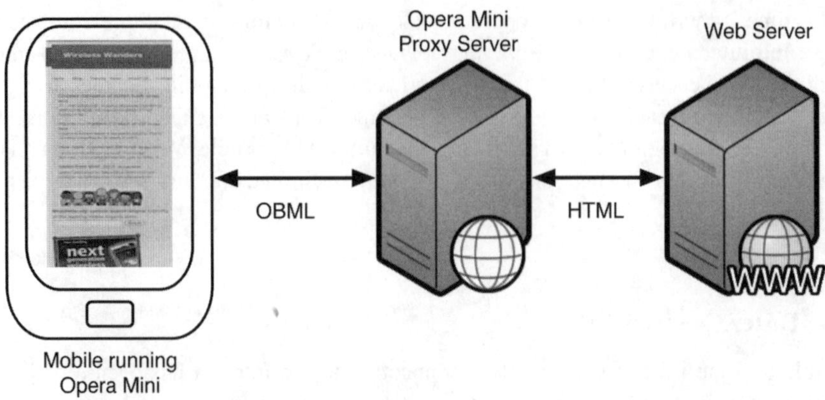

Figure 6.7 Opera Mini browser using a proxy to reformat web pages.

the browser accesses a website via a proxy server that converts the HTML into a proprietary format called Opera Binary Markup Language (OBML), as shown in Figure 6.7.

OBML produces a highly compressed version of the original HTML, enabling the page to be transferred quicker and also, just as importantly, to be rendered quicker by the browser, especially on devices with relatively limited processing power and memory resources, like many of the so-called "feature phones" that make the bulk of phones sold and used on the planet today. Additionally, the Opera Mini browser can deploy a technology called Small Screen Rendering (SSR) that reformats a Web page, where possible, to be a single column format that is easier to read on a device with a button-based interface, which lacks the navigational dexterity of a multi-touch interface with its pan and zoom features.

On regular browsers on feature phones, the optimal interaction is the up-down scrolling, which SSR deploys to maximum ergonomic effect. With SSR, various navigational or space-hogging visual elements, like long lists, are automatically collapsed in order to ensure that most of the screen is used to render actual content rather than the relatively superfluous "visual scaffolding" found on most websites (e.g. banners, menus, footers etc.)

There are other proxy solutions on the market, such as the SkyFire browser, which is best known for its ability to reformat video content that requires the embedded Adobe Flash player to view. Most phones do not support flash and mobile browsers, which can become a problem for viewing video given that Flash is used to present much of the video content found on websites. That said, even with the faster connection speeds of modern mobile broadband services, video is often a laboured affair with choppy playback and other irritations. Video is also problematic for network operators because it can rapidly consume a lot of the available bandwidth in a cell, adversely affecting the mobile Internet performance of other users sharing the cell. And, as we enter an era of limited data bundles, we should expect to see a change in users' sensitivity towards watching mobile video, as it will rapidly eat up their data bundles.

What we have seen in the post iPhone era is the shift from mobile devices that were internet-capable to devices that are truly internet-centric. We don't question the usability of accessing the Web via a smartphone like the iPhone or a Samsung Galaxy (Android).

6.5.5 HTML5

6.5.5.1 HTML5 Origins – It's all About Apps!

HTML5 is the fifth major revision of the language used to describe Web pages – Hypertext Markup Language (HTML). At the time of writing, the specification, managed by W3C, is still in draft, meaning that it is still a work in progress. As you might know, HTML was originally designed by Tim Berners-Lee to describe the layout of scientific papers. However, how many of us read scientific papers online? Not many. The use of the Web has blossomed into countless applications including the very important shift from a publishing medium to an applications medium. We don't just want to read content in our browsers, we want to access applications like online banking, shopping, social networking, word-processing and many more. HTML5 embraces a lot of features that make it much more amenable to creating a wide range of rich browser-based applications.

Don't forget, with HTML5, we are specifically talking about a browser technology. It does not encompass the wider Web OS story that we have explored, with CRUD, Web services, open APIs and all those other exciting areas of the modern Web story, many of which I discussed in Chapter 2. The Web 2.0 Services Ecosystem, How it Works and Why, and which have evolved separately to the HTML5 story. Ideas like social graphs, JSON interchange and RESTful APIs have nothing, or little, to do with browsers and HTML5 – they all work quite well with existing browsers and HTML 4. Nonetheless, the desire to create even more compelling user experiences in browsers, especially to bring us back to the notion of a universal client (i.e. the browser) that can provide a front-end to most, or many, of our applications.

The origins of HTML5 actually belong outside of the W3C. In fact, there was a meeting of various Web browser vendors and interested parties to ask the question as to whether or not the W3C should developer extensions to the current HTML and CSS technologies that would address various Web *application* requirements. The vote at the meeting rejected this stance, leading to a splinter group called the WHAT Working Group – Web Hypertext *Applications* Technology.

For about two years, the WHAT group wandered off and did its own thing, actually producing a substantial amount of work to gather and solidify how Web browsers might support applications, which they published as a body of work called "Web Applications 1.0." The W3C weren't necessarily ignoring WHAT, but they had gone off to focus on solidifying the core language, especially many of the issues that had cropped up as a result of turning HTML into XHTML (a story that I covered in my previous book). After those two years in the wilderness, the WHAT group was brought back into the fold. Web Applications 1.0 was renamed HTML5, and here we are, about to review what that piece of work is all about.

Let's begin with a description of HTML5's key features and then follow with a few of the questions that I frequently encounter from interested parties.

6.5.5.2 HTML5 – The Headline Features That Make it Special

6.5.5.2.1 Built-in Graphics Canvas

HTML5 defines a new element called `<canvas>` that defines an area in the webpage that can be used for rendering 2D graphs, game graphics or other visual images on the fly, somewhat akin to the way that Flash is used today to embed rich graphics into a Web page. To use the canvas, HTML5 defines an API that can be called by the JavaScript running in the page. The API includes drawing primitives to allow the creation of shapes, paths, gradients and transforms that allow not only drawing, but also animation to take place. Combined with the power of JavaScript processing of input events and its general-purpose language capabilities, it is possible to use canvas to create games and other interactive graphically rich features in a Web page without the need for a plug-in.

Canvas doesn't support 3D graphics, but that is an obvious and likely extension, at which point we might well expect to see the browser challenging the native applications' dominance over games. There are experimental builds of the Opera browser (Opera being at the forefront of defining and implement modern browser technologies and standards) that support a 3D canvas. There is also an experimental 3D add-on for Firefox.

6.5.5.2.2 Video Without Plug-Ins

One of the exciting features of HTML5 is its native support for playing video via the use of the `<video>` tag. Again, this is supposed to mean native support without the need for plug-ins like Apple's QuickTime or Adobe's Flash. Video support, by extension, means audio support, or perhaps we should revert to the unfashionable term "multimedia." Ironically, the original multimedia vision that accompanied the invention of the CD-ROM has never been more possible than with HTML5, with its ability to support just about any interaction mode and media type that most authoring and production tools could potentially spit out.

That said, video is a wide set of technologies. It's easy to get confused between AVI files and MP4 files and so on. Strictly speaking, these aren't really video files per se, but video container files. The video is housed inside the container and it's the compression technology (e.g. H.264) that is often the issue, because unless a device can decode a particular compression format, then it doesn't matter what the container is, the video won't render to the screen. So, just because HTML5 browsers will support video, don't think that this means you can now watch Flash videos without a Flash plug-in. If the browser doesn't support the Flash video encoding format, then it still won't play.

6.5.5.2.3 Local Data Storage

Local data storage is an exciting feature of HTML5, especially when we look back at the first entry in the table that compares native with Web apps, which compared application responsiveness. With most browsers today, any data needed for display inside a Web page has to be pulled down from the server, either via a complete HTTP GET request for a new page (see Section 2.2 Beneath the Hood of Web 2.0: CRUD, MVC and REST) or via the data-chunking method of AJAX. With local storage, data can now be stored on the local device and accessed via a JavaScript API. There's no local storage tag element, like the `<video>` or `<canvas>` tags above. The local storage is all accessed via the JavaScript APIs.

You may have heard rumours that local storage was removed from HTML5, which might have you concerned that it's not going to be supported by modern browsers. This isn't true. What happened is that the HTML5 spec became so large and unwieldy, that the working group had to relegate a few features to separate specifications, like local storage, in order to speed up the completion of the baseline HTML5 spec. Another rumour, or myth, is the idea that there will be numerous local storage techniques competing for developer's attention, including Google Gears, which you might have heard about. Again, this isn't the case.

Most browser vendors are committed to the HTML5 standard for local storage. However, local storage is *not* the same as a local database that will support a complex interface like Structured Query Language (SQL). Some vendors have built solutions that incorporate more SQL-like interfaces and there are some emergent and competing visions in this space, such as the Indexed Database API, usually simplified to IndexDB. This competes with another vision, called Web SQL Database, which has already been implemented in some browsers, such as Safari, Chrome, Opera and iPhone (Safari). However, it seems likely that IndexDB will be incorporated into HTML5.

The same principles of cross-site restrictions apply to the local storage as apply to posting from Web forms. If an application from site acme.com stores data locally, then the browser should only allow scripts running in applications published from acme.com to access that data. A site running from ackme.com won't be able to access the locally stored data from acme.com, even though the data all belongs to the one user and is stored by the one browser. This is to prevent unwarranted access across sites, which could well be used maliciously.

The benefits of local storage are that a Web app doesn't need to issue a GET cycle to the server to load its data. For example, let's say we implement an email application in the browser using HTML5, then we can do things like open the inbox using the data fetched during the last session (which had been previously stored using the local store), giving immediate access to data whilst the Web application fetches any new emails in the background. Additionally, a user could create a new email and then store this in the local store if there is no connection available, such as when out of coverage or flying on a plane. Developers of HTML5 games will also use be able to use local storage to save game state so that when you launch the game the next time, it remembers where you were.

Local storage only applies to storing values (in so-called key-value pairs) from within an application. In other words, it's for storing application data, not the application itself. However, as you can imagine, if we really want offline applications to work, like the games and email examples, then we need to think about how to store the actual applications locally on the device without the need to go fetch the apps from a Web server. This is the subject of the next HTML5 feature: offline Web applications.

6.5.5.2.4 Offline Web Applications

I know what you're thinking already – what's the point of an offline Web application? Isn't that an oxymoron? How can a Web app be anything but online, those two terms being synonymous? Well, offline doesn't mean that we do away with the Web. The Web is still where our application resides, running on the Web server, quite likely atop of a LAMP stack (see Section 2.2 LAMP and beyond: Web frameworks and middleware). But once we've gone online and downloaded our app, which might consist of a variety of pages and related resources, like CSS files, image files, JavaScript files, and so on, HTML5 offline cache gives us the chance to store all those assets locally so that we can access and grab them ultra-fast the next time we need them, without having to go back to the Web server.

Combined with the feature of local storage, we can also prime our application with an initial set of data. In other words, returning to our email example, we can load an email client in the browser and then prime it with some data (like the last emails fetched to the inbox) with a level of responsiveness that challenges the native application experience.

Once the developer indicates to the browser which resources it wants to store offline, then it's up to the developer how to exploit this feature. The browser will attempt to cache the items requested whenever it can, plus it provides an event model to the developer to signal when the browser has a connection, or not. This might allow the developer to go fetch some data and do some updates as soon as the device goes online again. A nifty feature of the application cache is that is also allows the developer to indicate another set of files that can be used in a fallback scenario when it hasn't been possible to cache the required files properly. Imagine a file called "whoops_offline_gone_wrong.html" that loads in this case and gives some meaningful and contingent user experience, should things have gone wrong.

6.5.5.2.5 Geo-Location API

A topic that I covered in the previous discussions of iOS and Android, most smartphones and an increasing number of "feature phones" incorporate some kind of location-finding capability that is exposed via the OS APIs (e.g. via the application framework libraries). As we have seen with the iTunes app store, ever since the location APIs became available, lots of iPhone and iPad applications have incorporated location-enabled features. This has happened so widely that it is no longer useful to talk of location-based applications, as we used to in the early days of location finding. Location is simply a common enabler across a wide set of applications, from shopping apps to photography apps. That being the case, it seems obvious that a lot of Web applications could equally benefit from gaining access to location, which is the intent behind the HTML5 geo-location API, which is an API that is exposed via the JavaScript environment.

Just as with all location features on mobile devices, privacy remains a concern and HTML is no exception in handling this via the usual mechanism of user intervention. The geo-location API is quite clear in specifying that a browser should not send location information to websites without the explicit permission of the user. In other words, as per common privacy practises in the mobile industry, the user remains in control.

The Geo-location API is already supported across a wide number of browsers, including Firefox, Safari, Chrome and iPhone/Android derivatives. The API allows, where possible, selection of either low accuracy finding mechanisms, such as cell ID, or, if available, higher accuracy options like GPS.

There are a number of alternative APIs supported by other browsers, including the OMTP BONDI implementation that is part of the telco WAC initiative. The BONDI standards define a model for controlling access permissions to browser APIs, so that telcos can manage these features, just as they insisted on a similar control mechanism within the security model applied to J2ME applications. It isn't clear how this standard is going to make its way into devices, but has the support of lots of operators.

6.5.5.2.6 Background Processing with Web Workers

If you have followed the evolution of iOS, then you've probably come across the idea of multi-tasking and background processing, especially as it was an area of great controversy and criticism with iOS. Indeed, for iOS 4.0, Apple made a lot of noise about their support for multi-tasking. So what is multi-tasking and why do we need it, especially on mobiles?

Figure 6.8 shows the problem in general, which we will expand to cover the browser case in a moment. Here we show the user running a game ("Racer game") on her mobile. Whilst it is running, it grabs the screen and the user interface (e.g. buttons or touch screen). It is the job of the underlying OS to make sure that all UI interactions are assigned to the game and

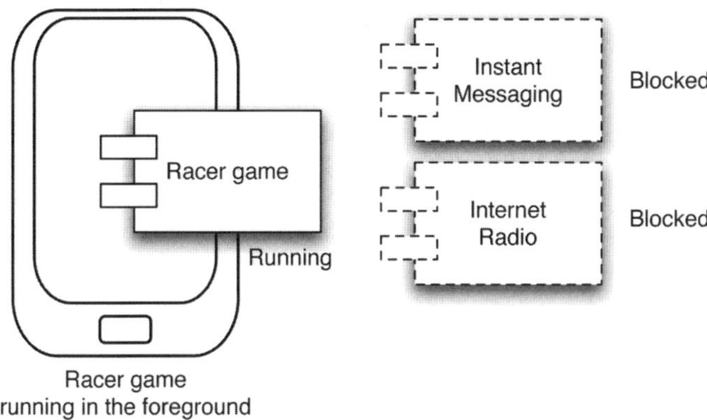

Figure 6.8 Foreground and background tasks on a mobile.

not sent to other applications. Meanwhile, other applications that the user might want to run, like an Instant Messaging client or Internet Radio, are blocked from running because all of the processor resources are dedicated to running the game. This presents the user with a problem, or an inconvenience. What if one of her friends wants to send her an IM message? It isn't possible because the IM app isn't running.[26] That makes sense because while she's playing the game, she can't be sending IMs at the same time, notwithstanding that lots of folk probably could play and message at the same time (and often do on their laptops).

How can we solve this problem? The answer is with a mechanism called multi-tasking, which allows the foreground application to keep running and using the screen resources whilst letting the other applications run in the "background." This means that the background applications continue to get enough time on the processor[27] to allow them to carry out some tasks that make sense without having to take control of the screen. For example, an IM app can continue to remain logged in and indicate an "Available" presence state. If a message comes in, then the app can still receive the message because it can run its "message receiver" task in the background, but simply not display the message. Instead, it can optionally signal to the OS that it has something that it would like to bring to the user's attention.

One way of notifying the user, as we see with iOS (now that it supports multi-tasking), is to provide an API that allows the background application to push a notification to the screen that will sit over the top of the current display. In our example, this might mean a notification like "New message from Bob," if Bob sends an IM while the game is in play. The user can click on the link to go check the message and respond, bringing the IM app to the foreground whilst suspending the game, putting it in background (or completely suspended) mode.

Similarly, for applications like Internet radio, that only require the speaker to function (to play an audio stream), it is possible to allow the radio app to process the incoming audio stream, decode it and then play it to the speaker, either mixing it with the game sounds or

[26] Worse still, if it isn't running, then the IM system will probably show that she isn't even logged in and available to receive messages.

[27] Without interrupting the foreground application because processors are so quick that they can continually swap tasks without the user noticing.

giving it complete control over the speaker. Of course, these kinds of idiosyncrasies are what make multi-tasking tricky on mobile devices – these edge cases, like providing shared access to the audio output API, need to be considered and then ironed out.

Now that you hopefully appreciate what background processing and multi-tasking is about, we can return to our HTML5 story and the inclusion of Web Workers in the spec. A Web browser is a single application that runs on the OS, so it doesn't necessarily understand the concept of multi-tasking, notwithstanding that we could well imagine supporting the scenario of "tabbed browsing" whereby we have several sites open at once. However, that is not what we're talking about because tabbed browser sessions are all independent – there is no concept of foreground and background, nor any interaction between them.

Web Workers provide a way for browsers to run chunks of JavaScript in the background. This is more akin to what developers call "threads" whereby they can ask their program to do several things at once, split into foreground and background modes. The classic reason for this mode of programming is to allow complex, and therefore time-consuming, tasks to be offloaded into the background so as not to lock up the user interface, preventing it from reacting to inputs. Another example might be an application that needs to regularly poll a server for data updates, like information being streamed from a sensor (see Section 7.2 Sensor-net: Mobiles as 6th sense devices), whilst allowing the user to analyze the data and perform local manipulations of the data.

6.5.5.2.7 Semantic Web Support via Microformats and Extensibility

One of the grand themes of this book has been the evolution of the Web towards something beyond just a interconnection of pages to an interconnection of things, like people (see Section 3.4 Future Web: "People OS?"). Indeed, we might well ask if there are several Webs, like the Social Web, the Sensor Web (see Section 7.2 Sensor-net: Mobiles as 6th sense devices), and so on, leading us to think of the InterWeb idea that has been touted by various Web thought-leaders.

Earlier, I discussed the idea of adding annotations to data (see Section 3.2 Internet of Things – A Primer), which eventually lead to a discussion of micro-formats (see Section 2.6 It's all about people: Social Computing), which are tiny nuggets of semantic data that we can embed in Web pages in order to notify machines (rather than people) about what the pages contain, especially in relation to commonly embedded data like contacts, addresses, locations and so on.

Think of the need for micro-formats and semantic data as the need for new HTML elements like `<person>` to described a person, and `<vitals>` to describe a set of vital signs for health and wellness applications (see Section 7.2 Sensor-net: Mobiles as 6th sense devices). However, we don't have these elements in HTML5 and they probably don't belong in the HTML5 specification, unless you want to argue for them (which is kind of how standards bodies work).

Well, it turns out that we can add these kinds of elements without having to shout and argue for them. Those nice guys at the HTML5 meetings have given us a way to extend existing elements in order to add the kind of semantic data that could give those elements additional context-specific meaning within our application or application domain (e.g. Healthcare). Adding microdata works by declaring a microdata property within an existing HTML element. Let's say we have a block of HTML that looks like this:

```
<h1>Paul Golding</h1>
<lu>
  <li>65 bpm</li>
  <li>138 bpm</li>
</lu>
```

This HTML snippet would produce my name followed by two lines of text that indicate my rest and maximum heart rate during an exercise regime at the gym. However, while this might be recognizable just by looking at the text, especially now that I have told you, it isn't so obvious to an application that might want to ingest the data, such as a fitness monitor system. What we can do is to extend the elements like so:

```
<h1 itemprop="athlete">Paul Golding</h1>
<lu>
  <li itemprop="heartrate">65 bpm</li>
  <li itemprop="heartrate">138 bpm</li>
</lu>
```

These "itemprop" extensions will be ignored by the browser when formatting and displaying the data, but could use processed by any application that recognizes these properties and can do something with them. The properties themselves are defined in a separate vocabulary file that is referenced elsewhere in the HTML file and can sit anywhere accessible on the Web (via a URL).

This extensible feature of HTML5 makes it ideal for the Internet of Things. We can now pass around data in our Web OS that is annotated and more readily digestible by all kinds of software services running in the cloud and its associated sensor clouds. The annotations are like parallel footnotes or sidebars to the main data streams.

6.5.5.2.8 Device Element

This is an interesting development that is closely related to the ability to embed multimedia formats directly into HTML5 apps. The device element represents a device selector that allows the user to grant access to a Web page to capture input from a device, such as a microphone or camera, or potentially some other sensor (see Section 7.2 Sensor-net: Mobiles as 6th sense devices). The input data can then be streamed using something like Websockets, which I described in Real-time Web and Twitter. Websockets was originally part of the HTML5 spec, but was siphoned away from the main spec in order to streamline the HTML5 specification work. The device element is a great feature of HTML5 as it prepares the browser for becoming an interface for sensors on the device.

6.5.5.3 Does HTML5 Fix Everything?

A lot of folks in the mobile community have heard the rumours that HTML5 is a panacea to all mobile application woes, that it will become the only tool that developers need to build mobile apps, putting an end to the native versus Web debate, and so on. This simply isn't the case. HTML5 adds new features that improve the mobile Web experience considerably, some of which can give native-like experiences, such as the ability to cache websites and data locally so that mobile Web apps load instantly, just like native apps. This much is true.

However, most of the native versus Web comparisons in our table earlier will remain the same until HTML5 matures. Meanwhile, native OS APIs will continue to expand in capability as more innovative features are added to mobile devices. There is the inherent tension between innovation – the need to push new and differentiated features – and standardization of APIs in a body of work like HTML5. However, it is clear that the richness of the mobile Web will substantially increase thanks to HTML5, once all of its features are available across a wide set of browsers and devices.

One word of warning about HTML5 is that the success of a standard is often related to the ecosystem of tools and developers supporting it. Clearly, as far as developers go, you couldn't get a bigger audience. It's safe to say that most developers are comfortable with HTML and website development, even if it's not their day job. There is a plethora of design, development and testing tools for building websites and a very rich ecosystem of support, knowledge and deployment environments. HTML5 will be in good company!

However, the tools caveat is with some of the new features, especially the graphics canvas. What puts Adobe Flash far ahead of the crowd in terms of support for graphics rendering in browsers is the extremely comprehensive toolset on offer from Adobe, like the Creative Suite, now in its 5th major release. HTML5 needs the support of leading tools vendors, like Adobe, in order to become as successful as it promises to be. I wouldn't be surprised to find Google, and other proponents of HTML5 and open Web standards, releasing design tools free of charge. Let's see!

6.5.5.4 Do I Need a New Browser?

No. HTML5 isn't a single set of unified features that all need to be implemented at once in order for a browser and website to support HTML5. Indeed, various elements of HTML5 have been making their way into browsers for some time. This is what companies like Apple mean when they say that they are committed to HTML5 in their Safari browser. Safari already has some HTML5 features, like an API that enables mobile location to be detected by the script running inside the browser.

HTML5 is also backwards compatible. Many of the great features of HTML 4, which itself was strides ahead of HTML 3, will still work. HTML 4 sites will render in HTML5 browsers. This is one of the great beauties of HTML. It has been an evolving specification with a large integral thread that remains at its core. Many of the original elements to define document markup, like the heading tags `<h1>`, `<h2>` and so on, still work in today's browsers. It is still possible, if you can find them, to render early HTML 1.0 web pages in today's browsers.

6.5.6 Widgets

6.5.6.1 Introduction to Widgets

You have probably heard about widgets and wonder where they fit into the overall scheme of mobile Web and mobile applications. Some say that widgets are "native Web apps." The problem with discussing widgets has always been the absence of any universal definition.

There are so many application environments claiming to support widgets, all of them using different technologies and approaches. If you're a Mac user, like me, then you'll be used to widgets (if you ever use them) as part of the OS X Dashboard, these tiny mono-purpose applications that hover over the screen pushing out weather updates, stock updates and other info-nuggets. Some of them offer interactivity, like the dictionary and calculator widgets. But let's face it, widgets on the desktop are almost a lost cause, a kind of curiosity and an idea looking for an application.

Perhaps that application is mobile. There seems to be a fit with the single-function single-window format. After all, many of the applications in the iPhone and Android app stores are designed this way. But wait a minute! Doesn't that mean that native apps and widgets are really the same thing, but with different names? Well, yes, almost. Seen from the user perspective, they are the same, in almost every way. They're downloadable, installable and runnable applications. The key difference, from a technical perspective, is that a widget usually requires a container to run. For example, Yahoo has its Blueprint widgets solution, but this requires the user to run a Yahoo application on the phone in order to then see and run any widgets.

Amongst the main widget vendors and solutions are:

- Nokia WidSets (Java based)
- Nokia WRT (Web based)
- Opera widgets (Web based)
- Motorola WebUI (Web based)
- Yahoo Blueprint (Java based)
- Sun Java ODP (Java based)

Where a widget requires Java, this means that a special container application is required to run first in order to display the widgets. This has had very limited success with Nokia Widsets, one of the earliest widgets frameworks on the market. Yahoo has similar limited success and the Sun solution (now Oracle) has failed to gain any traction, requiring developers to learn a new flavour of Java to create the widgets. For a comprehensive comparison of the above widget frameworks (plus a few others), I recommend consulting the well-researched study by VisionMobile.

Let's focus on the widget environments that are Web based because these are the most relevant to our connected services story and lead us to another story, which is the emergent Wholesale Applications Community, a multi-operator multi-vendor initiative aimed at creating a unified applications environment and ecosystem across all operators and handsets.

6.5.6.2 Web Widgets and the Mobile Web

In our discussion of HTML5, you might have noticed that the idea of storing a Web application locally, via the cache (and using local data storage) feels like we're talking about enabling local Web apps installed on the device. If, via the phone's home screen, we provide a short-cut to such a local app that causes the app to launch in the browser instantly with all its assets immediately visible (from the cache), then isn't that the app experience that users are familiar with? How does it differ from launching any other app? Also, we could, if we liked, invoke the app in the so-called browser "chromeless" mode (which means without the address bar and other UI scaffolding that comes with a browser), making it look even more like an app and less like a browser session.

This notion of locally installed Web apps is exactly the intention behind Web widgets. This is the vision behind the various Web widgets frameworks mentioned in the above list. In all cases, the applications are created using Web languages – HTML, CSS and JavaScript. There are only a few key differences between a Web widget and an ordinary Web application:

1. Web widgets are intended to be stored locally on the mobile (just like HTML5 local cache).
2. Web widgets can define a minimized version, which is usually a very cut-down UI intended to run on the home screen, like an Twitter client that only displays the latest Tweet, for example (similar to the Windows Mobile 7 idea).
3. Web widgets support the concept of being installed, unlike a mobile website. Therefore, they fit the paradigm of an app store model, including its "app-store economy" (i.e. users are prepared to pay for things they can download and install[28]).
4. Web widgets are able to access native features of the phone via JavaScript APIs, like the address book, the camera, and so on.

In other words, the idea of Web widgets is to enable developers to create apps for mobiles that in every way mimic the native app experience (app store mechanics, installation, access to phone features) but can be authored using standard Web languages, tools and methodologies. But that's not all! In case you've missed the obvious point, Web widgets use the browser (or it's inner core at least – the so-called "Web runtime"[29]), which means, on the surface, that Web widgets will run cross-platform (i.e. across different operating system platforms) because they will run wherever there's a Web browser.

It is an enticing promise, especially to operators who saw the opportunity to reclaim some of the ground lost to Apple, Google and other centres of gravity in the mobile applications economy. Enter then, the Wholesale Applications Community and OMTP BONDI, both attempts by various operators and their cohorts to make Web widgets a mass-market reality.

Taking point 4 from the list above, a key component of a Web widget is gaining access to the underlying phone operations via APIs. This is how native applications work, as we've examined in some depth in the previous two sections (see Section 6.3 Key Platforms: iOS and Section 6.4 Key Platforms: Android). However, JavaScript APIs are somewhat different to native APIs because they follow a different model and intent. JavaScript APIs exist to enable a Web programmer to manipulate the contents of the current document inside the browser. To do so, the document is represented into the JavaScript environment as something called the Document Object Model (DOM). The DOM is a list of all the elements inside the current document, such as all the `<h1>`, `<p>`, `<lu>`, `<li>` tags. The DOM API provides programmatic access to these elements. A quick example might make this clear. Let's say I have snippet of HTML code like this:

[28] This is not only because of the precedence set by the download economy of the Web, but because it is felt that users feel a sense of ownership when they download something, as opposed to visiting a website. Owning something has value.

[29] Hence the WRT in Nokia WRT widgets.

```
<p id="message">Your health is fine!</p>
```

This is a paragraph tag that a browser will see and render the contents ("Your health is fine!") to the browser display. However, let's say that I wanted to change that text dynamically based on some context, like taking readings from health or wellness sensors networked with the mobile (see Section 7.2 Sensor-net: Mobiles as 6th sense devices). JavaScript can modify this tag directly from within the page using the DOM API to extract the tag via its ID ("message") and then setting the HTML inside the tag (called the inner HTML). This might be done using the following JavaScript snippet:

```
document.getElementByID('message').innerHTML = "Your health
has problems!"
```

The `document` is an object (JavaScript being an object-orientated language) that I send a command to (called a method), called `getElementByID` with the added field of `message`. This invokes a piece of code inside the browser that supports the getElementByID API. You can imagine that it searches through the HTML until it finds the tag element with the appropriate ID (selected by `message`). Finally, I set a property of the element called `innerHTML` and this causes the API to replace the HTML inside the tag with the string provided to the API - "`Your health has problems!`"

Whether you fully understood this JavaScript API example or not, you will hopefully appreciate that we can use JavaScript to modify the contents of the current Web document. We can also use JavaScript to respond to events, like the press of a button or the hovering of a mouse over an element. What Web widgets do is to extend the JavaScript APIs beyond just the DOM and into the device itself, so that we can do things like this in JavaScript:

```
// Get the text message to and message fields from the page
phoneNumber = document.getElementByID("phoneNumber").value;
messageText = document.getElementByID("messageTxt").value;

// Set the text message to and message fields
txtMessage.To = phoneNumber;
txtMessage.BodyText = messageText;

// Now send the message dude!
myPhone.IMessaging.Send(txtMessage)
```

The first two lines grab some values from input boxes on the page, to get a phone number and some message text, ready to send a text message. The next two lines of code set the values of a JavaScript object called txtMessage and then, on the final line, we send the contents as a text message using an API called IMessaging, which is one of the APIs provided by the Nokia WRT widgets framework. As you can see, here we are using JavaScript inside a Web page to do something on the phone – send a text message. We can also access other phone features via the Nokia WRT JavaScript APIs, such as:

1. Calendar API
2. Contacts API
3. Landmarks API

4. Location API
5. Logging API
6. Media Management API
7. Messaging API
8. Sensors API
9. System Information API

Perhaps you're beginning to get excited about all these powerful APIs available via the Web browser. Not as comprehensive as the native APIs, but headed in the right direction nonetheless. But maybe you've got a nagging doubt here too. Are Nokia WRT APIs the same as Opera APIs? If not, then so much for the promise of cross-platform development with Web widgets. Sadly, they are not the same. However, recognizing the problem of fragmentation, various bodies came into being to try and tackle the problem, if not avoid it altogether. There are three distinct standards efforts of note in the Web widgets world:

1. W3C widgets standard, most notably the Devices and Protocols Working Group (DAP).
2. OMTP BONDI.
3. Wholesale Applications Community (WAC).

6.5.6.3 W3C Widgets

There are a number of specifications in various stages of development relating to Web widgets. These have primarily focused on the basic scaffolding issues of how to package widgets for distribution and installation. There are also issues like digital signing, updating and resource control (to limit what widgets can and can't do via internal APIs).

Of most interest to our current consideration of JavaScript APIs, there is a W3C working group to look at the definition and standardization of device APIs. As stated on its homepage, the mission of the Device APIs and Policy (DAP) Working Group is to create client-side APIs that enable the development of Web Applications and Web Widgets that interact with devices services such as Calendar, Contacts, Camera, etc. Additionally, the group will produce a framework for the expression of security policies that govern access to security-critical APIs. This last part is deemed essential by operators who have adopted industry-wide best practises to ensure that users are always in control of using any resources on their mobiles that relate to either privacy or charging. The telco industry remains very sensitive about the issues of abuse.

The APIs being considered by the DAP all have the following starting points:

- Nokia's calendar API
- Nokia's camera API
- Nokia's contacts API
- Nokia's messaging API
- Nokia's System Info API
- Nokia's DeviceException Interface
- BONDI 1.1 APIs (including Application Launcher, Messaging, User Interaction, File System, Gallery, Device Status, Application Configuration Camera, Communication Log, Contact, Calendar, Task)
- BONDI Architecture and Security 1.11 Approved Release

As you can see, these have come from existing widgets projects, like Nokia's WRT and OMTP's BONDI. This is a positive indication that the vendors are trying to push for a common set of standards for the device JavaScript APIs.

Don't expect all of the various "standards" efforts, even in the W3C, to necessarily align. Don't forget that standards bodies are just a bunch of people with a common intent, such as enabling mobile Web widgets to gain traction, but each with their own particular agenda, usually commercial. Each of these commercial agendas is likely to be slightly different, informed by different commercial strategies and capabilities.

If a particular vendor has a keen strategic need to push for local file access in widgets, then that vendor will raise his or her voice at the standards meetings and try to influence the adoption of file access. Later on, this interest might overlap or coincide with interest from another vendor, perhaps in a different (but related) working group. This might eventually cause those two efforts to merge, or for one to dissolve. Additionally, vendors will continue to innovate and pursue their commercial strategies irrespective of standards activities. This could be for any number of reasons, not all of them sensible or particularly well engineered. Don't fall for the myth that big companies and clever people know what the heck they're doing! I'm just saying all this in case you go seeking a unified and consistent view across standards efforts and don't find it. It also absolves me of the duty of having to try and tie all these threads together for you – because I can't!

All I can say here is that there is an effort by the W3C to write a standard for accessing device APIs. Whether or not vendors sign up to the standard and implement it is an entirely different matter. This is not the same arena as HTML5 where there is a wide consensus, comparatively speaking, to define and implement a common standard.

6.5.6.4 OMTP BONDI

The Open Mobile Terminal Project is a cross-vendor and cross-operator initiative set up to push for common standards relating specifically to the use of mobile applications. Founded in June 2004 by a group of eight mobile operators, the Open Mobile Terminal Platform (OMTP) was set up with the aim of simplifying the customer experience of mobile data services and improving mobile device security. This latter part of the OMTP mission seemed to be its focus, but the overall intent of the OMTP was to prevent fragmentation of key mobile applications enablers.

This intent gave rise to the BONDI project, when it became increasingly apparent in 2006/7 that the future direction and success of the mobile Web would depend on Web applications having access to the capabilities of the mobile device. Having seen the mess of MIDP, which was the previous cross-vendor cross-operator attempt to standardize on a single applications environment (native in Java), the OMTP wanted to avoid a repeat disaster. BONDI, therefore, focused on defining a common set of JavaScript APIs and a unified security and access model to enable operators and vendors to exert some kind of control over API usage, ostensibly with consumer interests in mind.

Following the introduction of the Wholesale Application Community, both OMTP and BONDI merged with WAC in July 2010, so I'd prefer to continue the story by switching tracks to a brief exploration of the WAC.

6.5.6.5 Wholesale Applications Community

The Wholesale Applications Community (WAC) is an open, global alliance formed from the world's leading telecoms operators. WAC is intended to "unite a fragmented applications marketplace and create an open industry platform that benefits the entire ecosystem, including applications developers, handset manufacturers, OS owners, network operators and end users." Just for historical completeness, WAC is an evolution of the Joint Innovation Lab (JIL) started by Vodafone and China Mobile (the two biggest operators in the world), along with Verizon Wireless (US). JIL has now become the WAC with the introduction of a lot more partners to the community. The focus of WAC is still on standardizing a set of APIs for access native device capabilities from a Web app.

Where possible, WAC aims to use existing W3C standards, like the Web Widgets standards work mentioned earlier. In some cases, this is not surprising because BONDI had already proceeded along this path, feeding some of their work into the W3C. In other areas, where an API already exists, like the Geo-Location API of HTML5, the WAC will supposedly prefer to adopt such APIs.

The WAC efforts also include the GSMA OneAPI standards work, which is supposed to create a unified set of network APIs to enable NaaS-type services (see Section 9.1 Opportunity? Network as a Service). It seems obvious that where an API exists to access Network services, then it will require not only standardization at the network level but at the device level if these APIs are to be made accessible via mobile clients in addition to Web applications (running on Web servers). However, for the time being, the network dependencies are an optional feature of the WAC specifications.

Let's briefly review the APIs in the current WAC standard:

1. **AccelerometerInfo** – Gives access to x,y,z motion vectors via an onboard accelerometer (see Section 7.2 Sensor-net: Mobiles as 6th sense devices).
2. **AddressBookItem** – Access to address book and contact records, including a number of standardized fields like name, email, company, address, homePhone, officePhone, mobilePhone etc.
3. **ApplicationTypes** – Allows definition of common application types so that they can be invoked accordingly, such as alarm, browser, calculator, calendar, camera, contacts etc.
4. **AudioPlayer** – A subset of the multimedia API, specifically to control audio playback and to check audio playback status.
5. **Camera** – To control the camera and take pictures.
6. **Device** – Returns info about the device, such as what apps are present, the device's position and system status information (like memory, cpu etc.).
7. **Messaging** – A unified API to control and access the messaging capabilities of the device (email, MMS and SMS).
8. **Multimedia** – A container API to control multimedia and camera operations.
9. **Widget** – An API to control and detect lifecycle functions associated with widgets, like whether a widget is maximized or minimized in the display.

In its design and intent, the WAC is a good idea because it seeks to define a common standard for Web widgets. Combined with the incredible and maturing capabilities of mobile Web browsers and run-times, advanced even further by HTML5, we have the ingredients for an enticing mobile applications platform that could well occupy a dominant position in the connected services story. However, we have to be cautious for the following reasons:

1. We have been here before with MIDP! Operators had exactly the same intentions to unify the world of mobile applications, except in Java (or its micro edition J2ME). It didn't succeed in preventing fragmentation. After all, it is not operators who control the mobile applications world, as great as their influence is. There are other interested parties with their own commercial agendas, not entirely aligned. More problematic is the idea that operators can collaborate successfully in the fields of technological innovation and co-ordination. It is hard to find any success stories. The core mobile-industry technologies of GSM and related cellular standards mostly came from technology vendors, like Motorola, Ericsson, Nokia and Siemens.
2. Not everyone wants WAC to succeed. Where is Google? Where is Yahoo? Where are the big Internet players who have led the innovation on the Web for the past decade? They are nowhere to be seen in the WAC initiative, even though Google is penetrating vast chunks of the mobile market with its Android OS. Where is Apple? Of course, they are all absent because that's seemingly part of the strategic intent of WAC – to provide an industry counterpoint or counterweight to the Web-venture domination of the connected services world.

6.5.7 Is That a Phone in My Browser?

When it comes to accessing the native capabilities of a phone from within a mobile Web app, making a phone call is an obvious step. In fact, it doesn't really require a new API as this feature has always been present. Most mobile browsers, if they recognize a phone number in a Web page, will enable a number to be clicked in order to place a call, usually prompting the user to confirm that this is what was intended. However, with the advent of VOIP and SIP-based calling, it's possible to go a stage further, which is to implement an entire phone within a browser!

The idea of soft phones is not new. These are applications that you install on a desktop machine and can then use to make VOIP phone calls, just like the Skype client. However, these are native applications. Now, with frameworks like phono.SDK, it's possible to implement a soft phone almost entirely with Web languages running inside the browser. Most of the phono.SDK software is an extension to the popular JQuery JavaScript library. The only part that is not native browser language is a Flash plug-in that's used to gain access to the audio path, especially the microphone. However, as I mentioned earlier, the `<device>` tag in HTML5, once implemented in modern browsers, promised to enable native access to the microphone from within JavaScript.

6.5.8 Mobile Web First?

I began this section of the book with a review of the evolving "Native versus Web" debate. I hope that by now you have become aware of some of the developments in mobile Web that are shaping the answer to this debate, if, indeed, it's really the right debate to have. With the emergence of HTML5 and so-called Web Widgets, we begin to see a much more powerful mobile Web environment, which is beginning to blur the lines between what we think of as native versus Web.

Another aspect of the debate that we hear in developer circles is the reaction to the very term "mobile Web," with some pundits claiming that there is no such thing. What they often mean is that with modern smartphones, like the iPhone, it's possible to access existing websites with relative ease. Therefore, they argue, there is no need to create sites that are specifically optimized or designed for display in a mobile browser. There is only one Web, they say. However, this is yet another question and debate that looks at the world somewhat back to front. Starting, as we should, with the user experience, we have to ask the question "what do we want our user experience to be?" What are our objectives on behalf of users? What are our own business and strategic objectives?

If, like many brands and sites, you want to make surfing your website from a large number of mobile browsers as painless as possible for your users, then making them navigate an unwieldy site that only looks good at a scale well beyond that of mobile screen sizes, is not being very helpful. It simply ignores many of the principles of good design practises for mobile devices, as explained in detail by the likes of Barbara Ballard, Brian Fling and Bryan Rieger, one of which is economy of effort.

No matter what you want your users to do with your app, you've got to make it easy for them, especially on mobile devices where tolerance to awkward design and task inefficiency is low. That's why a common design pattern for mobile Web is to re-create a mobile-optimized version of the site, usually with a subset of the functions that are possible on the "desktop" version. Facebook, Yahoo, Google, LinkedIn and all the Web giants take this approach.

However, we've heard noise that some thinkers in the industry are saying that mobile should come first. What do they mean? Well, leaving aside the strategic rumblings of Google whilst hinting at their strategy, the general wisdom of this approach is that, given the high and increasing numbers of mobile Web users, it is no longer an option to leave designing a site for mobile as an afterthought, or to leave it at all. And, given advances in the flexibility of modern Web coding and design techniques, it ought to be possible to design a site from the outset that will cope with mobile access, giving this priority in the design decisions and then layering in features to take advantage of non-mobile browsers without breaking the mobile experience. This is an argument that has been skilfully presented by Bryan Rieger and inspired by Luke Wroblewski.

7

Augmented Web

- There is no doubt that Augmented Reality (AR) services are going to occupy an important place in our digital lives.
- We interact regularly with both the digital and the physical world. Using one to augment the other is a natural progression and mobile platforms are the natural intersection points.
- The emergent "Augmented Web" is probably where the mobile Web was a few years ago, but with the rapidly increasing penetration of smartphones, we can expect to see faster take-off.
- There are many proprietary AR platforms vying for dominance, but it seems likely that the HTML5 browser will become the universal client for all webs, including the Augmented Web. The technology is capable of supporting AR via the browser, but we need standards for describing augmented data.
- Sensors are going to become the next frontier of the Web, enabled by convergence with mobile platforms and cloud-computing.
- Cloud-computing is a key enabler for sensor-based services because it provides the computing apparatus to process vast amounts of real-time data gathered from grids of sensors.

Connected Services: A Guide to the Internet Technologies Shaping the Future of Mobile Services and Operators, First Edition. Paul Golding.
© 2011 John Wiley & Sons, Ltd. Published 2011 by John Wiley & Sons, Ltd.

Figure 7.1 Augmented reality view via VR specs.

7.1 Real or Virtual Worlds?

7.1.1 Introduction

Let'sreturn to our consideration of the Web OS where we need to think seriously about exactly what "being on the Web" means in the post 2.0 age. First, let's recap a bit of "cyberspace" history. Do you remember those days when we thought that Virtual Reality was just around the corner? We would be wearing headsets (see Figure 7.1), walking through artificial worlds, rather than real ones, something like the holodeck on Star Trek perhaps. This is certainly what I believed back in the early 1990s when I was researching aspects of mobile virtual reality for Motorola Research and the prestigious University of Southampton's Mobile Multimedia Lab.

Perhaps the Holodeck isn't the first thing that comes to your mind when I mention virtual worlds. Most people, assuming they're not youngsters (who invariably don't know what virtual worlds are, even if they use them often[1]), will usually mention SecondLife, one of the most talked about virtual world services in the last decade. But, as we shall see, most of the current interest and activity in virtual worlds lies elsewhere.

My research interest, back in 1994, was not in virtual worlds, but in what we now call augmented reality. I was exploring how to compress and send real-time 3D models to a mobile user, allowing the models to be matched and aligned to a recognizable point in space identifiable through a wearable headset, as shown in Figure 7.1. A common use case that I would mention back then, still applicable today perhaps, is the road-side mechanic who needs to remedy a car problem safely and quickly. Rather than consult with various technical manuals, he would be able to don his VR headset and look at the car engine parts in both the physical and virtual realms, following 3D technical guides for mending parts in order to get the car back on the road quickly and safely.

However, what we didn't envisage back then was the arrival of a different type of viewing device. By the inclusion of a camera lens, and a powerful graphics processor, we have the ultimate *augmented reality platform* – the mobile phone!

7.1.2 Augmented Reality

Using the camera and a positional (or motion) detector in the mobile, it is possible to align points of interest in the digital world with actual image data seen through the viewfinder, as

[1] This isn't a comment about their technical ignorance, but an observation that kids are actually so familiar with virtual worlds, that they don't have a special name for them.

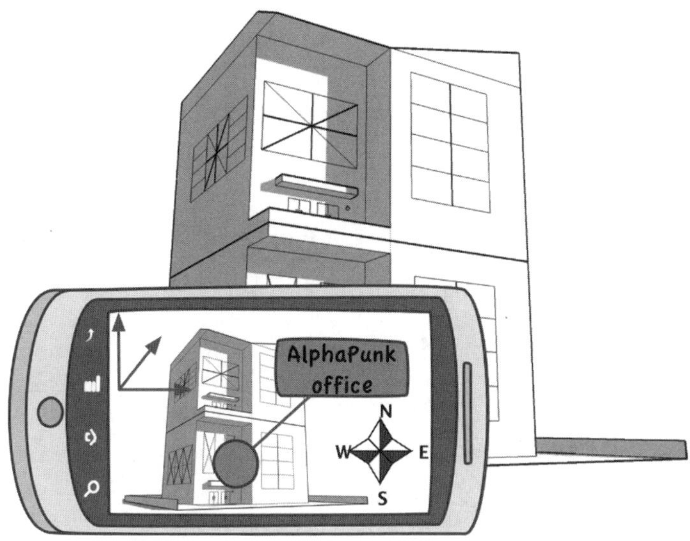

Figure 7.2 Overlaying digital information on a camera image in real-time.

shown in Figure 7.2 where a view of an office block has been augmented with an overlay to tell us who works in this office block.

In its crudest form, the digital augmentation of the image is entirely coincidental with what's seen in the viewfinder, using positional information from the mobile GPS and motion sensors (see Section 7.2 Sensor-net: Mobiles as 6th sense devices) to interpret where the camera is headed and what it might therefore be picking up through the lens. There is no attempt to detect what's in the image and then match this to patterns in an image database in the cloud. However, image detection is possible and can be used to provide better positional accuracy and object identification in the camera image.

Augmented reality requires more than just a viewing device. It requires a database of positionally-encoded ("geo-coded") information that can be matched to the "portal" view of the camera image, as determined by the positional sensor information provided by the device. The database must be capable of returning data points that are located in space within a sector that is coincidental with the view of the camera based on its position, as shown in Figure 7.3.

The backend is where opportunities exist to create platforms for augmented reality services. Thus far, a number of start-ups and established Web ventures have launched AR platforms, such as:

1. Layar
2. Insqribe
3. Juanaio
4. Wikitude
5. Google Goggles

Let's take a closer look at Layar, to see how it works. Layar requires the user to install a Layar AR Browser, which is a native application. Thus far, it is available for Android and iOS platforms. The browser works similar to the diagram and ideas described above, giving the

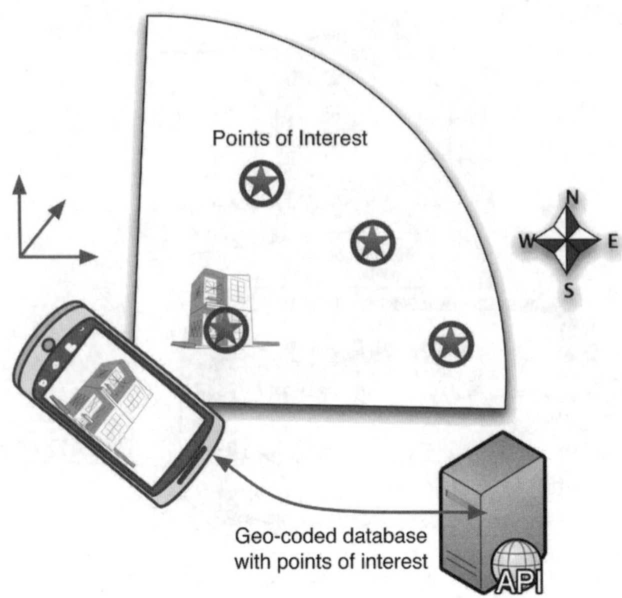

Figure 7.3 Backend database returns points of interest coincidental with mobile view.

user a viewfinder with an overlay that indicates the presence and direction of various points of interest coincident with the orientation of the device, as reconstructed in Figure 7.4.

As the diagram shows, the current view through the camera is augmented by a grid overlay to indicate perspective, with dots further up the grid being further away from the current position. These dots are points of interest within the current information layer, the idea being that the user can select from a number of layers, such as underground stations, restaurants, or any custom layer that a content provider wants to publish.

The Layar platform is open for any developer to create a layer and then publish it for consumption by Layar users. Clearly, the guys at Layar are trying to create the ingredients for an emergent ecosystem of layers that will form a kind of "augmented Web" in our web of webs that make up the Web OS. By attracting enough users to the Layar platform, they might succeed in building a big enough ecosystem with substantial network effects, big enough to dominate the "augmented Web," in the way that Google dominate the "search Web."

The idea has some merit, even if it has yet to gain much traction. We might be at a similar point to where the Web was back in the early 1990s. Don't forget that even Microsoft thought that the Web was a fad. Here we have a new type of browsing experience – browsing the real world. On the Web, we have pages that are located by Uniform Resource Indicators (URIs), like http://paulswebstuff.com. We "point" our browsers at one of these URIs and we get some content. This method of accessing hyperlinked content is now very mature and still evolving rapidly, like the advent of HTMl5 (see Section 6.5 Key Platform: The Mobile Web). We don't yet have the same standard for using a camera to surf the "Augmented Web." Indeed, we don't really have any standards for the "Augmented Web."

With HTML5, we will soon have the ability to interface with video directly in the browser. This is aimed at media playback applications, but there will also be APIs in the browser to

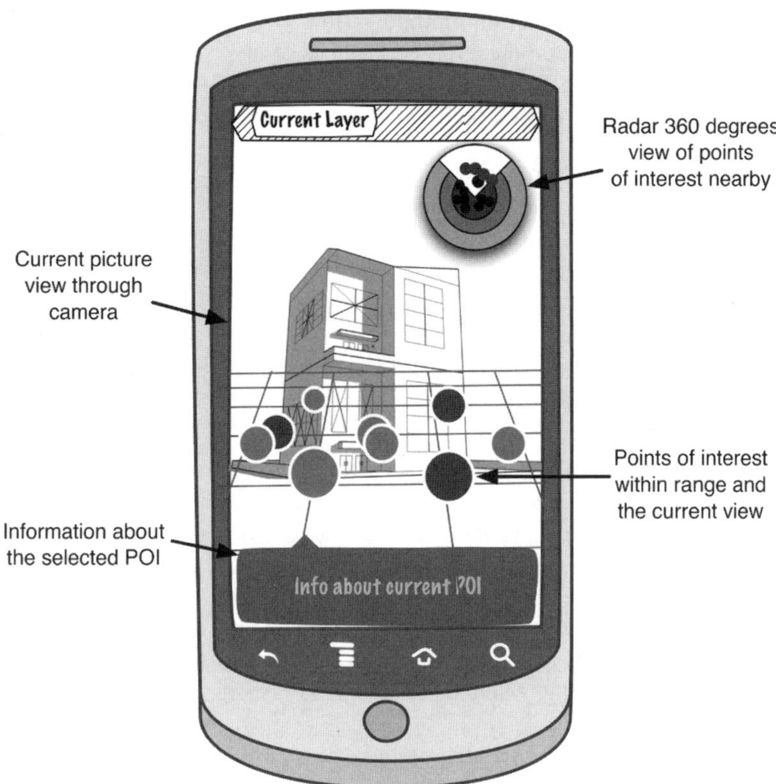

Figure 7.4 Representation of how the Layar browser works.

interface with the camera and the various sensors available, like motion, compass and gyro sensors. Additionally, HTML5 will offer us the capability of 2D graphics (and eventually 3D) drawn natively into the display using the `<canvas>` element. It is therefore easily foreseeable that the browser itself will be capable of augmented reality applications without the need for a dedicated AR browser. This could well turn out to be the "killer application" for HTML5 in the mobile browser.

The idea of creating standards for the emergent Augmented Web was explored at a W3C Workshop (Augmented Reality on the Web) in June 2010. That workshop published its findings, the major one being that it was time to form a working group within the W3C to pursue the development of standards for the Augmented Web. At the workshop, a number of issues were examined, which summarize the challenges for creating a set of standards:

1. **AR content markup & format** – How to indicate content so that it can be interpreted by an AR browser, just as we have HTML for the Web. A number of formats were proposed at the workshop, including Augmented Reality Markup Language (ARML), which has been developed by Mobilizy, the guys behind Wikitude (a similar service to Layar).

2. **AR content transport/interaction method** – How to deliver AR content to the browser and how to define and support the required methods of interaction with the content. What does interaction mean in the AR context?
3. **Representing 3D interactive AR content** – How should 3D content be represented in the browser?
4. **Device capability access APIs** – Which APIs on the device should be exposed to the browser in order to support AR applications?
5. **AR data formats** – POI (images, text, 3D models, URIs) – how should various AR assets be described? What data formats should be used (e.g. XML schemas, JSON etc.)
6. **AR data service API** – What APIs are required at the server level to support open AR platforms?
7. **Security & privacy** – What are the security and privacy issues and how might they be addressed for AR services?

In addition to the types of AR described above, there is also what I call "Tactile AR," which involves the ability to interact with digital objects within the augmented view, often triggered by some kind of marker, like a distinct barcode or pattern that is easily detected through pattern-recognition techniques in image processing. As a result of finding a marker (or some other pattern) in the image, an AR service can trigger certain responses, including augmentation with 3D interactive objects, like an image of a toy that might hover above a toy package found in a shop. This close-proximity and interactive AR has all kinds of exciting applications, including retailing, where there are already emergent interests from brands and major retailers.

At the OpenMIC barcamps in the UK, during a session to explore the future of AR (and virtual worlds), I coined the phrase "3rd-eye" to refer to the mobile camera as providing an entirely new way of viewing the world, with a whole range of cultural impacts and questions:

1. **Authenticity** – With entirely new ways to discover information about the world, how are we going to monitor and authenticate augmented data that might crop up anywhere, even above our homes? Do we need an "authenticity" capability in our Augmented Web OS? Is this an opportunity for telcos?
2. **Disintermediation of advertising space** – Will we still need advertising boards or will they be replaced by our mobile screens? Imagine, we could even overlay an existing billboard with entirely new content. What are the legal and ethical considerations for this possibility?
3. **Disintermediation of architecture and design** – The way we experience and navigate the physical world will no longer be as the architects and planners necessarily intended. We will have new insights and considerations when looking at objects through our "augmented reality." We can even peer into buildings and objects, or look back in time.
4. **Elasticity of destination** – We have seen that mobiles have made time more elastic. It's not so critical to arrange an exact meeting time and place when we can liaise nearer the event via our mobiles, saying "I'm going to be 15 minutes late." The same is bound to be true with AR. There's no need to arrange the place if we can simply point our cameras and find the nearest place that meets a certain criteria.

7.1.3 *Proof-of-Presence or "check-in" Services*

Who hasn't heard of FourSquare, Gowalla or now Facebook's Places? These are all variations on the same theme, which is a service that requires the user to say where he or she is, or has been, often using the check-in metaphor when arriving at a certain location. Again, what starts out like a fad seems to be gaining enough momentum that we might expect the check-in model to become another enabler of the Web OS. The question, yet again, as with all these various aspects of the Web OS that we have been exploring in this chapter, is what position this enabler is going to take in the Web OS. Will a few players dominate this space, or will we see the emergence of Web standards to allow any service to benefit from check-in mechanics?

I think that both are likely to happen. We certainly need the latter because it seems likely that check-in mechanics are useful to a wide range of online services. The so-called "Gamification" of these services (like Foursquare awarding of badges to make check-ins competitive) is a very vogue topic in start-up communities and investor communities, but I don't think that these aspects are as horizontally scalable as the check-in idea, at least for now. It is currently difficult to imagine what a "Gamification" layer or enabler would look like in the Web OS and who might provide it, even though it is the subject of much thought and speculation in the start-up community.

An extension of the check-in model is the proof-of-presence model, which requires a mechanism to verify that the user was actually present at the place. This could be done in a variety of ways, including:

1. Connection to a localized wireless hub (Bluetooth or WiFi[2]).
2. NFC – Nearfield Communications ("contactless" technology), as used for mobile payments.
3. Various photographic techniques, including QR codes.

The latter is all the more interesting when combined with the user of an AR browser, which might well become the standard way of using a proof-of-presence (POP) or check-in service. With any of these check-in and POP models, the notion of trust crops up again. How do we know that this person really did appear at this point in space? How do we know that this object (shop, restaurant, club) really exists at this point? Who says so? Who has authenticated the data? This is yet another opportunity for a "trusted provider," which might well be a telco in conjunction with a mapping company.

7.1.4 *Summary – Virtual is Just Another Layer in the Web OS*

It is obvious that AR services are going to occupy a place in our digital lives. We interact regularly with both the digital and the physical world. Using one to augment the other is a natural progression. The mobile is also the natural intersection point. The emergent "Augmented Web" is probably where the mobile Web was a few years ago, but with the rapidly increasing penetration of smartphones, we can expect to see faster take-off.

As you can tell, there are plenty of issues to be explored before we can finalize a set of standards for interoperable AR services, but the Augmented Web looks like it is going to be a

[2] I was once commissioned to conduct a feasibility study for how to build wireless proximity services into entire shopping malls.

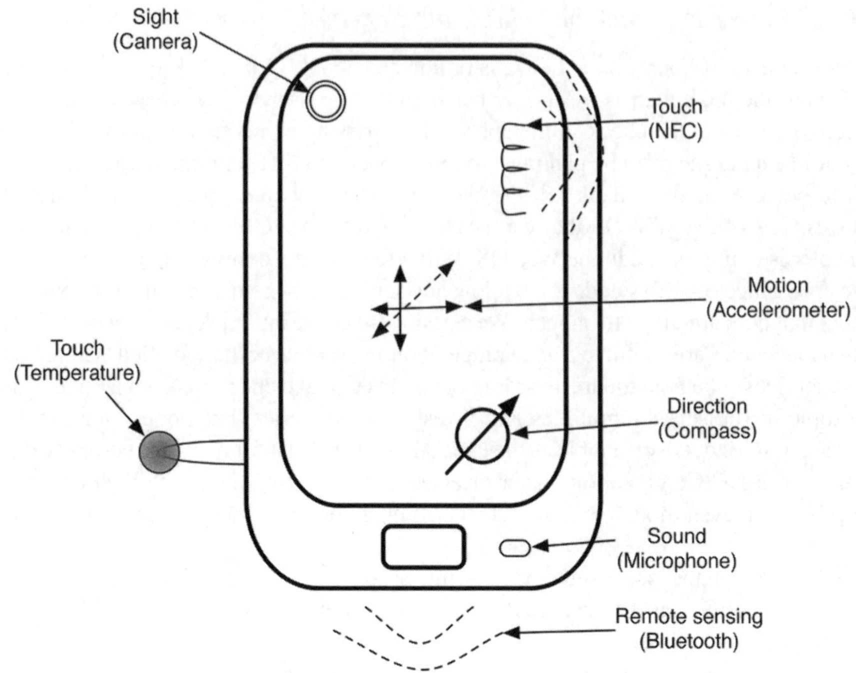

Figure 7.5 Sensor capabilities of a modern smartphone.

major part of the Web OS. It is still early days. We have yet to see the emergence of compelling standards and the tipping point, but it's only a matter of when, not if. There are still plenty of opportunities for various players in the emergent value-chain. Some of the components of AR services, like verified location, security, privacy, verified social graphs, are all areas where telcos could play a role, if they can move quickly enough and find the business models that fit with their strategic intent. Whichever way you look at it, the Augmented Web is very much going to be part of our connected services story.

7.2 Sensor-Net: Mobiles as Sixth-Sense Devices

One of the fascinating developments in smartphone technology is the inclusion of sensors that can sense a variety of aspects of the physical world, making smartphones even smarter, as shown in Figure 7.5. Combine this capability with the very powerful processors that we find in modern smartphones, and we have to start to wonder whether or not we entering a new phase in the evolution of mobile.

Sensors aren't just for mobiles. There is a gradual increase in the number and application of sensors in all kinds of fields, from sports to medicine. Many of these remote sensors will be wireless themselves and contain some kind of embedded microprocessor to add intelligence. Inevitably, the smartphone will become a control point, or hub, for some of these external sensors, either in nearby proximity, such as wearable sensors on the body, or remotely, such as motion detectors in the home.

I have already discussed the emergence of some sensor-driven applications, such as augmented reality, but in this section I will focus in more depth on the various sensor technologies and capabilities that are found in smartphones. I will deal first with the types of sensors and sensor applications that we find in current smartphones. I will then explore emergent and future sensor applications and possibilities, given that this seems like a major opportunity for new applications and services.

7.2.1 Current Sensor Applications in Smartphones

7.2.1.1 Motion Detection

Although not the pioneer of including sensors in phones – it was the Japanese who did it first – the iPhone has shown us the potential for including sensors in a smartphone. I'm not talking about the multi-touch interface, which is a kind of sensor, but the inclusion of an accelerometer to detect small-scale movement, like tilt, of the phone (with GPS able to sense larger scale movement). The accelerometer can detect both motion and orientation, which is one of its most common uses on the iPhone and smartphones generally, allowing the screen display orientation to adapt with the rotation of the device, flipping between landscape and portrait modes.

The motion sensor can pass its data through to application via an API on the iOS and Android platforms. This allows developers to include motion sending in their applications, which can be applied in a number of ways, limited mostly by the imagination of the developer. Some examples include:

1. Pedometer applications that can detect the steps of the user, whether walking, jogging or running.
2. Camera applications that only fire the shutter when the device is still enough (under low light conditions).
3. Scrolling via tilt – using the tilt of the device to scroll an article on the screen. Higher tilt results in quicker scrolling.
4. Artistic purposes – moving the phone around to create artistic "light trails" on the screen in harmony with the motion of the device.
5. Image manipulation – changing some element of a display in accordance with motion via some algorithm, which might be random seeding of an image generator.
6. Random number generator – creating a number in relation to the movement of the device to create a unique seed for a random number generator used to encrypt communications.
7. Games – plenty of games have already taken advantage of the motion sensing, from steering cars to moving balls around a maze.
8. Remote control of other devices, such as robots and Lego Mindstorm (via wireless connectivity).

Motion sensing using an accelerometer can be improved by the use of a gyroscope, allowing better accuracy in all degrees of freedom. The first device to include a gyroscope was the iPhone 4. This gives the iPhone a number of additional sensing capabilities:

1. Three-axis motion detection, including angular velocity (i.e. how quickly the device is being rotated).

2. Pitch, roll and yaw detection, which means the tilt back and forth, sideways and up and down.
3. Rotation about gravity, which is the ability to detect the phones position and movement in relation to true up/down relative to the Earth's gravitational field.
4. Six-axis motion detection by combining the gyro with the accelerometer. This means not only the tilt in relation to the vertical, horizontal, but the movement in those directions (back and forth along each axis) too.

Of course, we can, as I have done in previous books, predict all kinds of applications for the use of this technology. However, we should put on our "platform thinking" caps and remember that the best opportunity lies in providing this capability to developers and enabling them to innovate with the technology. This is exactly the approach taken by Apple by incorporating all of the sensor data in a set of APIs (called CoreMotion APIs), allowing developers to incorporate advanced positional and motion detection into their applications. Thus far, the biggest category of application by far has been gaming.

7.2.1.2 Direction

Similar to motion sensing, some devices include sensors to measure direction, like an electronic compass. These sensors are magnetometers that can actually sense the Earth's magnet field. Thus far, magnetometers have only been shown in production phones to be useful for electronic compass applications. This in itself is interesting because direction is a useful quantity to sense, opening up entirely new application possibilities, like augmented reality. Additionally, researchers have shown how magnetometers can also be used to sense nearby objects, such as a finger wearing a magnet, in order to provide new types of device interaction.

7.2.1.3 Image Detection

The camera on a phone is also a type of sensor, able to sense light. Initially, use was limited to capturing images for photographic purposes, but with faster processors and better APIs, it has become possible for developers to build applications that process the images in real-time, including augmented reality whereby images in the viewfinder are matched to shapes in a shape library, such as the outlines of buildings or landmarks.

Google has put this technique to good use with their Google Goggles application, which is a kind of visual search. Googles can recognize text, such as items on a menu (and then translate them if the menu is in another language), landmarks, books (by their covers), contact info (e.g. Biz cards), artwork, wine labels and even logos. It is also able to recognize people's faces, but this feature has thus far been omitted from the publicly available version, fearing that it might spark controversy.

7.2.1.4 Proximity Detection with NFC

NFC stands for near-field communication, which is a technique that allows two devices to exchange messages when they are in close proximity to each other, almost touching. It's the same technology that many of us have known for some time in the form of smart cards

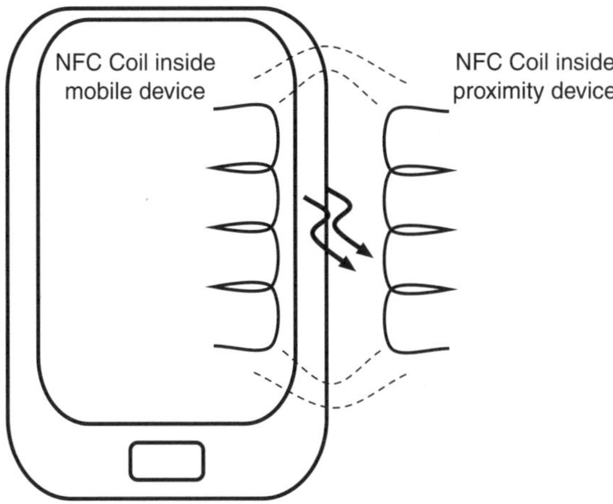

Figure 7.6 NFC communications via magnetic induction.

(or ID badges) that open doors in secure buildings. Unlike Bluetooth or other low-range wireless communications standards, NFC doesn't require the two devices to have some kind of prior "mating" beforehand. The whole point is that the "mating" and communication is effectively combined into one step, which is the touching of the two devices.

NFC devices communicate via magnetic induction, as shown in Figure 7.6. If you remember your school physics lessons, you might recall how electricity pulsed in a loop of wire causes a magnetic field to be induced around the wire. If another loop of wire is placed in close proximity, then the magnetic field from the first coil causes an electric current to be generated in the second coil. The electric signal in the second coil faithfully mimics the signal in the first coil thereby allowing a signal to pass between the two coils. A mobile would house one of the coils and an NFC device, such as a payment terminal, would house the other, as shown in the figure below. By bringing the two into close proximity, they can exchange electric signals, which is exactly the same principle by which mobiles communicate with the outside world.

7.2.1.5 Hearing – Aural Processing

The original sensor on the mobile phone is the microphone. We speak into it and our voices are digitized and then sent across the network using all kinds of amazing processing tricks to compress the voice. In fact, unknown to most of us, the sophisticated voice compression technology in the phone (called transcoding), is able to recognize various vocal sounds, like the voicing of vowels and consonants, in order to build a model of our voice box. It then reconstructs this model at the receiving end of the call and transmits the vocalization of sounds rather than the sounds themselves. In other words, rather than record our voices and send the recorded sounds, it models our voices and sends various codes to tell the receiving model how to reconstruct the sounds.

That same technology can be used in a variety of ways. For example, in the music industry, they use similar techniques to "listen" to music tracks and extract distinct patterns (called spectrograms) in the sounds. Using this technology, it is possible to fingerprint music. If those fingerprints are stored in a large database, it is possible to look up the names of tracks based on the input of fingerprints. This has been exploited by my favourite mobile application of all time, called Shazam.

If you haven't used Shazam before, then you are missing out on an amazing experience. Shazam allows the user to hold their mobile phone out towards a music source, such as the music blaring from the speakers in a coffee shop or store (places where I have used the application many times). The music is then passed to the cloud-based platform where the fingerprint recognition and matching takes place. Based on finding a fingerprint match in the vast music database, the Shazam service will then send back the name of the track, plus the recording artist.

Similar to Google Goggles, Google run a service called Voice Search. It uses speech recognition to detect keywords that are then sent back to the search engine, which conducts a search in the usual way, providing a speech interface for search. Of course, it won't be long before this capability is combined with other sensor information, such as position and direction, in order to add possibly useful context to the search.

7.2.2 Emergent and Future Sensor Applications in Smartphones

7.2.2.1 Spatial Image Processing and Motion Detection

The Xbox Kinect has demonstrated a whole new category of image detection applications based on advanced image processing. It is based on software technology that was reportedly developed by a British subsidiary of Microsoft Game Studios, called Rare, working with a device called a "Range Camera," developed by PrimeSense, an Israeli company. The range camera can sense and interpret 3D information in its field of view by projecting an array of dots from an infrared light. This projection of dots is known as "Structured light," which is the process of projecting a known pattern of dots onto a scene.

The way that these dots deform when glancing off of surfaces (like shining a torch beam along an wall) allows an infrared camera and associated processor to calculate the depth and surface information of the objects in the scene. This is the same process that you might have seen used to "scan" 3D objects into a computer, like bodies and faces when making films that require CGI. Supposedly, the Kinect is capable of simultaneously tracking up to six people, including two active players for motion analysis with a feature extraction of 20 joints (i.e. limbs) per player.

Now imagine what could be done if a similar technology were integrated into smartphones. Gesture control is one possibility, as used by the Kinect. This could also be used with multiple participants, perhaps a number of people gathering around a device placed on a table. This might provide a new kind of collaborative communication experience or be the basis for new types of interactive gaming involving groups of players. Perhaps even the inclusion of infra-red detection on a phone would enable a whole new category of applications, such as making it easier to detect people in the viewfinder, and so on. Again, the answer doesn't lie in you or I thinking of the applications. It lies in providing the technology and the APIs to access it, letting an army of developers dream up wonderful and new applications.

7.2.2.2 Advanced Aural Detection

With Shazam and Google's Voice Search, we are beginning to see new types of interaction via voice and sound generally. The ability for our mobiles to "hear" on our behalf, combined with the "brains" of cloud computing to interpret on our behalf has a number of interesting future directions. For example, it is possible to detect the emotional tone of a conversation. This could be used in a variety of ways:

1. **Lie detection** – Applications include negotiations, phone interviews and vocal-based reporting.[3]
2. **Stress analysis** – Applications include health, life-coaching and time-planning.
3. **Emotional Intelligence feedback** – The ability to indicate during our call how the caller and the callee are feeling about the topic, allowing modulation of the call to improve the communication in some way.

Sound processing can also be used to recognize voices in order to authenticate the person speaking. This has applications in security and related fields. The technology has already been used for remote banking applications in a number of countries, although with varying degrees of success. However, in most cases, the recognition took place at the receiving end of the call after the transmission of the voice with subsequent loss of clarity and fidelity.

With the power of modern smartphones, it is possible to process the speech locally where the quality is far higher. We have already seen the porting of various speech recognition applications to smartphones, such as the Dragon Dictation applications from Nuance, proving that smartphones are more than capable of providing sufficient processing power locally.

Sound processing can also be used to detect ambient sounds and noise. This is already used by some smartphones using an additional microphone input (e.g. iPhone 4.0) in order to eliminate the background noise from the conversation. However, there is no reason why this capability couldn't be used for other purposes. For example, we might envisage a service that enables people recognition based on their voice inputs. This could be used in an ambient fashion at a party or meeting whereby the various voices at the gathering place are gradually picked up by the phone and then converted into contact details – it would be like collecting business cards by voice.

7.2.2.3 Remote Sensors for Health

7.2.2.3.1 The Scope for M-Health

One of the biggest costs and challenges in any society is health care. The costs are only increasing as we live longer and discover new technologies to extend the range and effectiveness of treatments.

- More than 1 billion people in the world are thought to be overweight, with at least 300 million of those are considered as clinically obese. Without action, more than 1.5 billion people are expected to be overweight by 2015.[4]

---

[3] For example, we might ask an employee to regularly phone a number in order to report on progress with a project or task.

[4] World Health Organization. (2005) Ten Facts About Chronic Disease. Retrieved May 19, 2006, from http://www.who.int/features/factfiles/chp/07_en.html.

- Over 600 million people worldwide have chronic diseases, and the spending on chronic diseases is expected to increase dramatically. For example, in the US alone, spending is expected to rocket from $500 billion a year to $685 billion by 2020.[5]
- Globally, the number of persons 60 and older was 600 million in 2000. It is expected to double to 1.2 billion by 2025.[6]

These statistics are compelling, suggesting that there are huge opportunities in addressing these challenges and problems. As we become more aware of the importance of health, many of us seek new ways to enhance our health without having to suffer the inconvenience of visiting doctors and clinics. As we get older and live longer, we seek to find ways to maintain our freedom and independence for as long as possible, trying to delay the need for extensive care, including the need for care institutions.

One of the promises of modern technology is the ability to maintain healthier lives. Within that context, it seems that sensing of various bodily functions is an important technological advancement that will enable new categories of health-care options, with abundant opportunities for service providers to run these health sensor networks.

At the moment, it is not clear who will win in this race. There are a number of players with an interest, including:

1. Healthcare companies of various types who can provide the health treatment services.
2. Governmental health services, who are also suppliers of healthcare.
3. Telcos who can provide the network infrastructure for a network of sensors.
4. Various lifestyle brands and companies who want to access the market, such as sports brands and gymnasiums.
5. Equipment manufacturers who can make the sensors and the various embedded electronics that provide intelligence to the monitoring.
6. Cloud-computing and other "Big Data" providers who can provide the infrastructure for the centralized collection and processing of the sensor data.

There are probably other players that I am missing. Also, some of the players are seeking to move into adjacent categories. For example, telcos might not be content just to provide the network for shifting of sensor data from one place to another. They might want to provide additional services, including direct provision of lifestyle services. I use the word "lifestyle" deliberately because there are vastly different restrictions and opportunities when offering a service that claims to be a lifestyle service as opposed to a healthcare service where there are the additional liabilities of compliance with a number of potentially complex legal and regulatory requirements.

Whatever the eventual "augmented health" market ends up looking like, it is clear that mobile and wireless technology has a key role to play. Many see the mobile itself as being the hub for a variety of remote sensing applications. After all, it has a lot going for it – it's personal, always on and always with its user. It can be used in the home, office or anywhere

[5] Scaling Mount Proteome to Bring Down Chronic Disease. The Pfizer Journal®, Global Edition Volume 1I, Number 2, 2001, 4–9.

[6] World Health Organization. (2006, February 13). The world is fast ageing - have we noticed? Retrieved May 19, 2006, from http://www.who.int/ageing/en.

else, providing round-the-clock access to services. Therefore, many are ready to declare that "remote health," or "E-health," is actually "M-Health."

7.2.2.3.2 Bluetooth Low Energy – a Key Enabler for m-Health?

We have yet to see the emergence of sensors embedded directly into smartphones that would facilitate m-health applications. In remote monitoring of health, there seem to be a number of categories of detection:

1. Bodily functions – for example, blood pressure, heart function, blood glucose levels, oxygen levels, respiration etc.
2. Environmental factors – for example, House temperature, movement around the home, safe and habitual use of appliances etc.
3. "Lifestyle" factors – for example, Diet, stress levels, mood, sports activities and progress, etc.

Mobiles can play a role in all of these, although we should not think only of mobiles as the devices that we also make calls with. For example, there is likely to be a role for a multi-functional "home hub" device, which could be an Android-powered tablet, or similar, that provides communications services and monitoring services via an effective user interface that is usable by folk of all ages and abilities. That device would provide access to a range of remote sensors in the home, including bodily sensors that are worn and used in a variety of ways.

Whether via the smartphone or a home hub, it seems that a common method of monitoring is likely to be a remote wireless connection. A key technology for enabling remote sensing is the recently defined Bluetooth Low Energy standard. This is similar to existing Bluetooth technology in that it is intended as a means to connect peripheral devices over so-called Wireless Personal Area Networks (PANs) and with relatively limited transfer rates (compared with WiFi and other WLAN technologies). Characteristics of Bluetooth Low Power are:

- Ultra-low peak, average and idle mode power consumption.
- Ability to run for years on standard, coin-cell batteries, as found in watches and useful for a number of always-on sensing devices.
- Low cost in order to allow widespread integration into a range of products, providing the much-needed tipping point for mass integration of sensors into products.
- Multi-vendor interoperability, just like Bluetooth itself, allowing devices from different vendors to connect with each other (similar to inter-operability of Bluetooth headsets and phones).
- Enhanced range, allowing connections over 200 metres.

Bluetooth Low Energy has been designed specifically with remote sensing in mind. It is a remarkable technology that would enable sensors to be embedded in all kinds of products at extremely low cost. These could include training shoes, running vests, home appliances, drug bottles and just about any device you can imagine where a sensor might be useful.

The current Bluetooth standard includes the concept of profiles. These are particular categories of usage that determine a certain expected mode of operation from a Bluetooth peripheral. A good example is the use of a peripheral as a voice-telephony device, like a headset or car hands-free system. It is expected that similar profiles will emerge for Bluetooth Lower Energy. Two profiles that have already been set in motion are:

1. Sports profiles – with the main focus on vitals monitoring, but within a sporting context.
2. Health care profiles – with the main focus also on vitals monitoring, but for patient care.

The health-care profiles are being considered by the Continua Health Alliance, which is a non-profit standardization body set up to explore and define standards for personal connected health solutions, which includes m-health and sensor types of application. They have identified a number of service categories where remote sensors and Bluetooth Low Energy would have a role, including:

Independent living with age – applications include:

- Assistance with daily activities and monitoring of those tasks.
- Reminders to carry out routine medical tasks, such as taking medicines.
- Prompts to carry our certain types of activity, such as visit to clinic.
- Monitoring for early-warning signs of health deterioration or concerns.
- Dietary assistance and advice.
- Emergency services callout.
- Real-time alerts and communication in a range of circumstances in and out of the home.

That last point brings two key dimensions to our attention. First, the dimension of real-time, which is an attribute of connected service design that we have explored often throughout this book (see – Section 5.1 Real-time web and Twitter). The need to gather, process and react to data in real-time requires new types of service architecture that will inevitably combine various elements of cloud-computing, "Big Data," smartphones and the real-time Web. The second dimension is mobility. It might seem obvious, I hope, but independent living should not mean being confined to the home. This is another reason why mobile technology is key to the healthcare domain.

Health and Wellness – applications include:

- Extending healthcare out of the clinic and into the home, office and leisure worlds. Always-on healthcare is going to become an important aspect of our lives in the next decade.
- Carrying out initial testing and triage through the use of remote sensors and remote communications technologies.
- Scheduling appointments – this doesn't mean access to online calendars, but the use of intelligent sensors to figure out when we need to book a dental appointment, eye appointment and so on.
- Weight loss and diet monitoring – with increasing pressure on our lives to eat well, exercise and avoid unhealthy diets, the number one enemy is our ability to forget or take the easy path. The judicious use of assisting technologies ought to be able to address some of these shortcomings, helping us to eat well and on time.
- Fitness – there are myriad ways to monitor fitness. The introduction of low-power sensors into clothing and equipment will inevitable mean the advent of "smart clothing," with training vests, shoes and other garments telling us how well we're doing, monitoring progress, bodily stress and so on, helping to keep us fit and to prevent injury.

- Long-term data collection and trend monitoring – as we begin to collect increasing amounts of data from a host of sensors, we can store all of that data and start monitoring for trends and anomalies.

Again, this last point ought to trigger certain ideas in your mind, like the possible application for cloud-based services that can collect vast amounts of sensor data and layer in various processing and monitoring services. Doesn't that sound like a platform opportunity? It certainly is! Just as we have a giant network of Web servers to handle myriad Web applications, we will need a similar cloud-based network of "Sensor servers" upon which to build various sensor-based services.

It isn't obvious who should be building and providing that infrastructure, apart from the obvious big IT players of today, like IBM and HP. However, we are seeing new players, like McLaren Applied Technologies, who are extending their expertise in real-time sensor management (on racing cars) to other domains, like the real-time management of ground vehicles on an airport apron, or the real-time monitoring of athlete performance in professional sports teams and games.

McLaren Applied Technologies has seen a number of similar opportunities, leading them to believe that a unified cloud-based architecture is the way forward, allowing vast amounts of data collection and processing in real-time. Again, it is this real-time aspect that seems critical, combined with some kind of "Big Data" capability that can find patterns in the data quickly and effectively.

As Figure 7.7 shows, real-time cloud architectures will be required for collecting data from potentially billions of smart sensors in distributed sensor grids. That data will need to be stored in real-time and then processed in real-time. Today, such architectures simply don't exist, or are hard to find. We see examples of scalable real-time processing in the financial industry and – where else – in telcos with their massive rating and charging platforms.

We don't just want to store and process streams of sensor data. We want to do something with the outputs. We want to provide real-time output mechanisms of various kinds. Doctors

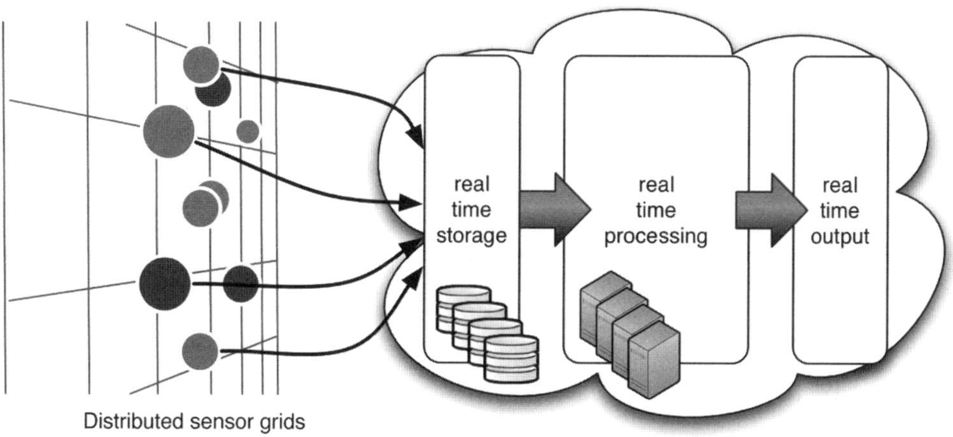

Figure 7.7 Cloud-based platforms for processing sensor data.

and nurses need to know when something is going wrong amongst the million sensors worn by their patients. But knowing that something is wrong is not enough. They will need accurate and useful information presented quickly to them in an appropriate format, which might need to change as circumstances change, moving from big screen to the mobile screen with ease.

We are seeing the emergence of real-time capabilities on the Web (see Section 5.1 Real-time web and Twitter). However, these are not generalized architectures and platforms that can be deployed across a wide range of sensor-based applications at massive scale. Clearly, here lies an opportunity!

7.2.2.4 Remote Sensors for Machines

We aren't interested only in sensors that detect bodily functions. Most sensors in the "Sensor Net" are likely to be attached to physical objects, particularly machines. However, for our discussion of device platforms, the question is how will these sensors interact with mobile devices? One such example is the connection of mobile devices to cars, allowing interaction with a range of car systems, including the engine management system with its array of sensors.

The displacement of in-car systems with smartphone systems is already well underway with navigation devices where Personal Navigation Devices (PNDs), like those made by TomTom and Garmin, are being replaced by Android and iOS smartphones. This is causing considerable disruption to the PND market. However, there is a further move towards the smartphone becoming a kind of communications hub in the car, providing the driver (and passengers) with access to all kinds of services, including:

1. Video/audio playback via in-car entertainment system.
2. Gaming services.
3. Home automation – for example, setting various home appliances in readiness for the car arrival.
4. Advanced navigation – combining in-car navigation with feedback from car sensors, such as road conditions, fuel consumption, brake wear, cruise-control, type pressure, etc.
5. Safety and security services – always-on monitoring of sensors, road conditions, passenger wellness (e.g. tiredness), traffic patterns to advise and manage safety of the passengers.
6. Road and vehicle monitoring – the smartphone becomes the central hub for sensor data collection and processing, including sharing with other drivers.
7. Advanced communication services – voice detection, concierge, message alerts etc, whilst implementing various safety procedures to prevent distraction.

There are already various standardization attempts to integrate open device platforms into automotive platforms, including the GENIVI alliance, which is a non-profit group driving the adoption of a In-Vehicle Infotainment (IVI) reference platform. Thus far, they have chosen the Linux-based Meego operating system to be the platform for the reference implementation. However, the GENIVI platform is all about decoupling the client device (e.g. smartphone) from the automotive platform.

Once the smartphone becomes an integral part of the driving experience, we can combine this with the power of the cloud to provide all kinds of service opportunities. Again, the provision of a services and connectivity layer might see automotive companies teaming up with telcos to provide new "auto-cloud" platforms upon which these new services can be delivered.

7.2.3 Sensor Net – Is This Web 3.0?

We have looked at concrete examples of sensor applications, such as Google Goggles, that are cloud-based. They are also "Big Data," applications, reliant upon vast amounts of data crunching to index images and embedded patterns in order to provide a near real-time response to a search query. Indeed, with the local processing power of a smartphone, which enables initial image processing to take place locally (such as edge detection, character recognition and so on), the combination of smartphone with cloud processing is an enticing combination that will emerge to be the dominant mode of mobile computing and the pattern of many future services. This is why it is vital for telcos to take a strategic view of cloud computing (see Section 8.1 What is Cloud Computing?) and "Big Data," (see Section 4.1 What is Big Data and where did it come from?). These are major forces in the evolution of connected services.

8

Cloud Computing, Saas and PaaS

```
{
  "id": "fluffy",
  "name": "Paul Golding",
  "first_name": "Paul",
  "last_name": "Golding",
  "link": "http://somewhere.in.the.cloud"
}
```

- Cloud-computing is one of the key enablers of connected services.
- A key attribute of cloud-computing is elasticity – the ability to provide scalable and economic amounts of computing power on demand.
- An ecosystem of new ventures has emerged around cloud-computing, including entirely new types of service and platform offerings.
- There are a variety of infrastructural components and services that combine to enable and support cloud-computing.
- Software as a Service (SaaS) is a major paradigm of the modern Web, enabling entirely new industries and ventures to emerge, many of them strategic threats to traditional incumbents, including telcos.
- Platform as a Service (PaaS) has had a dramatic effect on lowering the barriers to entry for new Web ventures, often enabling them to provide new types of services previously beyond their reach. PaaS services are now being extended to new areas, like "cloud telephony," which are leading to examples of major service innovation in traditional industries (e.g. Telco).
- Telcos must develop meaningful strategies for cloud-computing, PaaS and SaaS, both as providers and consumers of these technologies.

Connected Services: A Guide to the Internet Technologies Shaping the Future of Mobile Services and Operators, First Edition. Paul Golding.
© 2011 John Wiley & Sons, Ltd. Published 2011 by John Wiley & Sons, Ltd.

8.1 What is Cloud Computing?

8.1.1 More Than Just a Fluffy Phrase

Cloud Computing is the new cool. It's hard to go for long in any conversation about the Web before someone mentions "The Cloud." It seems that everything these days is in "The Cloud." It's the phrase of the day!

Everyone seems to have their own idea of what "The Cloud" really means. A frequent, though incorrect, use of the term is as a synonym and metaphor for The Web, or, more aptly, Web 2.0. As I've described elsewhere, and is familiar to most of us by now, many software services have moved onto the Web. For example, we can now store our pictures online, perhaps using a service like Flickr. This notion of services moving away from our local machines to ones that are distant is easy to describe as "moving to the cloud." In this sense, many services are now running in "the cloud" and, therefore, are more than likely using "cloud computing," right? Perhaps, but this isn't how IT folk think of Cloud Computing, which is more of a technique and set of technologies, rather than a metaphor.

IT provides services to its users. These services are delivered by running various software components on top of IT infrastructure, which is mostly servers and various interconnecting apparatus. In the IT world, IT folk need ever increasing amounts of computing resources to deliver IT services. Traditionally, these resources are bought and paid for by the IT department and become business assets. Cloud Computing changes all that. With Cloud Computing, IT resources that reside in remote data centres, accessible via the Internet, can be *rented* on a pay-for-what-you-use basis. The key attributes of Cloud Computing are:

1. ON DEMAND – It is available "as needed", meaning that the resources are only paid for when they're being used.
2. ELASTIC – Resource can be scaled up and down, as required, only paying for what gets used.
3. ONLINE – The resources are always available and can be instantly deployed and used for IT services.
4. MANAGED – This is often overlooked, but a Cloud service, at least the infrastructure, is managed by someone else (i.e. the Cloud vendor), not by your own IT staff.

The term Cloud Computing is often used interchangeably with Grid Computing. Again, no one owns the definition, so it is difficult to be authoritative about the comparative meanings. However, the ideas are similar, because of the parallel with buying electricity from "The Grid." Cloud Computing is very similar in principle, turning computing resources into utility services.

If there is a difference, then Grid Computing is still a term that tends to get used in academic circles to mean linking disparate (often loosely coupled) computers to form a single computing resource to tackle a single problem. A fairly well known example is the use of lots of home computers by the SETI[1] project. Home PC owners were able to hand over spare computing resource on their PCs (e.g. When not being used at night) to enable massive data-crunching that was beyond the available local computing resources of the SETI project.

[1] Search for Extraterrestrial Intelligence.

Figure 8.1 Cloud computing.

Grid Computing is usually concerned with the ability to throw massive amounts of parallel computing resources at a single data-crunching problem, such as pattern searches in data. This is distinct from deploying a particular software service (e.g. an e-commerce platform or massively social game) that needs to remain available to a large set of users in a scalable fashion.

8.1.2 Open and Commodity: Key Enablers for Cloud Computing

Internet-driven economics has shaped Cloud Computing and made it possible. Aggressive commoditization is at the heart of Cloud Computing. The availability of low-cost hardware and software infrastructure, including connectivity, has led to the cloud computing revolution. A massive amount of CPU power is one thing, but a key driver has been the aggressive erosion of storage costs.

Coupled with hardware price erosion is the growing widespread availability of open-source software at zero cost! This is a massive factor in the explosion of Web services generally, so not surprisingly it is also a key enabler for cloud computing, which has thus far been driven by companies seeking to provide Web services, as shown in Figure 8.1, that will scale easily and cost effectively, so-called "Web scale."

Of course, this doesn't mean that cloud computing is confined only to open source software stacks. There are plenty of enterprise applications that need cloud computing and which rely on licensed software, such as Oracle database technology or Windows operating systems platforms. These too are available via cloud computing platforms, but it was the open source free-of-charge software that paved the way to cloud computing and what often makes it a viable proposition. After all, if a consumer wishes to scale their application from 40 to 4000 servers, then the incremental costs of software licensing might well be prohibitive in many cases, especially considering that cloud computing has been particularly attractive to scalable start-ups. Indeed, whole new entrepreneurial endeavours are possible because of the affordability of scalable computing that previously would have required millions of dollars of hard-to-get venture capital investment.

Another key enabler of cloud computing has been the growing trend of virtualization. This is where it is no longer necessary to confine a single server application to a single physical server machine. A server application, such as an email server, expects to run on a CPU with memory and a disk drive, abstracted by an operating system, such as Linux. Using virtualization software, it is possible for the underlying physical server to pretend that it is actually more than one server, each with its own operating system and dedicated hardware machine. These machines are virtual machines, as shown in Figure 8.2.

Virtualization has distinct cost advantages because it enables a single large and powerful server to act like several dedicated servers. To be clear, this arrangement is not the same as multi-tasking on a single operating system instance, such as running a Web server and database server together. Each virtual server acts as if it is a completely standalone server, even though it is running alongside other virtual servers on the same hardware box. In this way, for example,

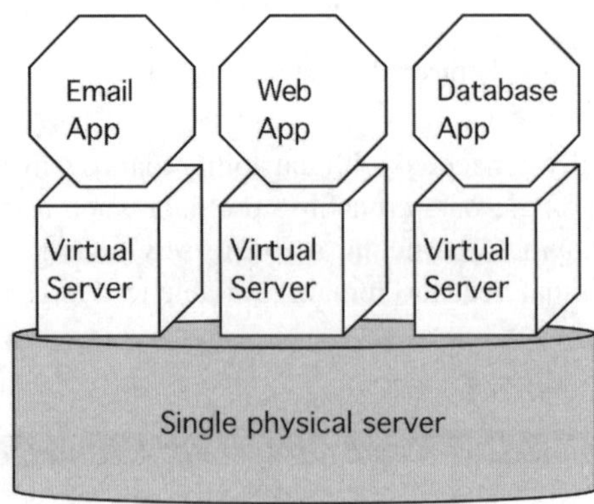

Figure 8.2 Virtual servers.

each virtual server is completely isolated from the others. When a superuser logs on to a virtual server, he or she is the only superuser and has complete control over his or her server instance. This would not be the case in a shared server instance.

8.1.3 Public or Private Cloud?

The earliest examples of cloud computing were public clouds, which are elastic computing resources provided offsite and over the Internet by an external company, like Amazon or GoGrid. Public clouds continue to be a rapidly growing business opportunity.

Naturally, there are some concerns with the use of public clouds, such as security of data. Many companies that would like to take advantage of cloud computing are concerned by the idea of sharing a common infrastructure with other companies. They are also concerned about the level of protection and service on offer. Thus far, public cloud computing providers, such as Amazon, seem reluctant to offer the types of Service Level Agreement (SLA) that large enterprise customers would ordinarily expect from an enterprise IT supplier, such as from a large systems integrator, like IBM.

There are a variety of approaches to the security problem. For example, Amazon now offers a Virtual Private Network arrangement for segmenting part of its cloud in a cryptographic sense, so that a cluster of servers is ring-fenced using robust encryption apparatus. This seems to be making the transition to cloud computing more acceptable to some enterprise consumers of the services, notwithstanding other SLA concerns, particularly concerning levels of availability.

Another solution to the security concerns is to deploy a private cloud. This doesn't mean that it sits locally on the consumer's premises. It can still be hosted in a remote data centre (which typically it will be), but this is probably a dedicated arrangement. In other words, the machines are made exclusively available to the consumer, but still with a degree of on-demand provisioning.

A private cloud would seem to negate the economics of cloud computing, such as the on-demand pricing. This is largely true. After all, if a company decides that it needs 10,000 servers to provide an internal cloud, then the up-front capital costs of the machines still have to be met. However, the concept of a private cloud seems to be more about emulating the on-demand infrastructure capabilities of cloud computing, such as scalable and flexible server image deployments.

The internal cloud model does seem to offer some distinct advantages. These include high degrees of flexibility and agility with the deployment of IT resources for internal uses. These advantages seem to be more important in companies where there are high degrees of software innovation, such as Yahoo, who has invested heavily in providing its own internal cloud service called Sherpa.

Yahoo has a large number of software projects either in production or development. Bringing these to market very quickly and in a scalable fashion is essential for maintaining a competitive edge and for sustaining acceptable levels of performance for its high numbers of website visitors. Yahoo is clear that the use of an internal cloud enables innovation. It allows for rapid development, test and deployment in a unified environment. In other words, by eliminating the complexities of IT infrastructure through the use of a highly automated cloud platform, innovators can focus more of their efforts on converting ideas into cool services without the distraction of various IT infrastructure headaches.

Part of Yahoo's internal cloud strategy is not just to provide an internal IaaS, but to take it a step further to a set of software services running atop, such as a massive shared key-value data store (Sherpa) and a common load-balancing infrastructure (YAK). The advantages of the Yahoo private cloud are that a software team can immediately access a useful set of common software infrastructure components that are already optimized to take services all the way out to Yahoo users in a scalable fashion.

8.1.4 Key Use Cases

It seems fairly obvious that any computing task that needs to scale across lots of servers is a potential candidate for the cloud. This is generally true, but some applications are ideal candidates for cloud deployment, including:

8.1.4.1 Scalable Websites

A common problem with websites is that demand can fluctuate. For example, small websites can suffer from what's sometimes known as the "Digg Problem," which is when the site gets featured ("Dug") by Digg (or some other high-traffic news site), causing a spike in interest and traffic. Popular Facebook apps can suffer the same strain. A website confined to fixed infrastructure will not be able to scale and will therefore crash, causing potentially irrecoverable damage to the service's reputation. A website hosted by a cloud provider will be able to scale on demand. It takes potentially a few minutes to deploy extra servers to take the extra load.

Deploying to the cloud has become a fairly standard practise for start-ups once they are ready to unleash their wares on the world. Beforehand, they might well test everything on a single server. Indeed, in the world of "Lean start-ups," it is recommended to use as few resources as possible to get an initial idea into the hands and minds of some users. Nonetheless, when the time comes to drive traffic to the site, then scalability is important. Cloud computing has been a great advantage to many Web start-ups as they scale up to cope with increasing demand.

8.1.4.2 Test and Development

It is increasingly commonplace for Web start-ups and Web projects generally (whether in a start-up or not) to follow agile development processes. Once an idea is unleashed into the world, it becomes important to stay agile in order to maintain a competitive advantage and to keep users engaged. Agility is only as good as the slowest part of the process, which historically could well have been the awkward hassles with deploying infrastructure to ramp up development and testing.

With cloud computing, development and testing can proceed and scale at a greater rate of knots than before. For example, if the developers need to try out sharding[2] on a new storage solution, then it's much easier in the cloud. In the cloud, there's virtually no difference in effort between sharding across two servers or ten. It's easy to bring up ten server instances for a day to try out the sharding and gain confidence that it works before handing off to operations.

[2] Sharding is the process of adding additional server instances to a storage solution in order to scale the overall storage space in a horizontal fashion.

8.1.4.3 Batch Processing (Grid Computing)

There are lots of processes in software that don't need to run all of the time, but when they do, they require lots of processing power. Examples include:

A. Converting media files from one format to another.
B. Statistical analysis of a large data set (e.g. correlation of vast numbers of test results).
C. Pattern matching and searching in a large data set (e.g. fraud detection).
D. Billing reconciliation – checking large numbers of billing records against financial transaction records.

In all these cases, the size of the job, or the required speed, could benefit from the temporary deployment of substantial amounts of computing power. This is much easier to do with cloud computing. It is just as easy to run up 1000 server instances as it is 10. Indeed, depending on how amenable the job is to being mapped to a parallel processing arrangement, it is possible to save money by the deployment of extra servers because of the substantial savings in time, assuming that the time advantage can be exploited by the enterprise. For example, a job to process 10,000 records might take 250 hours on four server instances, but take only 1 hour on 10,000 instances as shown in Figure 8.3. All kinds of interesting business opportunities arise when a process can be accelerated this much, allowing certain processes to become more real-time.

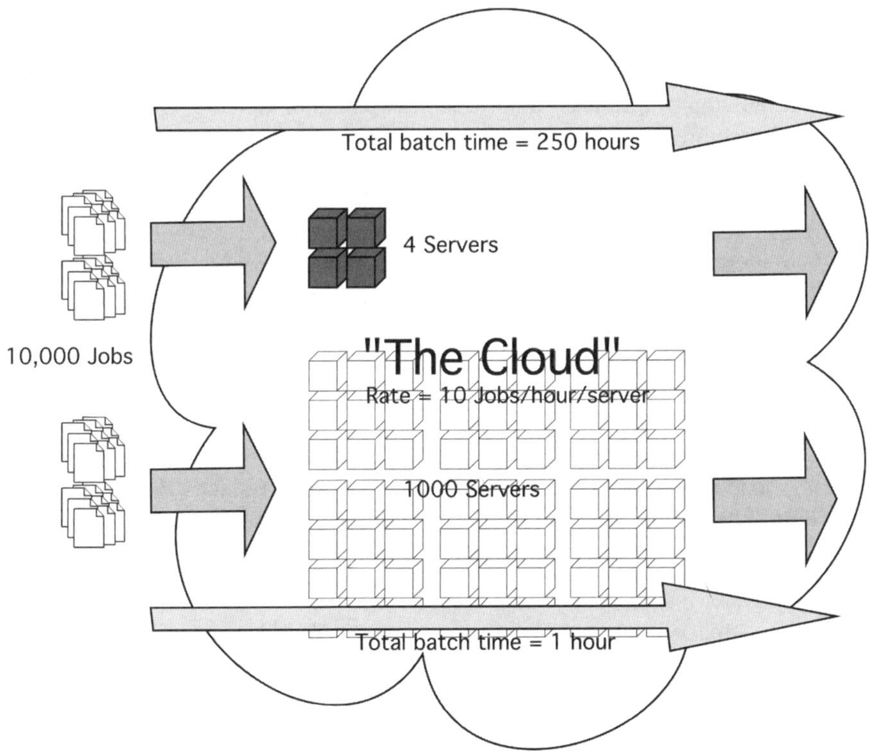

Figure 8.3 Speeding up processes with grid computing.

An insurance company that can process claims in real-time can adjust their premiums more rapidly, enabling them to increase profits. A pharmaceutical company that can process test results in real time can substantially accelerate the arduous task of searching for new chemical compounds for the next big drug, thus speeding up the time to revenue recognition from expensive research. A media company can digitize their back catalogue of content much more quickly and cost effectively, opening up new sources of revenue. These are just a few examples to show that when various batch-processing tasks can be accelerated, new business cases will often arise to justify the move to cloud computing.

8.2 On-Demand: Cloud Computing Infrastructure

8.2.1 The Infrastructure Level: Servers, Images and Templates

At the lowest layer is the physical hardware itself, the smallest unit of which is really the CPU, but which almost always gets packaged in the form of a server. It is servers sitting in data centres that form the base unit of cloud computing services.

At the lowest level, these servers can be rented out as raw hardware. The consumer loads an "image" onto the hardware, meaning a complete set of software including the operating system and any other software needed by the consumer to carry out the task required. For example, it might be that the consumer is running a particular accounting application and needs to increase the resource during end-of-month financial closures or at other times of data-intensive processing. The accounting package might need MySQL server for data storage, Apache Web server to serve up data to clients and PHP to run the accounting package. All of this might run atop of a Linux variant, such as Ubuntu or CentOS. Of course, particular versions will be required for each software element in the stack. This entire software stack is bundled into an image (a single installable file with associated installation scripts) so that it can be deployed instantly on a server running in the cloud.

The configuration of server images is a specialized technical activity best carried out by those with strong server administration experience and skills. Unsurprisingly therefore, in the nascent cloud-computing economy, providers have emerged to provide pre-built images, such as BitNami. They provide a wide range of pre-configured images that are ready to deploy straight to the cloud, such as Amazon's EC2 service and GoGrid.

You might be wondering how an image gets installed on a particular server in the cloud. Each cloud computing provider, such as Amazon's EC2, provides administration tools and infrastructure to enable instant deployment of a particular image. In the case of EC2, Amazon already has an extensive library of images, including those from BitNami. These can be readily deployed to EC2 hardware in the Amazon Web Services data centres.

Examples of images available from Bitnami include:

Web Infrastructure. These are server images for Web developers who want to host scalable Web solutions, such as:

A. **Django Stack** – This is a Web framework for Python programmers, but includes the underlying Apache Web Server, MySQL database, PostgreSQL database, Python language runtimes and SQLite, a development database solution. This stack is bundled with Ubuntu operating system.

B. **LAMP Stack** – This is the popular Web framework discussed in 2.3.1 Introducing LAMP, which includes Apache Web Server, MySQL database and PHP language runtimes. This stack is bundled with Ubuntu operating system.
C. **Ruby Stack** – An increasingly popular Web framework for Ruby developers, which includes Apache Web Server, MySQL database, Ruby on Rails Web framework and Ruby language runtimes. This stack is bundled with Ubuntu operating system.

Blog Infrastructure. These are server images for webmasters who wish to host scalable blog solutions, such as:

A. **Wordpress** – Apache Web Server to server up the Wordpress instances, MySQL database to store the blogs and PHP language runtime to run the Wordpress code.

Content Management Services (CMS) Infrastructure. These are solution stacks to support the various popular and sophisticated open-source CMS systems, such as:

A. **Drupal** – A stack of Apache Web Server, MySQL database and PHP language runtimes, supporting the Drupal CMS framework, all running atop Ubuntu OS.
B. **Joomla** – A stack of Apache Web Server, MySQL database and PHP language runtimes, supporting the Joomla CMS framework, all running atop Ubuntu OS.

And the list of Bitnami stacks goes on, including: chat forums, poll management, project management, bug tracking, version control, wikis, e-commerce, e-Learning, photo sharing and so on.

Nearly all of these images are available for deployment in the Amazon EC2 cloud infrastructure. Amazon has already consumed the images into its image library so that they are ready for instant deployment by an Amazon Web Service's customer. And there is no limit to the number and combination of images that a customer can use. It is possible to run Web, blog, CMS and e-commerce instants all in one cloud computing account. These might well work together to form a single solution to the end-users, but that is entirely up to the consumer of these services. All that AWS does is to install the stack images on-demand and then give access to them via the Internet. The configuration of the stacks is down to the AWS customer.

Taking this a step further, the configuration of the stacks is clearly going to be important. For example, if a consumer wishes to run a scalable social game in the cloud, then this is going to require a number of instances that need to collaborate in some way. Let's consider an example of a scalable Web solution.

As shown in Figure 8.4, there are collections of various server images required to make a complete Web solution that will scale. Indeed, to scale this up, we could simply add more App Server instances, growing from the two shown to say ten. This might still work with the number of load balancers and database servers shown. At some point, we may need to change these too.

Clearly, each instance in this group is part of a wider set of instances that are working together to provide a complete solution. How do these instances know about each other? How do they remain isolated from other instances running in the EC2 cloud belonging to completely

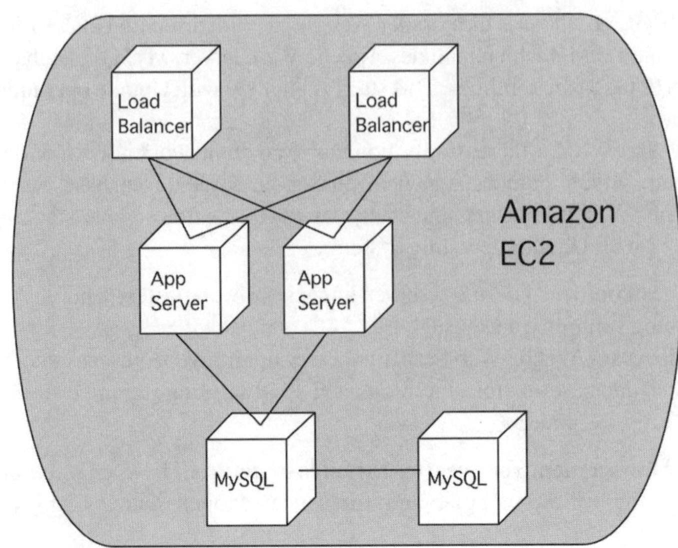

Figure 8.4 Example deployment on Amazon EC2 cloud.

separate applications? These are configuration issues. One approach to configuration is to deploy the instances and then log into the boxes to configure settings. Another approach is to include configuration scripts that run as part of the image deployment in the first place. This is a better method because the scripts can be saved and associated with images. They can also be versioned so that the deployment can revert ("rollback") to previous versions should new configurations go awry.

Configuration and deployment management can quickly become a challenge for large installations with lots of instances. As the solution scales, it is important to enable rapid deployment of new instances, especially as it is likely that new instances will be crucial to meet various service level agreement (SLA) policies that the consumer might have imposed on the design.

Again, unsurprisingly, various solutions have come into play to solve the configuration problem, including companies like RightScale who specialize in providing innovative solutions to manage entire cloud-based deployments. As shown in Figure 8.4, deployments can have various servers and connections between them within a cloud, or even between clouds (across disparate cloud service providers). Within such deployments, RightScale allows each instance to be pre-configured and controlled using a proprietary RightScale Template solution. A key component of a RightScale deployment is a *RightScript* that runs as part of the instance set-up, configuring the stack in a way that is specific to the overall application deployment.

RightScale templates allow the administrator to:

1. Dynamically configure servers at run-time.
2. Confidently launch and grow a predictable infrastructure.
3. Preserve cloud portability (where possible[3]).

[3] Based upon RightScale's commitment to more than one cloud provider, thus allowing their config scripts to work for multiple vendor solutions.

The RightScale template system can best be thought of as a cookie-cutter for images, including image version control. This is very useful for managing the cloud deployment, especially as it scales.

When just thinking about instance-deployment of images, whether template-assisted or not, this level of cloud computing service is often referred to as Infrastructure as a Service, or IaaS. It is worth noting that the instances don't necessarily run on dedicated servers; they may well run on virtual servers.

There are numerous cloud-configuration solutions available, with more coming online all the time. Some of these, like RightScale, are clearly aimed at the market-leading public cloud services, like Amazon Web Services. However, there is also an increasing interest in private cloud solutions, available within a large organization to serve its internal IT needs. The benefits of internal (private) clouds are still being debated and understood in the IT industry, but most of the advantages are related to efficient use of resources, including rapid and flexible deployment. However, just as with all IT projects, the devil is in the details of total cost of ownership and other key performance indicators.

8.2.2 *The Service Level: Storage, Queues, Load-Balancers ...*

8.2.2.1 Going Beyond CPUs

Thus far we've considered the notion of elastic computing where the basic unit of service is a server, or CPU, that is rented from the provider. It is left to the consumer to configure the servers with whatever software image is required to support the application being provided. However, it is possible to provide elastic services at the software service level. Note that these services are still generic in nature, not dedicated to any particular service domain like media, enterprise, or mobile. Let's explore a few of the interesting services, such as storage, task queues and load balancing.

8.2.2.2 Scalable Storage

One example of a scalable software service for generic cloud computing is a scalable database service, as shown in Figure 8.5.

With an elastic database, the application connects to the database and can then assume that the database will scale automatically as more data is added. The store might be a key-value store, a relational database or a document store. There are two approaches to providing the store:

1. **Proprietary store** based on the cloud provider's own technology and design – for example, Amazon's SimpleDB.
2. **Open solution store** – for example, Amazon's Relational Database Service (RDS), which uses MySQL, or MongoHQ's store, which uses MongoDB.

Let's briefly explore each of these examples:

1. **Amazon SimpleDB** – This is promoted as a highly available and scalable store suitable for non-relational data. The developer calls the SimpleDB service via Web-API requests and let's Amazon take care of how the data gets stored, where it gets stored and everything

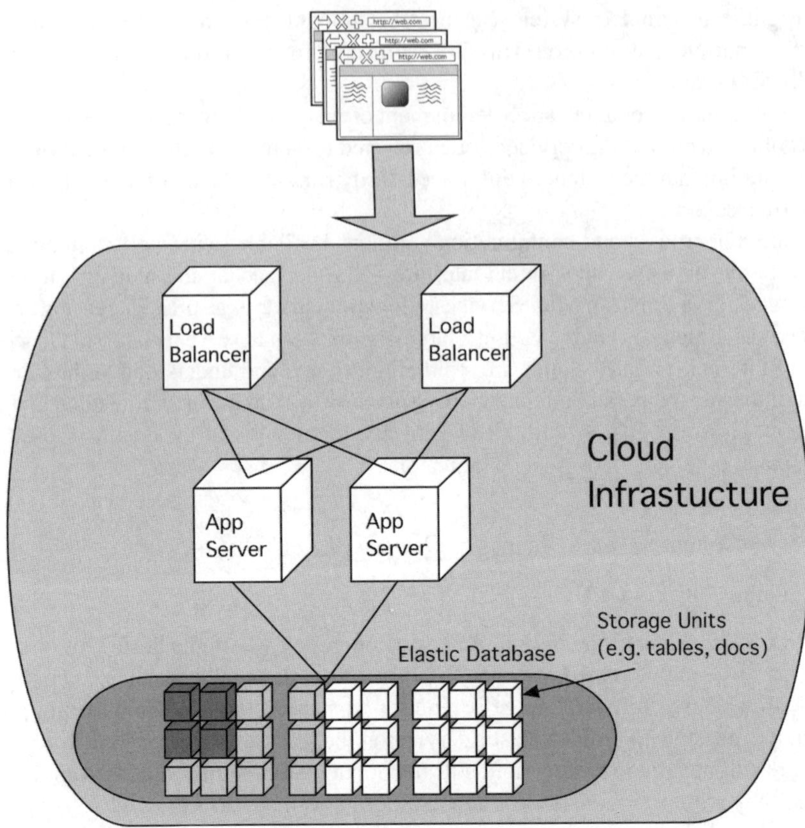

Figure 8.5 Elastic database.

else related to taking care of the data. Moreover, the developer doesn't have to think about configuration and expansion. The store scales automatically as data gets added. A few examples of ideal use cases are logging, online games and metadata indexing, although there are no limits to how the store gets used. So long as the developer has a need for a scalable store that is suited for key-value type constructs, then SimpleDB is a candidate.

A major advantage of using something like SimpleDB is that the developer is left free of all concerns about how to set up, configure, manage and grow the application store. These are traditionally heavyweight problems, especially if a service proves to be popular and needs to scale quickly. In this age of rapid innovation and the need to prove ideas quickly, changing direction often, any "low touch" solution is going to be attractive to developers. In this way, the developer focuses on the value-add of making the application compelling and refreshing it often, pivoting towards a viable product that resonates with users. All of the typically gritty work of managing infrastructure is handed off to Amazon.

The store is easily managed using the concept of domains, which we can think of as "tables." Thereafter, data is managed via a simple set of commands: **put** and **delete**. The store is highly available, distributed across several data centres, and very fast when accessed from an application that already resides in the Amazon EC2 cloud, allowing "near-LAN

latency" to be achieved, according to Amazon. In reality, for apps co-located in the same data centres as the store, this is likely to be true. This speed advantage is now being exploited by other cloud providers who are providing various value-added cloud solutions atop of Amazon EC2, such as Heroku's Ruby-hosting cloud (which I shall return to later).
2. **Amazon Relational Database Service (RDS)** – This solution is designed for developers who require the full features and power of a relational store, like MySQL, which is what RDS uses. Again, by using RDS, the developer is freed from the potentially significant burden of administration, configuration and operation of the storage layer. Again, all of the advantages of SimpleDB apply, especially the ability to scale the resources on demand.
3. **MongoHQ** – This is a service provided by Commonthread, allowing developers to gain access to a dedicated or shared instance of MongoDB without having the headache of installation, configuration and operation. It is relatively early days for services like this, which are based on emergent and immature database technologies (from the so-called No-SQL family). The service is similar in concept to SimpleDB, except that SimpleDB uses Amazon's own proprietary technology to enable a flexible and scalable deployment, whereas MongoHQ are exploiting the power of an open solution to scale automatically; at least that's the theory. MongoHQ have yet to deploy a solution that promises to scale via Mongo's own auto-sharding capabilities. It will be interesting to see how this new breed of start-up cloud providers, like MongoHQ, evolve in the next few years.

A fascinating feature of the MongoHQ solution is that it sits atop of the Amazon EC2 cloud computing platform. This is interesting because it shows how a value-chain is already emerging within the cloud services and provider market. It is also interesting because of the potential for a "network effect" in the supply chain. The more that services are built atop of EC2, the more likely it is that developers will be attracted to the platform, if only because of performance reasons and familiarity. We are beginning to see the emergence of a cloud ecosystem built around the EC2 platform.

8.2.2.3 Scalable Queues

The Amazon Simple Queue Service (SQS) is a cloud-hosted queue service. I discussed the operation and principle of queues in some depth in my previous book *Next Generation Wireless Applications*",[4] but it is worth a brief recap here. In the world of Web 2.0, we might be excused for tending to think of software as always being driven by a request from a Web browser. That happens a lot of course. However, there are all kinds of *events* that drive the execution of a software task: arrival of an email, arrival of a text message, uploading of an image for processing or a file for converting, and so on.

If we have enough time and CPU processing power to deal with events as they arrive at the system, then we might consider processing them in real-time. However, a number of things can thwart this approach:

1. The number of events is too large.
2. The deadline for processing the events is too short.

[4] See Section 13.3 "Message Handling using J2EE" on page 503–507.

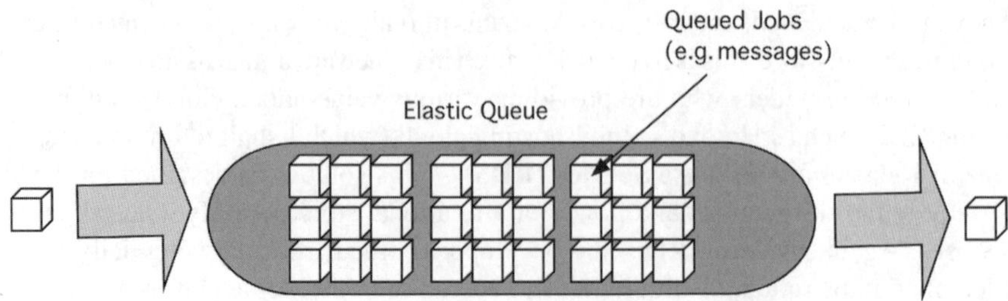

Figure 8.6 Elastic queue.

3. There is a surge in events, such as text messages sent during a TV show.
4. Something interrupts the software process flow, preventing the events from being processed for a while (e.g. while the software deals with an unexpected alarm event).

The ideal solution is to deploy a queue that can store all the incoming events and then dispatch them to the application as and when it's ready to process the next event, as shown in Figure 8.6.

The input events, whatever they might be, are added to the queue as they arrive. Whenever the software is ready to process a queued job, it removes the next job from the queue (in a first-in, first-out – FIFO – sequence).

It might be tempting to think that the queue service is trivial. This is true, especially for a small queue. However, keep in mind that some applications can have very complex architectures and large numbers of events, running into the millions. In these cases, being able to rely on a scalable queue service might well be attractive.

It is still early days for cloud-based queue systems. Some start-ups are emerging in this space, such as StormMQ, which not only promises on-demand performance, but the use of "open" standards, such as the emergent Advanced Message Queuing Protocol (AMQP). The use of standards is interesting and attractive because of the increased demand for disparate applications to collaborate via the Web. There is no reason why applications shouldn't collaborate via a message-passing paradigm, which has been happening reliably and successfully in the finance world for a long time.

8.3 On-Demand: Software as a Service

The previous section described cloud computing as mostly about turning raw computing power into an on-demand service available over the Internet – a kind of utility computing not dissimilar to the provision of electricity from a grid or water from a network of pipes. However, the trend is far more interesting than simply providing raw computing power (Infrastructure as a Service).

What many IT consumers are looking for is the provision of actual software services atop of this infrastructure, but still in a scalable fashion. Consumers want software services that are domain specific, like CRM, but available on a pay-per-use basis without the need to install any software on-premise. This is the so-called Software-as-a-Service model, or SaaS.

SaaS isn't a new idea. Even in the early days of Web 1.0, we saw the emergence of Application Service Providers, or ASPs. One example was e-government, where services like parking-fine payments were moved to the Web. The providers of vertical solutions that could only be accessed over the Web became known as ASPs. Many other industries followed suit, such as the travel booking and reservations chains that were some of the early adopters of Web-based service presentation.

What has changed is the relative ease with which SaaS can be used as deployment model for new and existing software services, thanks to the same internet-economics and technological enablers that have made cloud-computing possible. As you might expect, cloud computing is increasingly underpinning SaaS, enabling on-demand software services to scale with usage and subscriber increases.

SaaS is generally associated with business and enterprise software applications. Whereas the early ASP model tended to be niche vertical solutions from specialist suppliers (e.g. travel companies), there is a growing trend towards moving more and more everyday software solutions to the SaaS model. This is evident with the emergence of office productivity solutions, such as Google Apps and Microsoft's BPOS (Business Productivity Online Standard Suite).

Previously, Microsoft products were shipped as installable files on fixed media (e.g. CD-ROM), expected to be installed on local machines running Windows Server operating system. With BPOS, core enterprise applications, such as Microsoft Exchange, are now made available over the Internet via a Web Browser at a subscription rate of so many dollars per user per month. The entire BPOS suit comprises of Exchange (email), SharePoint (intranet and document management), Office Live Meeting (Web conferencing) and Office Communications (instant messaging and unified communications).

The Microsoft move from shippable product to on-demand product is a shift in two respects:

1. *The platform* has changed from being the Windows operating system to being the Web (enabled by cloud computing).
2. *The pricing* has changed from fixed annual licenses to pay-per-use monthly subscriptions that can be scaled up or down, as required.

It is these two components that define the essential characteristics of SaaS: *Web platform and per-usage pricing*. In the case of Microsoft, these are new patterns of software delivery and business. In the case of Google, as with other born-on-the-web ventures, this is fundamental to the core of the way the company does business – everything has always been on the Web.

8.3.1 Opening SaaS with APIs

Providing software services via the Web platform ties us back to our previous discussion about platforms (see Section 1.6 From Platforms to Ecosystems). I said that platforms provide the potential for new services to be built on top of its core services. This is possible with Web platforms via the existence of APIs. It is an increasingly common feature of SaaS to provide APIs that enable service extensions via independent software services accessing those APIs.

Google Docs, with its many APIs, is a better example here than Microsoft BPOS, which currently has no APIs.[5] In some ways, the existence of a rich set of APIs is one of the major

[5] The underlying software, such as Microsoft Exchange, has lots of APIS, but these are not currently exposed in BPOS via Web APIs.

features that makes Google Docs attractive despite the distinct lack of a rich user experience seen with native applications like Microsoft Office. The APIs provide ample possibilities for extending Google Apps services, or, perhaps more usefully, integrating them into existing enterprise workflows.

The APIs provided for Google Apps include:

1. *Gmail gadgets* – tiny programs that can be inserted into the Gmail user interface (UI).
2. *Gmail inbox feed* – a means to access the unread email messages.
3. *Calendar Data API* – allows external apps to create new events, edit or delete existing ones and search for particular events (e.g. Birthdays). This works across individual and group calendars.
4. *Calendar gadgets* – add functionality to the calendar UI.
5. *Contacts Data API* – gain access to a user's Gmail contacts.
6. *Documents List API* – managed Google Docs documents (see below).
7. *Sites Data API* – enables other services to access, publish and modify content with a Google Site ("intranet").
8. *Spreadsheet Data API* – allows manipulation of spreadsheet content. This is one of the most powerful APIs because of the widespread use of spreadsheets for many business-critical applications.

As an example of the sophistication of each of these APIs, let's dig a bit deeper into the Google Documents List API. This allows other Web applications to access and manipulate data stored with Google Documents. Here are some of the things that could be done using the List API:

1. *Document and Content Discovery* – the API can be used to retrieve documents that match keywords, categories and metadata (e.g. author).
2. *Sharing* – the API allows document access permissions to be modified, including sharing to individuals, group emails or across an entire Google Apps domain.
3. *Revisions* – review, download or publish a document's revision history.
4. *Download* – export any documents in a variety of formats, including PDF, RTF, DOC, XLS, PPT etc.
5. *Create/Upload* – create online backups of local word processor documents, spreadsheets, presentations and PDFs.

8.3.2 Using SaaS for an Ecosystem Strategy

It is clear that Google's strategy is to go far beyond just a platform. They wish to create an ecosystem around their Google Apps offering, as evidenced by the vast numbers of APIs and the support for developers to build and then offer extension products via the Google Apps Marketplace.

The marketplace is a growing directory of software services that work with the Google Apps services via API integration. The list of service categories in the marketplace is already quite extensive:

1. Accounting & Finance
2. Admin Tools

3. Calendar & Scheduling
4. Customer Management
5. Document Management
6. Productivity
7. Project Management
8. Sales & Marketing
9. Security & Compliance
10. Workflow

Each category has a growing number of services that integrate with Google Apps. Just to be clear, not all apps are merely extensions of the Google Apps platform. Some of them are entire services in their own right, like the calendaring extension called YouCanBookMe, which enables a user to expose his or her calendar to a public Web page such that visitors can book fixed time slots. This is great for tutors or professionals who work with hourly appointment slots. This is a relatively simple extension product. It is also free of charge, although there is an optional two dollars per month fee to remove the developer's logo.

Other apps are complete services that would work standalone, but work better for Google Apps users when integrated with Google Apps. An example of this is Insightly, which is a CRM and project management application. It integrates with the Calendar, Contacts and Docs parts of Google Apps. In this way users can maintain customer appointments, for example, in their existing Google Apps calendars rather than use the Insightly calendar.

8.3.3 Opportunities for Telcos

Other than fairly limited and conventional managed services, like Blackberry Enterprise server and online back-up, telcos are slow to adopt SaaS offerings in their service portfolios. Of course, one might argue that the opportunity is limited because operators are not in the business of providing software services anyway. This is a fair criticism and has to be considered. Telcos are good at operating networks and the various operations layers on top, including formidable marketing and support operations. It simply isn't in the telco DNA to think about software, never mind a SaaS-based offering.

Of course, the number one problem, besides DNA, is the lack of a suitable platform to offer SaaS services from. The IT stacks within telcos are fairly often pre-Web in architecture and operation, lacking in low-friction integration points and service creation support. Efforts have been made over the years to mitigate the high-friction problem through the use of orchestration technologies and methodologies like Service Orientated Architecture (SOA), but these can often struggle to deliver anything with enough scale and agility to support a substantial platform play.

Nonetheless, it seems a logical extension in other ways. Here's a list of assets that might lead a telco to consider SaaS as a plausible business extension, at least from an intrinsic point of view:

1. Large user base of SME and corporate customers who are in the market for other business support services.
2. Existing billing relationship with the base.

3. Support and trust relationship with the base.
4. Core place that communications occupies in the business toolkit.
5. Increasing importance of smartphones and other connected devices in the business toolkit.
6. Various underlying network enablers that could possibly provide key support functions for business services (e.g. telematics, authentication, location, billing etc.).
7. Existing, and often growing, online relationships with the base.
8. Great brand and potential network effect that can easily attract other partners.

Some services seem obvious candidates for consideration, such as anything to do with unified communications. Reselling Microsoft BPOS presents an interesting opportunity in this regard, offering value-add through the tight integration with mobile communications services. From there, CRM is only a relatively tiny step away, again with seamless integration of communications services.

Of course, none of these services are the natural purview of telcos, so a heavy degree of partnering and/or acquisition seems to be the only course of action likely to yield results. Whilst most large telcos have fairly decent R&D facilities, often buried in the parent company offices, mostly these have proven to be highly ineffective in delivery of useful services. In fact, rather contrary to their remit, they are very often a block to innovation because it is all too easy to suggest that a native R&D shop can build a service, when often they can't.

The "build versus buy" process in telcos is invariably highly skewed in favour of build because of a combination of naive technological ambition, corporate politics and crude financial measurements that don't put the right value on innovation in the first place. Not having enough confidence in the ability of their R&D shops to deliver, the various business units within the telco end up prevaricating about key business decisions that depend on local R&D efforts.

My view is that telco R&D shops should focus on "innovation through partnership" rather than the myriad "lab experiments" that can easily occupy thousands of employees in everything but fruitful research for the company. What I mean by *innovation through partnership* is using technology (rather than commerce) to make partnerships work in productive and integrated ways that add up to greater than the sum of the parts. This approach fits well with the use of modern Web platform technologies and methodologies.

For example, if telco business customers could use a single identity to log into a range of business services, then it is possible, at least in theory, to begin to offer harmonized services that will eventually evolve to compelling and seamless customer experiences. The focus here should be on the glue logic that enables the single sign-on approach. There are obvious technologies like OpenID and related projects (OAuth on the backend), but these need substantial support and input from telcos to work at scale in a reliable fashion that makes sense to consumers.

This is just one example of a technology that could enable "innovation partnering" and there are numerous similar invented-for-the-web technologies that could be readily adopted by telcos to seed and accelerate "ecosystem" development. However, returning to the DNA issue, these potentialities are seldom recognized, or not beyond an anecdotal level that is capable of being converted into a real strategy and execution plan.

It is my view that the efforts of operators to open their networks via APIs to become more "platform like" should focus mostly on services that enable the mechanics of business partnering to be as frictionless as possible. In this way, a telco can move towards offering SaaS services that are mostly delivered by partners, but which have enough "telco personalization" to be attractive to the telco's business base.

The name of the game is "personalization," but this has been greatly overlooked by operators during a period in which the Web has perfected (Web 2.0) and is busy re-perfecting (Web 3.0) personalization technologies and experiences. After all, the mainstay of much Web economics has been advertising, which simply isn't possible without high degrees of personalization and targeting.

8.4 On-Demand: Platform as a Service

Platform as a Service is a natural evolution of SaaS. As with all these emergent memes on the Web, there is no single or authoritative definition, but the name at least conveys the central idea, which is the ability to provide a platform on which to run something. Taking the familiar CRM domain as an example, the idea of accessing a CRM tool via a Web browser, paying a monthly subscription, is what SaaS promises. But what if I want to customize the CRM solution?

There are a few approaches to the problem. I might think about building my customization elsewhere on the Web and then trying to integrate it with the CRM SaaS solution. For example, I might import data from the SaaS solution into my customer CRM application. As with most SaaS solutions, I can probably achieve the data import using an API, as shown in Figure 8.7.

This approach will work and is a common enough pattern on the Web for building one application "on top" of another, as discussed in detail in the opening chapter. Indeed, if a SaaS provider can persuade enough application providers to link up with their solution in this way, then they are well on the way to establishing an ecosystem around their solution, especially if the other applications strongly complement the solution and each other, causing subscribers to sign up for a number of apps in the mix. When this happens, the ecosystem gains strength via direct network effect.

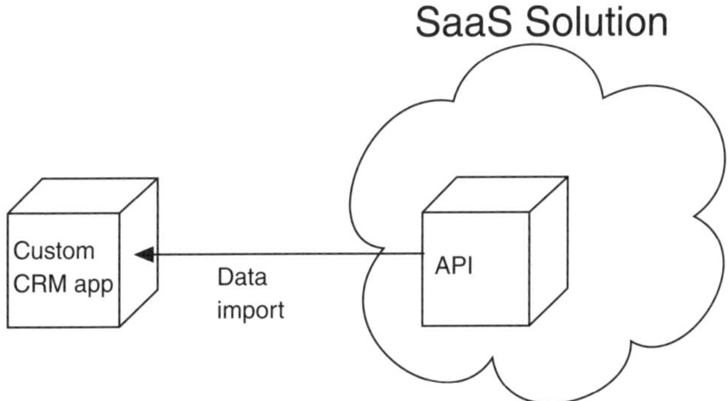

Figure 8.7 Data import from SaaS solution.

However, what some SaaS providers have done, recognizing the increasing strategic importance of platforms, is to provide the means by which a custom app can be built and hosted directly on top of the SaaS solution stack, as shown in Figure 8.8.

Figure 8.8 PaaS approach.

The best way to understand PaaS is to explore a few examples. For these, I will take one each from the business application, telco application and Web application domains.

8.4.1 Business PaaS – Force.com

Force.com is the PaaS offering from the company Salesforce.com, which is well known for its SaaS-based CRM applications, widely used throughout the business world, from SMEs to large corporations. The genesis of the force.com platform is fairly similar in concept to how Amazon evolved from an e-commerce business to a cloud computing company.

Originally, Salesforce.com built a SaaS-based CRM solution allowing businesses to outsource their CRM solutions to the Web, accessing the Salesforce.com application via the Web browser. It wasn't long before the need for custom CRM apps arose, as shown in the top half of Figure 8.9. However, in re-factoring the original Salesforce.com platform to accommodate custom application extensions for customers, it became apparent that a lot of the platform services needed to run CRM would be needed by any business application. For example, the need for account management exists in all business applications – users need to log-in securely and manage their data.

Salesforce.com took the decision to create a more generic business application infrastructure that would be capable of not only running Salesforce.com CRM apps, but any business app. This new platform became known as Force.com and would enable Salesforce.com to develop their own CRM solution, but any business application that might sit alongside it. In fact, the new platform allowed any business application to be created, whether related to the use of the CRM solution, or not. And, since a platform exists for internal developers to build new biz apps, why not let any developer build apps on the same platform, enabling a brand new business model for Salesforce.com.

In Chapter 2, we explored the mainstay of the Web application world, which is the LAMP stack. When developing enterprise applications, the LAMP stack lacks many features that might be needed by a mission critical business app. For example, it lacks transaction management, which is the ability to "record" steps taken by a software process in case it all

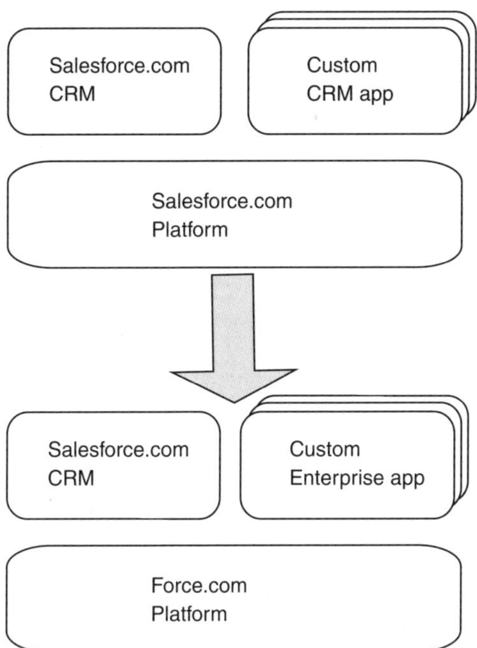

Figure 8.9 Salesforce.com evolution to a platform business.

has to be "undone" (i.e. the steps retraced and the state of the software and data set back to how it was before).

In my earlier book *Next Generation Wireless Applications*, I discussed in some detail the need for an "App Server," which is like a Web server but with lots of additional software services that support distributed enterprise applications (such as transaction management). An example of an app server technology is Java Enterprise Edition, which is widely used in enterprise environments.

Salesforce.com didn't do away with the need for app servers, but they discovered that a whole set of additional software services could be layered on top of app servers in order to support enterprise applications running in the cloud, as shown in Figure 8.10. One example

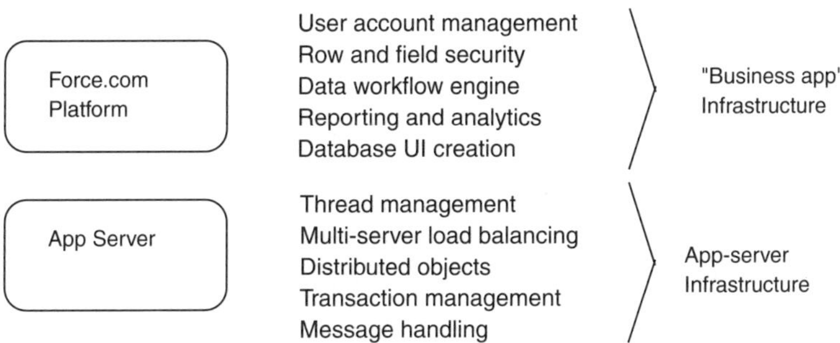

Figure 8.10 Force.com application server layering.

Figure 8.11 Form-centric UI common to biz apps.

is workflow technology, which is useful in many enterprise situations. Workflow is used to set a chain of software tasks in motion based on an input trigger event, such as an email from a customer. The sorts of workflow that spring from receiving a customer email might include flagging the email to an appropriate customer service agent, auto-responding with a boiler-plate email, and so on. Another triggered task might be to create a service incident report, should the email warrant it, and so on.

Workflow management isn't a standard software feature of a Java Enterprise Edition app server, or similar. Salesforce.com identified and developed a number of key "business infrastructure" services, like workflow, that could be abstracted to a set of common business software services that could be used by any business application running on the app server. Of course, then they took it a step further, which was to create an architecture that would automatically scale, thus fulfilling the promise of cloud computing. Force.com is a *business* cloud computing service.

Salesforce.com realized that many enterprise applications are database-centric. Of course, most Web apps utilize a database (remember the "M" in LAMP stack[6]), but this might not be obvious to the user. Take an application like Twitter. It uses massive database resources to store all those tweets, user accounts and all those lists of their followers. However, there is nothing obviously "table-like" in the user experience.

When we use a Web app that requires us to give our details, such as the billing and delivery information for checking out from an e-commerce site, the use of tabular information is more obvious, with all those fields to fill out in the form, as shown in Figure 8.11. It turns out that a lot of business applications are similar in structure and in their pattern of usage – lots of fields to fill out and results to save, modify, or delete. To create a database application with a form-centric user interface can take quite a bit of programming skill, not to say all of the additional overheads of turning the app into a secure app that runs in the cloud, accessible by a business, measurable by some kind of analytics.

The Force.com platform contains a whole raft of services, accessible via the Web browser, that enable a developer to create a form-centric database application that runs securely in the cloud with all the cloud computing benefits discussed earlier. Let's say that you wanted to create an expense-management tool for your team of 100 co-workers. No problem! Create a

[6] Stands for MySQL database, in case you forgot already.

database object with the appropriate fields (e.g. item, description, date of expense, amount receipted, reason, etc.) and then specify the type of data to be stored in each field (e.g. text string, date, currency etc.).

With nothing other than a bare LAMP stack, we'd have to build the HTML files and then program them in PHP. We'd have to program the database connections and so on. With Force.com, it's as simple as "pressing a button" in order to create the web forms and all the database access logic automatically. Moreover, it's possible to use the platform's built-in analytics and reporting tools to continually measure and present all kinds of metrics, such as how often expense claims are made, the typical amount, the amount per department, and so on. These analytics and reporting features are another example of "business infrastructure" that's commonly required by many enterprise apps. With Force.com, it's all part of the platform.

Force.com also includes a rich set of developer tools and related development resources. This is important for the development of rich apps that require deeper functionality or integration than is possible using the visual programming paradigm. Force.com has its own language, called Apex, which can be used to extend, enhance and enrich applications. This is done via an Integrated Development Environment (IDE), namely the Eclipse IDE, which is a very familiar tool to developers. Furthermore, as the Force.com is based on Java technology, it is also possible to program using Java, again via the Eclipse IDE. Java is a widespread technology in the enterprise world.

A further component of the Force.com PaaS is App Exchange, which is a kind of app store for Force.com apps. The model is similar to any other app store – developers can create apps on Force.com and then offer them to other users of the platform. Sharing and selling apps is a highly useful part of any platform strategy, as this enables a richer ecosystem to flourish via the network effect.

8.4.2 Telco 2.0 PaaS – Tropo.com

This is one of my favourite platforms because it takes a very (very) old idea, which is telephony, and brings it up to date using some cool Web 2.0 software ideas. Tropo is an example of a "telephony in the cloud" service, similar to Teleku[7] and Twilio. It enables developers to create telephony and messaging apps and then run them on a platform that can scale from one telephone call (and number) to potentially millions. For the uninitiated, it is worth reflecting first on just what an achievement this is. In a traditional telco environment, like the one we'd typically find inside an operator's data centre, most of the telephony and messaging VAS[8] infrastructure is locked inside of application silos.

Let's take voicemail as an example. A voicemail solution will probably have a lot of features, which I don't need to describe here. But let's say we want to try out a few new features, such as the ability to change the announcement message via Twitter. If that feature isn't part of the solution already, which it probably isn't, then in most cases it's very difficult to create such a feature. One can imagine returning to the vendor of the solution to ask if it can be done. An accommodating vendor might say, "Sure, but it'll cost you X," where X is most likely some large figure.

[7] Bought by Voxeo who own the Tropo platform/product.

[8] Value added services.

A less than friendly vendor might say that it isn't part of their roadmap and is unlikely to happen. They might point to some friendly (and expensive) systems integrator who'd be more than happy to build a bespoke Twitter interface for the voicemail box. Either way, I think we all know that the whole process is going to be prohibitively expensive and time-consuming, unlikely to survive the rigours of business justification. The numbers simply won't stack up, although no one really knows because without trying the service out, the subscriber interest is difficult to predict.

Now, before telling you how easy it might be to build such an app on a telco PaaS solution, it's worth the detour to consider your objections to the idea of a Twitter-controlled voicemail app. That's right – I'm saying that you're more than likely to object to the app,[9] wondering "Who on Earth would use such a thing?" That's a good question, but I want to point out that the solution here isn't about answering such questions. What if I can build the app so easily, cheaply and quickly that we can simply find out if it's attractive to users? That's right. We can try the idea out for real with a few thousand customers, and get their real feedback. In the final chapter, I shall return to the theme of building stuff quickly enough that it provides a solution to the problem of *how to innovate*, but let's continue for now with Telephony 2.0.

With a "programmable telephony stack," of the sort offered by Tropo.com, it's relatively easy to create a Twitter-based voicemail app. The entire voicemail app can also be created, all using standard Web 2.0 programming languages, such as PHP, Ruby and Javascript, familiar to most Web developers. And that's part of the appeal of a platform like Tropo – it allows Web developers to build telephony and messaging apps. Even better, it allows developers to mash-up the two, such as integrating a conference calling widget inside of a Web app used to manage projects. Imagine being able to see who's working on a particular project and then reaching out to them all with just one click of a button on the Web page – instant tele-conferencing without the hassles of access codes, PINs and so on.

With a Web app, the basic unit of input is a request for a Web resource, such as clicking on a URL link. With Tropo, the basic unit of input is a phone call (to one of the numbers handled by the platform) or a message, which could be a text message, an instant message, or a tweet (from Twitter).

By writing a *Tropo Script*, which runs on the platform, as shown in Figure 8.12, software can be used to control what happens when a call or text message arrives. And, because it's software, anything is possible! Let's start with the basics, such as answering the call, which is a single line of code:

```
answer()
```

Of course, that's not very interesting yet, as it doesn't do much, but we could follow up with an announcement (and then hang up):

```
answer();
say("Welcome to the future of telephony!");
hangup();
```

[9] Don't worry – most people object to the idea, or anything radically new at all, so you're not unusual. Just remember that when I was designing digital cellular telephones (2G), it was almost impossible to find anyone who said that they'd buy one. "Why?" was the typical response. The same was true when I built one of the earliest text-messaging application gateways. "Who's going to use that?" And on it goes.

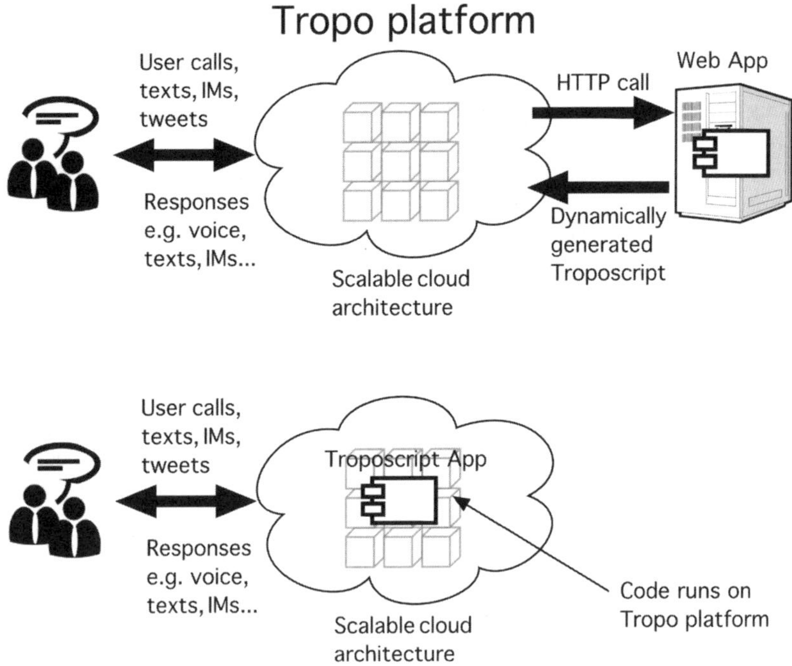

Figure 8.12 Tropo platform architecture.

Okay, that isn't very useful either, but it's the bare bones of a telephony application in the making. What's significant is that it takes just a few clicks to deploy this application to the cloud-based platform where it can be put to use immediately, getting feedback from users. And it probably took less than a minute to write. Now, because this is software, we can play with some ideas, like replacing that welcome message with my latest tweet, or with a tweet that contains the hash tag "#say" (or any hash tag we like). In this way, my callers could get an instant vocal update of my latest status via Twitter, as shown in Figure 8.13.

Twitter-controlled voicemail might be gimmicky for some, but if Twitter is already a natural part of daily habits, which it is for me (admittedly on and off), then presence updates via Twitter make perfect sense. In fact, it's more natural than having to update my "call" status via a method that requires a new habit to be formed. Where possible, it's usually a good idea in product design to fit the product in with the user's natural way of working rather than expect the user to adopt a new habit.

I agree that an app that merely tells you my Twitter status via a dedicated number isn't going to be of much use. However, with just a few more lines of code I can turn this app into a voicemail app – the announcement is what you hear if I don't answer my phone, which is what I built as a hack at Twitter's Chirp conference in San Francisco. Customized and current announcements are often more useful than playing the same greeting over and over again. The problem is updating them often enough, or trying not to forget to refresh an out-of-date message. This is usually because it's just so damn hard to update the message. However, what if the user is already using Twitter to update status messages anyway, often frequent enough to be useful? Why not re-use those messages (or a filtered subset – for example, using a dedicated hash tag) to update the voicemail announcement?

I would prefer for my wife, or a colleague, to know why I can't take a call – "Hi, I'm in a meeting for the next 2 hours, so please leave me a message or send me a text." It's an untested theory, but I suspect that customized messages like this are more user friendly and more conducive to better communications: callers might feel more inclined to leave a message, or just simply happier to get the extra detail.

It's relatively trivial to add call screening in my app in order to play announcements that are customized specifically for my wife or a colleague – "Hey John, I'm tied up just now, but I'll get you that report this afternoon." Everyone else will get a standard greeting, which can still be customized and current of course – they get to hear that I'm in a meeting, but only John knows the details. I might use a particular hash tag – for example, #wife – to control tweets that are aimed at my wife, with Tropo picking these up to convert into announcements. The possibilities are seemingly endless.

If a user prefers IM or text messaging, then the Tropo platform will allow those message channels to be used instead of Twitter. It's fascinating to think of how often someone might change his or her voicemail message using IM, as it's relatively low-friction. Features could include:

1. Announcement control via text message, Twitter and IM.
2. Ability to change preset announcements via keywords – for example, "Away," "Meeting," "Holiday."
3. Ability to select pre-recorded announcements (i.e. actual voice recordings) for the [away | meeting | holiday] greetings.
4. Ability to set the greetings for the announcements using any text string sent via a message, which gets converted to speech.
5. Ability to record the greeting and then forward it as an email attachment.
6. Ability to record any of the preset greetings via a voice call to the voicemail number.
7. Ability to send a text message to the caller to ask them to call back later, should they fail to leave a voice message.

That's the beauty of a platform like Tropo – developers can try out new ideas as fast as they can think of them. Lots of telcos have fashionably adopted the "agile method" for IT projects, but their IT platforms just aren't up to the job. This is where the promise of platforms like Tropo lies. If the architecture is made-for-cloud, as it claims to be, then there's no reason why Tropo shouldn't scale to meet the demands of lots of users. Ideally, this is where we want to be with platforms like Tropo – "build, try, tune and deploy."

I have yet to see any credible demonstration of a large-scale telco platform built using scalable Web technologies, but there's no reason to doubt that it's possible, or that it's coming soon.

These are the sorts of commands (called "verbs") that can be run inside a Tropo script:

1. Answer – to answer the call.
2. Ask – to prompt the user for some input (via voice or key press).
3. Call – to set up and connect an outbound call.
4. Conference – to set up a conference bridge ("chat room").
5. Message – send a message to a recipient (IM, Tweet, SMS).

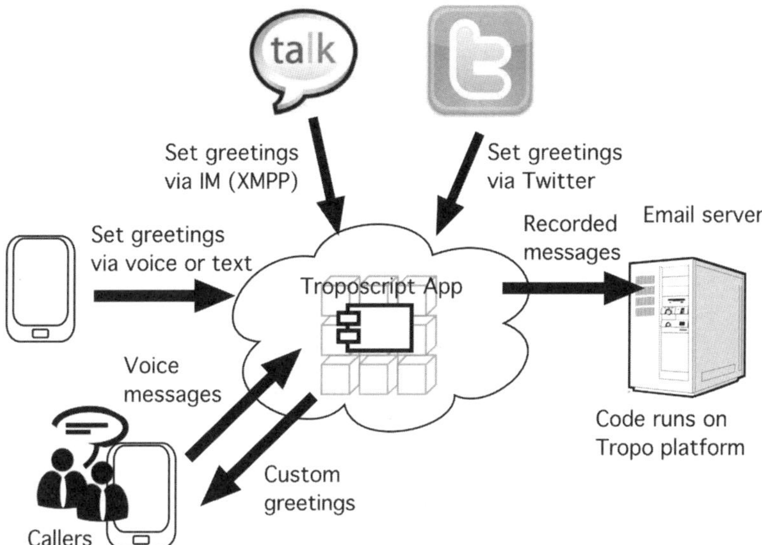

Figure 8.13 Web-connected voicemail.

6. Record – record the incoming call – good for voicemail apps.
7. Say – make an announcement – either an audio file or a text string using text-to-speech conversion.
8. Log – write a message to the debugger (see below).

There are plenty of other commands that can be called from the code, which, combined with the richness of language itself (e.g. Ruby with all its built-in libraries), gives the developer lots of power to build compelling apps and services. On top of that, the platform comes with a debugger that tracks and logs every operation as the code executes, enabling the developer to find mistakes when things go wrong. The debugger reports general execution errors plus any log entries that the developer includes by the use of the log command in the script.

The platform also supports a provisioning API, as shown in Figure 8.14. Among other things, this enables applications to be assigned to selected inbound numbers. This would allow a user to sign-up for a service via some self-serve portal, which then uses the provisioning API to associate the selected service with the user's number.

With a platform like Tropo, there are all kinds of integration opportunities with other platforms, such as legacy ones in the telco network. Such integration would need to be bespoke, but is probably necessary if a platform like Tropo is to run alongside an existing telco network. Integration points might include those shown in Figure 8.15.

8.4.3 Web 2.0 PaaS – Heroku.com

8.4.3.1 The Rails Way

It has become semi-fashionable to build Web apps using the Ruby programming language. Ruby programmers even have an affectionate name for themselves: Rubyists. I'm not sure that

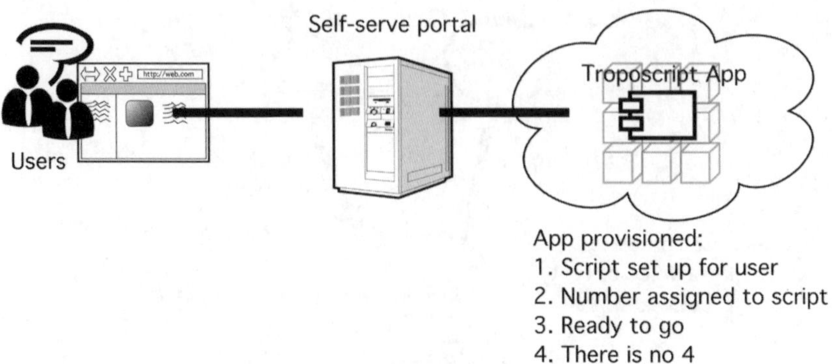

Figure 8.14 Provisioning a Tropo app via a portal.

Java programmers ever called themselves Java-ists, or anything else. I know I didn't! (And it kind of sounds naff.)

Where did this fervour for Ruby come from? One version of the story is that a bunch of guys at a company called 37 Signals built their own Web framework using Ruby. They called it *Ruby on Rails*. In one of those "tipping point" moments, possibly aided by the heavily publicized Twitter's use of Rails, giving it a degree of "it scales" credibility, and thanks to some infectious "watch me build an app in front of your eyes" demos by its creators, Rails took off. We're now at Rails 3.0, which was a major release in 2010, mirroring a fairly significant update of the Ruby language itself.

In a world dominated by the LAMP stack and its various cousins, it has been more challenging to find a reliable and scalable solution to hosting large-scale Web apps built using

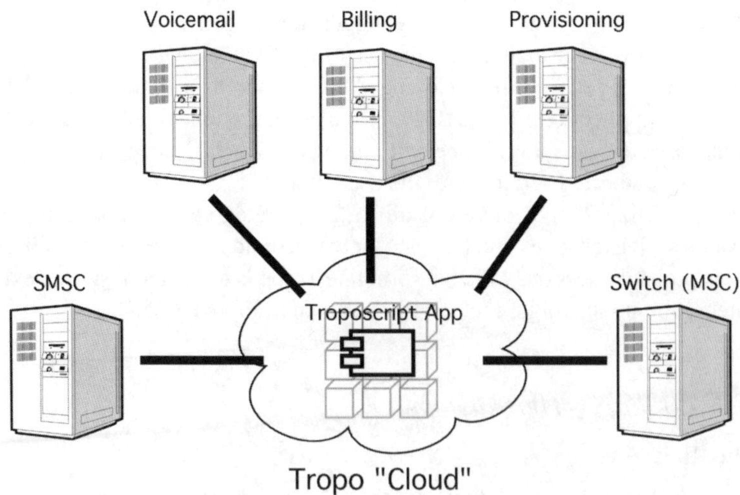

Figure 8.15 Hypothetical integration of Tropo into existing telco network.

Rails (or other Ruby frameworks[10]). As Rails has matured and the community around it has expanded, lots of open source tributary projects have fed the Rails ecosystem and increased the appeal of developing on Rails.

Just as there are lots of Rubyists, bless them, there are now plenty of Rails fans who love Rails. Even so, it is one thing to jump on the Rails wagon and get a project moving along the tracks, but quite another to "industrialize" that project, enabling it to scale and take a hammering from users. Heroku is possibly the solution.

8.4.3.2 The Heroku Way

Heroku is a service that allows Rails developers to host their Rails-based Web apps on a cloud-computing platform that will allow their apps to scale. It is a multi-tenanted platform, which means that any number of developer's Rails apps could be sharing the same server or servers. Multi-tenanted hosting enables the price entry point to be lower than if assigning dedicated servers to each app (or developer).

The major selling point of the Heroku platform is that a developer can host a Ruby Web app without having to worry about scaling. The developer doesn't need to configure a load balancer or a cluster of servers. All this is done automatically and somewhat transparently by the Heroku platform. What's more, the underlying platform is continually improved, thus allowing the developer to benefit from improvements in infrastructure performance and functionality without any effort. Deployment to the platform is very simple and doesn't require any unusual configuration of the Rails app.

8.4.3.3 Heroku Architecture

The architecture of Heroku looks like Figure 8.16.

Let's take a brief tour of the platform:

> **HTTP Reverse Proxy** – This layer handles all HTTP requests, whether from Web browsers, mobile apps or other websites (e.g. via API calls). This layer is entirely maintained by Heroku and works transparently – from the developer perspective, his or her Rails app receives HTTP requests directly from the client app. The developer is unaware of the load balancing and fail-over mechanisms that are working as a proxy for the app. Some of the processing-intensive HTTP tasks are offloaded to this layer, such as SSL[11] and gzip compression (where supported in the HTTP headers).
>
> The layer is highly tuned by the Heroku engineering team, who use the open source Nginx software as the foundation of the layer. Nginx is a free open-source HTTP server and reverse proxy, already hosting millions of sites worldwide.[12] A reverse proxy sits between source HTTP requests, usually from the Internet,

---

[10] There are a number of Ruby frameworks for building Web apps, such as Sinatra and Merb, but Rails still seems to dominate the scene, thanks to the network effect of lots of developers contributing to the wider project.

[11] Secure Sockets Layer.

[12] See wiki.nginx.org.

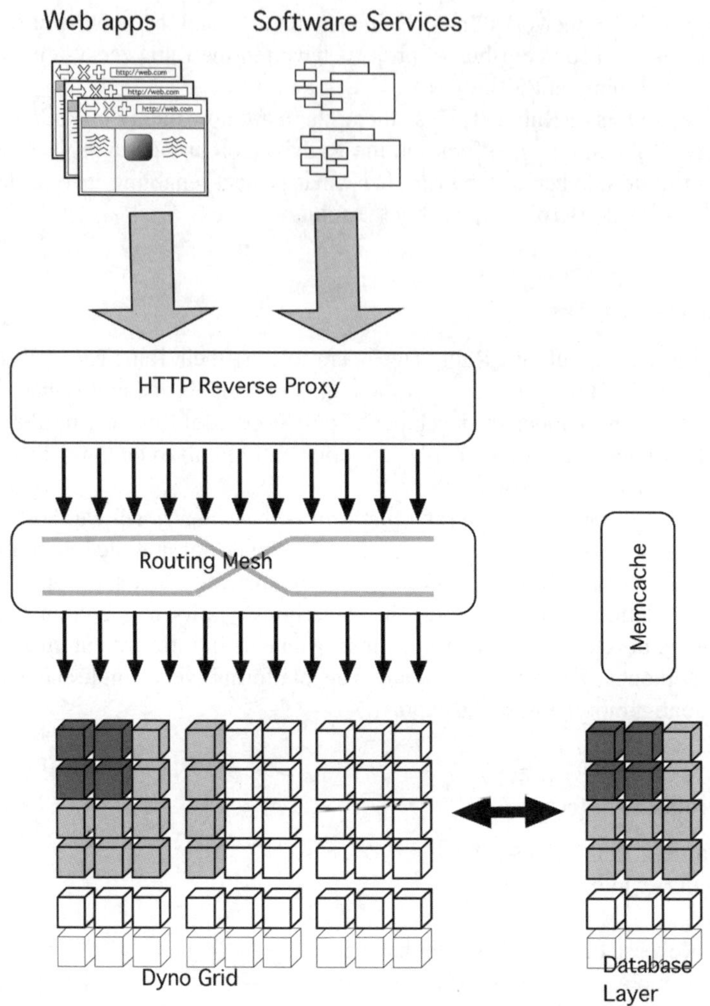

Figure 8.16 Heroku platform architecture for Ruby Web apps.

and the target application servers, which in our case are Ruby on Rails apps.[13] A reverse proxy takes care of security issues, SSL encryption and load-balancing, which means spreading the inbound requests across a cluster of servers running the same Rails app in order to dynamically adjust resources to match (scale) the traffic demand for the app.

Apache can also be used as a reverse-proxy (via pluggable modules), but Nginx is favoured in cloud environments where its high performance and architecture more readily lends itself to scalability. Remember, Apache was designed initially

[13] More accurately, it can host any Rack-compliant app, Rack being middleware that sits between a proxy and a Rails framework, like Rails, which is Rack-compliant.

for standalone Web servers. We could argue that in this new era of scalable apps, where multi-server is an expectation, Apache is falling behind the performance curve.

Unlike Apache, Nginx uses a novel architecture that avoids the use of threaded processing to handle requests, which is the traditional way of handling concurrent requests in a single program. Instead, it uses an event-drive architecture, which is not that unlike the multiplexor concept used in telecoms processing and switching. In many ways, we can think of the HTTP Reverse Proxy layer as a switching fabric atop of the Heroku platform.

HTTP Cache – Another high-performance layer in the Heroku stack, the cache layer can give a significant performance boost to an underlying Rails app. It is a simple idea. If an incoming request asks the Rails app for data that it has already (and probably recently) given, then it is quicker just to send out the same result again, stored in the cache, than ask the Ruby app to go produce the result all over again from scratch.

Heroku uses the open source Varnish caching app as the basis of the HTTP cache. Varnish uses some of the advanced features of the underlying operating system (Linux 2.6, FreeBSD 6/7 or Solaris 10) to achieve performance gains. Again, Varnish is more compatible with distributed cloud-computing architectures like Heroku, so it is a good fit. However, caching is not an entirely transparent and magic process.

The developer will need to be aware of various techniques to exploit the caching layer, otherwise opportunities for performance gains will be lost. It is possible in the Ruby code to set the HTTP header that specifies to the HTTP client whether or not the requested resource should be cached. A developer should think carefully about which outputs from the code should be cached. There are generally two scenarios to think about: static and dynamic caching.

Static caching – This is where the output from the code is unlikely to change ever, or for a long time. One example might be a 2D barcode generated for a particular URL. Once generated, which might take a few seconds of processing time, the barcode will not change (unless the encoding[14] method is updated). Therefore, there is no need to ever run the code again to generate the barcode, should it be requested by any future HTTP requests.

It might be tempting to save the barcode image to a file store and then have it dished up from the store. At its crudest level, assuming a naive (and non-cached architecture) this is not scalable because the image is likely to sit on one store, thus forcing all requests for this image to go down the same routing pathway between the proxy layer and the file store. At the very least, the image should be copied across multiple servers. Better still, it can be stored in the cache, which is not only faster (because of its tight integration with the underlying operating and file system) but is also faster because it negates the need for an additional fetching cycle from a server.

[14] The encoding is the algorithm that says how to generate an image for the barcode based on the input data.

Static caching is effective for any image or file content (e.g. MP3 files, PDFs etc.) although in the case of Heroku, which is hosted on top of Amazon EC2, it is recommended to offload any static assets like these to Amazon file services, like S3, which are better optimized for serving static content. Another example of static caching might be a URL shortener, like bit.ly, where, similar to barcodes, once the mapping has been calculated between a long URL and a short URL, the mapping is not going to change (assuming the shortening algorithm doesn't change).

Dynamic caching – This is where content can be cached for a short while, but is likely to be updated soon. In this case, we want the cache to serve the cached content only for as long as it is current. Once the content dished up by the code changes, then we need the cache to request the code to run again and return the updated content. If the developer is aware of a reasonable period for caching the content, then this can be set in the HTTP "Cache-control" header. This can be done easily via the Ruby code, typically in a controller responsible for dealing with the request:

```
class MyController < ApplicationController
  def index
   response.headers['Cache-Control'] =
                             'public, max-age=300'
   render :text = > "Rendered at #{Time.now}"
  end
end
```

The 'max-age = 300' setting tells thc IITTP user agent that the content should be cached for 300 seconds (5 minutes). It might not be possible for the developer to guarantee that this is always correct, but it might be acceptable most of the time. The developer really has to think about the consequences of serving potentially stale data from the cache. If it's unlikely to cause any problems for the consumer of the data, then it might be acceptable for the occasional misalignment of cache data with dynamic data.

There are various strategies to manage dynamic caching refresh, but the key point here is that the caching layer is available to any app running on the Heroku platform without the developer having to install and configure it. This is good news for most developers.

Routing Mesh – This layer is custom-coded by the Heroku team using the "telco grade" language Erlang, which is a language written from the ground-up to build highly reliable, fault tolerant distributed systems (of the kind found aplenty in any telco IT environment). The routing layer performs intelligent routing of inbound requests to the underlying Rails-app cluster, which we shall explore in the next section.

The routine mesh not only enables automatic scaling, but supports fault tolerance by routing traffic around any underlying components that seem to have lost function, or that are functionally poorly or erratically. The beauty of the mesh is that it can spot suspect components beneath it and make requests for new ones to be dispatched in their place, automatically inserting these into the mesh.

Dyno Grid – As the Heroku guys say – "this is where the action happens." In other words, this is where the Ruby Web app (e.g. Rails or Rack app) actually resides and runs, sitting inside of a wrapper. A Dyno also represents a chargeable element in the Heroku pricing and billing scheme, with developers paying per Dyno as and when they're needed, starting small and then scaling upwards as the service takes off. Dynos are priced in CPU hours.

A Ruby Web app will invariably have dependencies, such as third-party code components (typically packaged as Ruby Gems) and database connections. The app, along with its dependencies and configuration is compiled into a self-contained and read-only "blob" of code, called a "Slug." These can then easily be deployed to new instances of Dynos, thus enabling rapid and dynamic scaling of Ruby Web apps on the Heroku platform.

The whole Dyno Grid is itself scalable, running atop of Amazon EC2. Its size will vary depending on the overall load that the Heroku platform is experiencing at any one time, remembering that there are lots of apps spread across the grid in a multi-tenanted fashion. Where, how and when a particular Dyno will run on a particular server in the underlying EC2 cluster is completely hidden from the developer. In fact, there is no need to know the details. The reason for the grid size variation is economics. It is Heroku attempting to optimize the use of the underlying EC2 resources in order to constantly optimize costs.

The Dyno stack is shown in Figure 8.17.

The layers of the stack are:

Posix Environment – The Dyno Grid runs on Debian Linux, but there is a mapping of a particular Dyno to the Posix enviroment via a dedicated unix user per Dyno. In this way, the management of permissions (i.e. which Dyno can access which

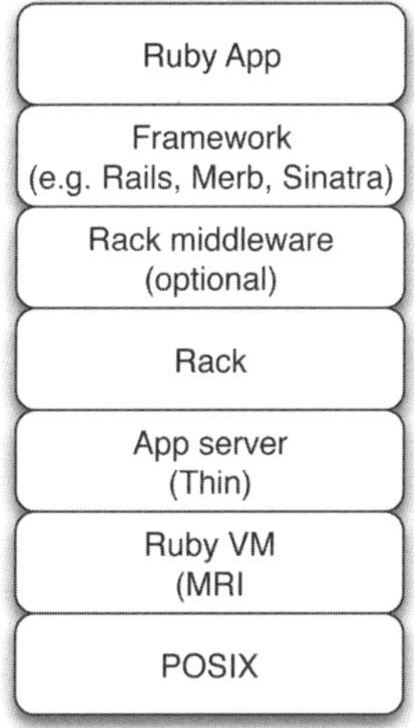

Figure 8.17 Dyno stack on Heroku.

resources) is taken care of by the well-proven unix permissions sub-system. For example, if a Dyno is assigned a unix user "dyno123," then we can use this user in the allocation of database resources, creating database tables with the username "dyno123." Therefore, if another Dyno runs on the same server, but with user "dyno888," then with the right use of permissions, there's no way that "dyno888" can access any databases that belong to "dyno123," and vice versa. This approach is nothing new. It is exactly how most app servers (e.g. Java 2 Enterprise Edition) will solve the problem of security and authentication by relying on the mature underlying security apparatus of the operating system.

Ruby Virtual Machine (VM) – Ruby is an interpreted language, so it requires a virtual machine to convert the Ruby constructs into code that will run natively on unix.

App Server and Rack – In many ways, with the Dyno stack, Heroku have effectively designed their own "app server," although there is no reliable definition of an app server. In this case, Heroku refer to one part of the Dyno stack as providing the "app server" function, although it is really only a Web server in the conventional sense. Here it is useful to appreciate that a Ruby Web app is unable to run standalone as a Web app because it lacks critical components, such as the ability to handle HTTP requests. This is usually given to another component to run. In fact, it's better to think of things the other way around – the app server, in this case called *Thin*, is what loads the Ruby app. It is then the Thin server that handles requests and dispatches these to the appropriate piece of Ruby code in the Ruby Web app. Think of this like a dialogue in a play:

- *Action: HTTP request received by Thin.*
- *Thin: "Whoa, judging by the URL, this looks like it belongs to a Ruby app called 'DukeMe.' Better go get that sucker and give it the HTTP contents. Hey DukeMe, wakeup, time to do your stuff.:*
- *Action: Thin uses Rack to go talk with the Rails framework, DukeMe flavoured.*
- *DukeMe (A Ruby app): "Whoa! Thanks Thin (and you Rack). Yep, I'll take that HTTP request and do something with it. Let me see. Oh yep, that URL needs my 'Zipper' controller to do its thing. I'm gonna wake that sucker up and give it what it needs from the HTTP contents."*
- *Action: DukeMe dispatches the Zipper controller and the outputs propagate back up through the Rails framework, via Rack, out to Thin. Thin will go take care of pushing the content back out over HTTP.*

Middleware – This is optional, and to be installed by the developer using the Heroku app management framework. Rack allows slabs of software to sit between the incoming HTTP requests and the underlying Ruby Web framework, whether that's Rails or one of the others, like Sinatra or Merb. For example, let's say that we wanted to add a special HTTP header field for all requests dispatched to our app, this might best be done via a Rack middleware to act as a filter that changes the requests. Or, let's say we wanted to introduce an OpenID authentication layer to our HTTP requests, we could do this using middleware totally outside of our application. The list of possibilities is probably endless. There are certainly lots of Rack middleware options out there. Rack middleware options are best for implementing stuff that ought to happen at the HTTP level and that really belongs within that general "HTTP/Web infrastructure" domain, not to the specifics of a particular Ruby Web app.

Your App – Finally, we arrive at our app, which is any Ruby Web app that works with Rack. The most popular expectation seems to be that developers will be building and deploying Rails apps. However, the app could equally be a Sinatra app, or Merb. For ultra-fast testing of a Web idea, it's also possible to deploy an app using the Camping "micro-framework."

Then again, you could go ahead and deploy your own Rack app – in other words, just avoid any framework altogether. In effect, you'd be building your own framework, but that's just the kind of think you could do on Heroku and still benefit from all of its cloud-computing power and all of the benefits that we have been reviewing in this section, from the HTTP Proxy down to the Dyno Grid. Your Rack framework would simply run inside a Dyno, as just described.

Heroku Add-Ons

The Heroku platform supports add-ons to allow developers to customize their apps even further. The idea behind add-ons is to allow other third-party infrastructure components to be exposed to apps running on the Heroku platform. The add-ons are provided by third parties on a monthly subscription fee basis. Many of them run on Amazon EC2, thus enabling high-performance connectivity (because Heroku also runs on EC2). The add-ons feature is a classical example of taking a platform model and trying to extend its appeal and reach via a platform business model whereby suppliers can sell services to Heroku developers via the Heroku platform. Let's explore a few of the current add-ons as examples:

1. **Websolr** – Apache Solr is a robust and well-proven search server for content indexing and search for web apps.
2. **Amazon RDS** – Allows a Dyno-based app to connect with the Amazon Relational Data Store service in the Amazon Web Services cloud.
3. **Apigee for Twitter** – This allows an app to access the Twitter REST APIs via OAuth whilst significantly improving the API rate limits.
4. **Bundles** – Allow the developer's app and its associated database to be captured and downloaded for offsite storage.
5. **Cron** – Allows unix cron jobs to be run, which are tasks that run periodically. For example, say an app wanted to poll another API to gain periodic updates to weather information. A cron job would allow these API requests to be scheduled and then run according to the schedule.
6. **CloudMailin** – Another great example of cloud infrastructure service, this add-on takes the pain of handling and processing incoming emails. The service will receive an email at a designated address (assigned to the app) and then pass on the contents via a HTTP POST, within milliseconds.

9

Operator Platform: Network as a Service

```
strategy = case platform
  when "closed" then "die"
  when "open" then "thrive"
  when "walled" then "ossify"
  else "do nothing"
end
```

- Network as a Service (NaaS) is where a telco exposes existing network enablers via APIs, usually associated with the core capabilities of the network. NaaS patterns and strategies are described in this chapter.
- NaaS APIs are typically transactional, such as sending a message, and seldom expose customer-generated content or customer-related content.
- The "customers" for NaaS are developers, not the typical end-users of telco services. Telcos often find it hard to adapt to the language and apparatus of supporting developers as customers.
- Telcos will often struggle to fulfil the mantra of "low-friction" access, which is essential for successful platform adoption.
- The developer community covers a wide spectrum of niche developer segments with varying motivations, all of which are described in detail in this chapter.
- Examples of NaaS include the OneAPI cross-telco initiative, although it is somewhat limited in scope.
- Other examples include the O2 Litmus community, which migrated to BlueVia, Telefonica's global NaaS platform, and has some unique commercial elements.
- Via their #Blue beta service, O2 exposes an message-streaming API, which is one of the more innovative examples of NaaS.

Connected Services: A Guide to the Internet Technologies Shaping the Future of Mobile Services and Operators, First Edition. Paul Golding.
© 2011 John Wiley & Sons, Ltd. Published 2011 by John Wiley & Sons, Ltd.

9.1 Opportunity? Network as a Service

9.1.1 What is Network as a Service (NaaS)?

NaaS is when an operator exposes some of its internal IT capabilities via APIs, allowing these capabilities to be used by external services. It is not a new idea. The earliest example of NaaS was exposing bulk-texting capabilities so that partners could send and receive text messages programmatically. The emergence of premium SMS short codes then played a large part in driving the interest in this capability. Many operators have made a small fortune by exposing the premium-rate SMS service via APIs. The typical go-to-market model has been to work with select partners who aggregate similar capabilities across all networks. Developers and brands then work with the aggregators to gain access to text-related services.

NaaS usually follows the pattern of exposing existing network enablers via APIs, usually associated with the core communications capabilities of the network. It isn't like SaaS, where a particular (vertical) service is offered via the browser for a subscription fee, such as a book-keeping service (e.g. QuickBooks online). Nor is it like PaaS. The idea of building and hosting applications on top of the network, such as with services like Twilio, is seldom supported.

9.1.2 Characteristics of NaaS APIs

Different operators have exposed a variety of capabilities via APIs. Often, these are *existing* capabilities and mostly transactional in nature, such as sending a text, requesting the location of a mobile, checking for prepay credit, and so on. These capabilities are consumed by other services, usually as enablers. For example, a location API can be used to enable location-tracking feature in a wider service, like employee safety or security tracking. Text messaging is used to enable notifications in all kinds of services, from bank transactions to limousine services. The list of notification uses is a long one.

Operator APIs seldom expose customer-generated content. The APIs are not like the rich Twitter APIs that expose lots of user content, more than enough to build a variety of other services. With Web 2.0, we have seen users become more aware of and interested in playing with and getting more value from their data via other services. After all, it's the user's content and who doesn't have an interest in his or her own content?

It has taken operators a long time to understand this, often leaving it too late to benefit from the opportunities available when exposing user content. For example, whereas operator networks have long included address books (built into any GSM phone) and social networks, it was left to Web ventures to exploit the "social connectivity" space, with the likes of Google, Plaxo, Yahoo and Facebook.

NaaS APIs often differ from the API model seen on the Web. For example, they often lack self-serve capabilities, instead requiring complex permissions and set-up procedures. This is changing slowly, with operators learning from the success of Web ventures who enthusiastically welcome developers to their APIs, making it as easy to connect as possible, extolling and following the "low friction" mantra.

Low friction access to APIs is essential. It isn't just a case of allowing self-service to the APIs, although this is an important first step. The mechanics of the API should be as simple as possible to consume by another service. These days, developers prefer the ease of RESTful APIs and data formats like JSON. Operators tend to favour the clunkier technologies of enterprise integration – SOAP/XML. Of course, ultimately the difference isn't that important from a technical and programming point of view, but it plays to the matter of "voice." By adopting a certain "language," operators can all too easily alienate their audience, looking stuffy and out-dated. Given their antagonistic track record of being closed to external (third party) innovation, just speaking the right "developer-friendly" language isn't always enough – operators are still treated somewhat with scepticism. Operator APIs need to be delivered in an environment that is developer friendly from head to toe.

9.1.3 Opportunity?

Exposing capabilities via APIs has become a key trend in Web ventures. It is increasingly unthinkable to leave out APIs for *any* venture born on the Web. The long-term success of this approach has yet to be understood, but the logic is sound. APIs are just another way of giving customers access to your services (and their data), and if that isn't of interest to operators (or any business), then I don't know what is. Of course, monetization and business models are still a challenge. However, the mistake that operators always make is to try and figure out the entire business model upfront. Telcos will often assign the wrong key performance indicators to their developer community efforts. Or, as I have seen often, internal politics associated with the project will dominate the program, causing "must have" features to be left out because of budget and timing compromises that are unrealistic.

It is almost impossible to know what the business model is without a feedback loop in the process. This point is lost on those who haven't yet understood that the job of a Web start-up *in its initial phase* is not to make money. I repeat: it is not to make money. The initial job is to find the business model that has the best chance of sustaining the future of the business. This search for the model is done by trying many different ideas out, as quickly and efficiently as possible via a series of iterations. In other words, the issue of dealing with the unknown (i.e. the business model) is tackled by adopting an approach that might reveal the answer (e.g. agile platforms), not by trying to find the answer itself ahead of time. The relatively low cost of building an initial product (not the cost of building a business) means that the search for the business model can take place in real software and product, not on the whiteboard where, all too often, people with the loudest opinions win.

A Web start-up accepts that much is unknown about the business model. However, rather than try to think of the solution, a Web start-up "builds" its way to the solution by trying things out with real customers, measuring success according to various metrics (which could simply be the number of active users) and then tweaking the underlying implementation to see how this impacts the metrics. As the metrics begin to converge on a useful pattern with a growing user base, the start-up can begin to try different models for charging and then see what works, what doesn't, playing around with the parameters, design and delivery of the service. This approach should be applied to any API strategy. Without exposing various capabilities and trying different ideas out, it is almost impossible to arrive at a useful NaaS strategy.

Given the strategic importance of exposing capabilities via APIs, it is clear that operators should be giving more emphasis to this activity than many currently do. Some operators, like Orange, are well ahead in the game, having released lots of APIs, although they appear to have failed in executing any meaningful refinement. Other operators have yet to take the first step, but O2 and Telefonica have made some interesting moves towards finding the business models (see the case study in the next section).

It is my view that the best tactic is to focus on specific subsets of capabilities and then work hard to expose them and evangelize the APIs as aggressively as possible. A good deal of the challenge is to engage the developers, so let's take a look at this particular challenge.

9.1.4 The "Customers" are Developers, not the Users!

Operators have got this wrong for so long. The audience for APIs is developers, not end-users. The same customer insights and instincts do not apply to developers. Operators have failed to understand the mindset, needs and composition of developer communities. Early attempts at building developer communities were abysmal, often weighed down by punitive processes and unrealistic business models that were unfamiliar and mostly unacceptable to developers. Operators often published a set of APIs and then sat back waiting for good things to happen. Unsurprisingly, nothing much happened, leading operators to draw half-baked conclusions about the relevance and importance of APIs. This diluted approach has led many operators to attempt API strategies more than once, often alienating the developer community even further – "Is this yet another attempt that is going to fade away?"

Just like any other "customer" or "user," the needs of developers should be researched and understood before delivering a product to them, even if it's an API. Operators used to take the opposite approach, asking "What do we want?" and then "How do we get developers to do it?" It's an unsurprising attitude. After all, operators run very successful businesses that generate a lot of revenue from a lot of customers. They did all this without developers. The assumption is that merely giving developers access to the "operator machine" should be enough for them. It is a high-handed approach that comes across exactly like that, as though telcos were doing developers are favour. Developers, just like any other users, don't like to be treated that way. Indeed, if operators were to apply the same "customer care" to users as they do to their developers, the marketing director would be fired within a day. It is interesting that operators have failed to notice the disparity. I guess that developers aren't really seen as customers.

Given the importance of looking after the developer-customer, we ought to ask two things:

1. Who are developers?
2. What do they want?

9.1.5 Who are Developers?

This question might seem to have an obvious answer. If you're not directly familiar with the process of making software, you'd be forgiven for thinking that developers are all those folk who create software. Well, that's true enough, but it's worth digging a little deeper. The process of creating software can be divided into a number of disciplines, one of them being coding, which is the actual job of writing the code, be it in Java, Ruby, Groovy, Scala, or something else (Erlang seems in vogue these days). Most developer programs are aimed at coders. There are other guys out there too, like software architects and product managers, all of whom might also write code (the best ones often can, and do) and be attracted to your wares. It's important to engage these folk and all those involved in the wider process of producing software products using your NaaS offering.

Historically, most operator developer programs were aimed at coders. The language and design of the developer program was mostly coder-centric. I remember attending some of the earliest developer forum sessions held in the UK by operators like O2 and Vodafone. I got involved with programs in the US and other parts of Europe too. At the time, I was also a member of the Microsoft developer program. In fact, it was called a partner program and, unsurprisingly, was much more about partnership. Microsoft strived to create win-win relationships at the technical and commercial level, much more than simply enabling coders to code. I pointed this out many times to operators back then, but it was somewhat lost in the noise of indifference. It still is in many instances, with developer programs often run by guys who have never touched a piece of software in their lives, nor worked in a software business. They think that appreciating and understanding something is the same thing.

So, developers are often coders – the guys writing the code – but an effective "developer" program should be aimed at more than just coders. It should be aimed at the businesses employing the coders, or, better still, aimed at supporting *the business of producing software*, not just the process of making software do stuff (like send a text message). This assumes that the operator is serious about the NaaS opportunity and wants to push it to its limits in terms of developing revenues, innovation and new opportunities for commercial success, as yet unknown.

Digging deeper yet, it's also important to understand that not all developers are the same. The "target market" (i.e. developers in your program) can be segmented, just like any other market. I have yet to see a convincing segmentation, but operators like Telefonica (led by O2 UK) have attempted to probe into this issue by the commissioning of large scale developer surveys in conjunction with leading analysts VisionMobile. I was involved in various aspects of the survey, so I am somewhat familiar with its intent and results, but you can read the entire results online.

What surveys like this reveal – although it has never been explicitly discussed – is that developers can be grouped by what they want to achieve, rather than a more conventional grouping, such as size of business. Sure, that grouping is very often useful. It is sometimes useful to differentiate between the so-called "long tail" of developers who are mostly "bedroom/garage developers" (one-man shops) trying to build an app, and the larger SME players who are already in the business of software and looking for mobile enablers via the NaaS platform, as shown in Figure 9.1.

The figure indicates a few key points:

1. The spectrum of developers is wide, from the one-man shop to the largest Fortune-50 enterprises, contrary to the common misconception amongst some senior telco stakeholders that most developers are bedroom geeks with unrealistic views of the "real (commercial) world." The great news is that open APIs are open to the entire spectrum, which can lead to a number of different types of opportunity at no extra cost to the NaaS provider.
2. Innovation tends to increase towards the smaller end of the spectrum with the smaller guys often producing amazing stuff just because they can (and want to show others). This is why it's important not to ignore this end of the spectrum.
3. Scale obviously increases towards the bigger end of the spectrum, but this can also provide a kind of funnel to take innovative ideas from the smaller end and then upscale them to the higher end.

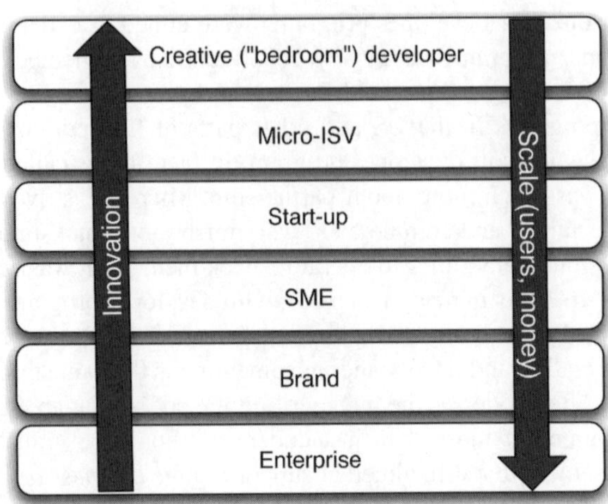

Figure 9.1 Spectrum of developers who might be interested in using NaaS.

Looking at the segmentation by what developers want, we have to ask the question "what do developers want?" This is the topic of the next section where hopefully the motivations for writing software will reveal the different ways of working with developers and the different opportunities. For example, if you're looking at the use of NaaS as a means to drive innovation, then you might take a different approach than if you were trying to generate significant revenue through the monetization of your NaaS platform. Let's explore further what developers want.

9.1.6 Ingredients for NaaS Success – What do Developers Want?

Having said that operators need to talk the language of developers and understand the different motivations for producing software, we should ask what that language is and what the motivations might be. As I elaborated in a talk I gave at the world's largest gathering of operators – Mobile World Congress 2010 – developers want power! That doesn't mean world-domination power, although that's certainly the aim of many an entrepreneurial developer. There are four types of power to consider when trying to empower developers, as shown in Figure 9.2:

9.1.6.1 Cool Power

Before going into more tangible details, I should probably qualify the use of the word "cool." Those not in the field of software will often fail to appreciate that it is not solely a commercial industry dedicated to churning out software products that make money. There is plenty of that going on and most developers want in on the commercial action. However, software is just like clay or paint – it's easy to pick it up and start creating something with it. This is a point that's missed by those who've never had the privilege and pleasure of writing code.

The low cost of entry into coding and the malleability of code gives us a world in which the software industry is also a *creative industry,* with lots of developers producing code purely for the sake of creating something playful, interesting and elegant, in the same way that

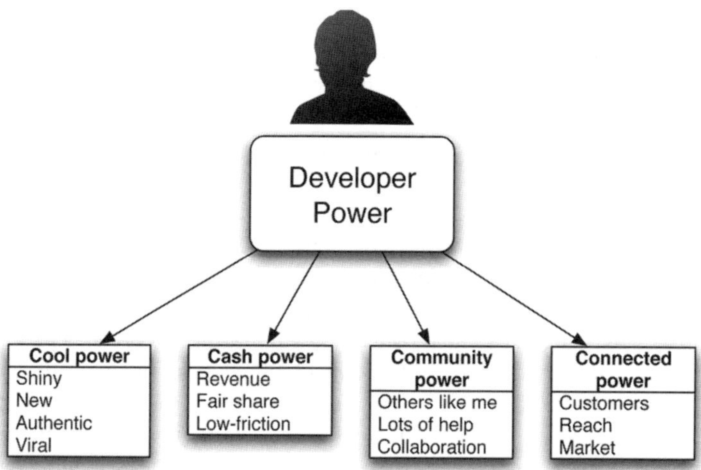

Figure 9.2 The power that developers seek from telcos.

artists engage with their artistic inclinations and talents. They might be creating software for themselves, for their own edification, or for others – the glory of producing cool software that is admired and re-used by others is a major motivation for a lot of developers. When a coder is just building stuff to be creative, that's when they're most interested in using cool shiny tech and showing it off to their friends (being cool).

Shiny – Given this insight about creativity, it is easy to see why the availability of shiny new technology attracts developers. Just like any other creative field, it has its explorers and pioneers who like to pick up new stuff and try it out. The field of software technology is no different from any other field. It is subject to fashion and trends. And where there are trends, there are trendsetters. And where there are trendsetters, there are lots and lots of sheep. Shiny new technology can – and does – have a major impact on attracting developers. It is no surprise that lots of developers flocked to the iPhone and Android platforms with their oodles of shiny tech. Many commentators and observers forget that playing with iPhone and Android software was irresistible long before the app stores appeared with their promise of a new gold rush. Now we think it's all about the money. It isn't. Cool and shiny matters. But there are other dimensions to coolness.

Operators will struggle to offer shiny stuff. They usually fall back to exposing existing enablers, which, by definition, aren't new and interesting. They will also prioritize exposure of capabilities that are obviously or immediately monetizable, or so it would seem. This seldom means shiny in the sense we're exploring here. Shiny to the moneymakers, yes, but we'll come to cash power in a minute.

In my view, having worked with so many operators, the best way to offer shiny stuff is to go get some shiny stuff, which often means investing in a new capability, technology or partnership, especially with a "shiny" company. Admittedly, this is difficult without a solid first-order business case that directly makes money. In the Internet world, there are so many players building stuff that's shiny and where the business case is second-order, such as simply trying to attract a user base that can eventually, it is assumed, be monetized in some way.

Of course, that's often not the case, so it is easy to sympathize with operators' lack of appetite for taking a similar approach. But it's not as if they can't afford some investment in Web platform development. However, the problem here is that even where an operator thinks that this might be attractive, they will totally fail to figure out the next steps in building a platform strategy. This is always because the right people to do so don't work in operators. They work in Web ventures.

Authenticity – Operators often recruit from within, using existing marketing and technical folk to design and deliver the developer community program. This is a mistake. One thing that developers can spot very quickly is whether or not the guys in the suits know what they're talking about. It's no use simply adopting the clothes of developers, such as ditching the suits and using Twitter to send out the same old marketing messages, just a lot pithier. Software is often fashionable, yes, but coders are attracted to the real deal, not pretence.

Openness – Whether operators like it or not, much of the developer world, especially in the age of the Web, is flooded with open this and open that. I'm not just thinking of open software here, which still makes many operators nervous and is an obvious manifestation of an open system. One of the greatest markers of the post-Web software age is the emergence of sites like GitHub, which goes by the tagline of "Social Coding." It is a website where coders can store their code online, which they do using a revision control system called Git (hence the name). The fascinating achievement of GitHub is how many coders post their code and make it publicly accessible – open – free for anyone to download and use, often creating what's called a "fork" of the code. A fork is the coder's own variation of the original code, but then uploaded back into the GitHub repository for others to use.

Code sharing takes place in so many places online, but don't mistake the sharing as simply a kind of hippy share-fest driven by love for one's fellow coders. There is an element of that, as found in threads of the anti-establishment bias of some early Internet pioneers, but it just makes sense to share code because you get back what you put in and much of the code that gets written isn't a trade secret anyway. This is a hang-up that many traditional corporate folk have about sharing anything. We are trained by the culture of "commercial proprietary secrets" and so on to think that any sharing is bad, that all knowledge is power. This is the walled garden, or castle moat, reflex. It's an attitude that has its place, but simply isn't how much of the modern world of software works on the Web. This disconnect – and dissonance – causes lots of potential problems for large corporations who want to get in on some of the developer action, but without the spirit of giving to get. It's perhaps better, in the world of developer programs, to assume that everything is open unless we can find a solid reason for it to be closed. That's anathema to usual corporate practice, which is quite the opposite: everything is closed unless we have a good reason to make it open (and only then under a number of restrictions).

9.1.6.2 Cash Power

The iPhone app store has clearly demonstrated that money attracts developers. Lots of developers turned their hands to creating iPhone apps in order to chase money in the app store: it's an exciting and enticing marketplace. At first, we heard some of the early success stories of developers striking gold, even with relatively inane and simple apps that produced bodily noises or flooded the screen with a single colour to convert an iPhone into a torch. Such applications were able to succeed in a new market where there was a heady combination of a large number of customers looking to buy cool new stuff in an uncrowded marketplace.

It sent developers into a frenzy, trying to think of which apps ideas were yet to be exploited and then rushing to fill any gaps in the store, raking in the profits.

The app store worked for a number of reasons. Firstly, it already had "customers," in that the store was not a new place, but simply an extension to the existing iTunes market. Secondly, and I think the most important lesson of all, the process of consuming apps was as low-friction as it gets, starting with discovery. The ability to simply hit a button (Apps icon on the iPhone) and immediately discover a catalogue of apps was, believe it or not, a new experience. Its success was shear ease of use. Equally low-friction was the buying process. Of course, thanks to the accompanying all-in data package, consumers had no fear of paying additional (and unknown) fees to download the apps, thus achieving a high level of pricing transparency. The low-friction process combined with low price points turned the phenomenon of buying and installing apps, of any kind, into a non-considered purchase for the first time ever. People began to talk of "ring-tone" apps, which meant the notion of "disposable" apps with a short lifespan – buy them, use them and then "throw" them away (i.e. delete from iPhone).

The lesson learned from the app store is that developers want to make money. It wasn't a new lesson. Desktop developers had long been used to the idea of making money from their apps, but the responsibility for doing so was entirely theirs. In the new world of app stores, the responsibility of selling the apps is handed over to the app store owner. They take the money and then share the revenue with the developers, such as the 70:30 split in the iTunes store, in favour of the developer. Of course, operators had been sharing revenue long before this, such as premium text message revenues. However, the revenue split was usually heavily weighted in favour of the operator and then the aggregator, with the developer picking up the smallest share.

Returning to the Telefonica developer survey, researched by VisionMobile, it firmly attests that developers want to make money. The research is based on a set of benchmarks and a survey across over 400 developers worldwide, divided into eight major platforms: iOS (iPhone), Android, Symbian, BlackBerry, Java ME, Windows Phone, Flash Lite, and mobile Web. The research probed deeply into the question as to how developers choose between these platforms. The results clearly showed that market penetration and revenue potential are the two leading reasons for choosing a platform. This shows that developers, while attracted to shiny cool stuff, are heavily guided by the economics of how they spend their time.

As Figure 9.3 shows, market penetration was chosen by a whopping 75 per cent of those surveyed. Revenue potential was the second most important motivator, cited by over half of respondents. What the figure also reveals is how these two factors are more significant to developers than any single technical reason for selecting a platform.

In his four-part blog article discussing the research findings, Andreas Constantinou, VisionMobile's director, says:

> The preference of marketing over technical reasons signifies a turn in the developer mindset. Developers no longer see programming fun as a sufficient reward in itself, but consider monetization opportunities as a primary priority. It seems that, mobile developers now have a sense of commercial pragmatism.

This is certainly true. Before the iPhone, it was very difficult to make any money from developing mobile applications, by which I mean apps that run on the devices. Developers were making money from mobile, but mostly in areas like mobile marketing and text-messaging aggregation that involved no development of device apps. However, I think we need to be a

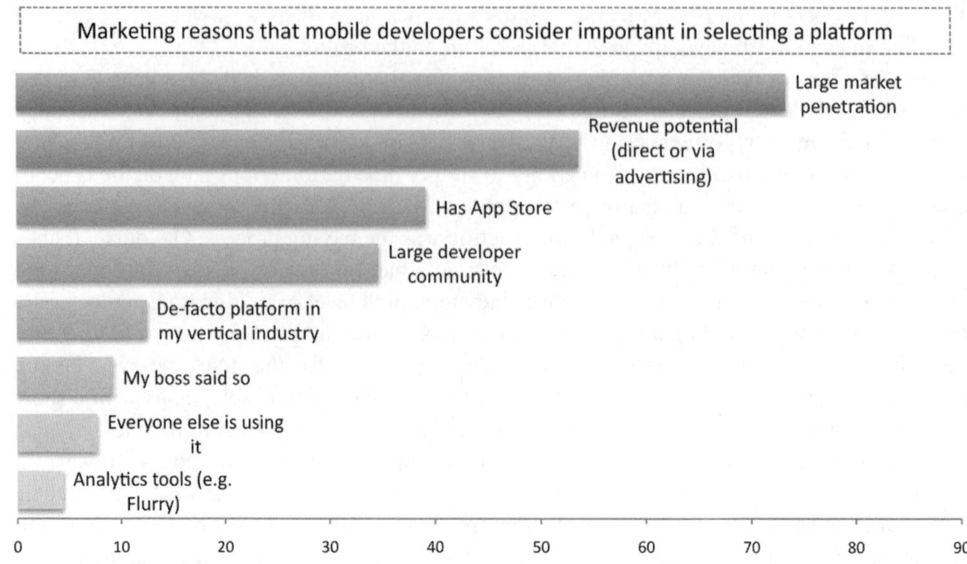

Figure 9.3 Reasons why developers choose a platform. Courtesy VisionMobile.

little cautious about placing all the emphasis on the "cash power" component of the four we are exploring in this section. There are a couple of caveats to consider.

If it were really *only* about making money from software, as the 75 per cent in the diagram seems to indicate, then most developers would be better off developing for other platforms, like the desktop and the Web. Most mobile developers won't make much money from the app store and it has been demonstrated on various blogs that the economics of mobile app development aren't very attractive for most developers in terms of average return on investment in a business endeavour (e.g. versus other software activities, or even versus working in a bar). Nonetheless, there is a growing interest in mobile apps and this is undoubtedly because mobile has grabbed our attention ever since the iPhone. This had more to do with possibility than profitability. The notion of just doing cool stuff because they can is still an important part of the overall motivation. Therefore, I caution our friends in the operator communities not to focus solely on the money issue. You still need to do, encourage or enable "cool" stuff.

Also, operators shouldn't focus their NaaS strategies only on mobile apps. Without doubt, most developers are still playing in the Web sandpit, building awesome services and apps with the greatest potential to scale. The Web is still a much lower-friction platform than mobile. Unsurprisingly, operators frequently miss this opportunity. However, I would hope that this book has given plenty of food for thought about the scope and power of the Web as *the* platform for connected services.

9.1.6.3 Connected Power

As Figure 9.3 revealed, developers are influenced by market penetration. Developers have been knocking on (mostly closed) operator doors for a long time and almost always for the same reason – access to customers! Developers want their apps to get used, whether for money or not. The large numbers of customers on operators' networks is enticing and still the number one asset that an operator has, along with their relationship with these customers. This is the kind of "connected power" that developers would like to access.

However, I would say that most developers unfamiliar with operators have an unrealistic view of exactly what operators can do for them in terms of market reach for their wares. The fundamental blocker here is that most operators simply don't have a channel for selling mobile apps, or any type of apps. Visit any operator website today and you'll struggle to find much evidence that mobile apps matter to operators. Where mobile apps do appear is on the mobile portals ("deck") that most operators offer to their customers.

However, the process for deck placement is usually that the operator selects applications from partners and then promotes these on the deck. This process is usually a fairly heavyweight commercial one driven by the marketing department. There will typically be a very small team looking for partners and promoting their apps on the deck. Naturally, the focus is on low-hanging fruit with predictable revenues and measurable customer satisfaction based on existing customer insights. In such an environment, it's almost impossible for most developers to stand a chance of getting their app selected for the deck, especially if it's innovative. This can be very frustrating for developers who simply don't appreciate the realities of the machine they are dealing with. I have seen so many times how a mobile apps start-up will struggle to get any traction with an operator. I have been in that position many times before. I've also been on the other side (which is why I often give start-ups a chance).

In the post-iPhone world, operators are realizing that the solution to the problem is to build their own app stores. However, this isn't so easy in practice. An app store is not merely a marketplace for selling apps. As we explored in Chapter 6 about device platforms, an app store is usually intimately wedded to a platform, like iTunes is to iPhone and Android Marketplace is to Android.

A point to ponder here is how the VisionMobile survey has shown unequivocally that iPhone (iOS) and Android are the number one platforms of interest to developers. This is curious when we consider that in any market, these two platforms are nowhere near the number one in terms of device penetration. In fact, Mobile Java (J2ME) and Symbian would be number one, which is a fact not unnoticed by GetJar, an independent app store (mostly for Java apps[1]). However, the facts are that most device owners don't download apps, except on the two platforms iOS and Android. A report by Ovum shows that iPhone has 69 per cent of all downloads while Symbian has 9 per cent of all downloads. The report also reveals that over half of all downloads in 2009 originated from the US, where a high penetration of iOS/Android device users accounts for the trend.

If we look at the meaning of these stats for iOS and Android downloads dominance, we are led to the insight, as pointed out by VisionMobile, that it's the addressable market that matters, not just the installed base. Of course, addressable market has always been the driving concern for any serious marketer and entrepreneur, showing again that mobile developers are getting real about the money game. Returning to our concern for "Connected Power," we might conclude that an operator doesn't have so much to offer after all in terms of access to the market, given that the addressable part is already being addressed elsewhere, namely in

[1] The Jar is GetJar originally refers to the technical name for a Java download, called a Java ARchive file, or JAR file.

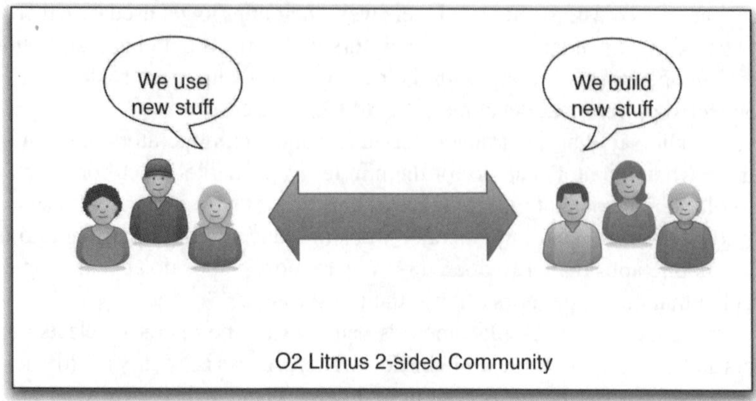

Figure 9.4 O2 Litmus 2-sided community.

the iTunes and Google stores. It is difficult to see how an operator would succeed with an app store play without any leverage over the underlying technical platforms.

The solution here, as operators see it, is to get into the platform game, and the name of that game is widgets, or Web apps, let's say. It has some kind of logic, given that operators do have a good deal of success with their mobile portals, which are entirely Web based (except for various, relatively minor and unsuccessful "on-device portal" attempts[2]). However, the issue of widgets as a platform is complicated. Moreover, the most unlikely part of the whole idea is that operators can collaborate to do anything in the platform space. They tried this with Mobile Java. That's exactly what the Mobile Internet Device Profile (MIDP) was supposed to be about. In J2ME, a Profile is simply an agreed subset and manifestation of the underlying technology. However, the idea failed for a number of reasons, too many to discuss here. The point is that there is little or no evidence that any similar attempt could possibly succeed. Nonetheless, operators are pushing ahead with the Wholesale Applications Community (WAC) idea to develop a common platform standard based on Web-standards widgets, a technology that I explored in Section 6.5 The Mobile Web Platform.

Whilst we are waiting for operators to make what they can of WAC, we can still ask what they can do to help developers reach the market in some way. As I commented in my 2010 presentation to the Mobile World Congress, I think that O2 Litmus can point the way here. What I liked most about the O2 Litmus program in its original inception was how it managed to recruit a number of O2's customers into the mix, people who expressed a willingness and enthusiasm to try new stuff, as shown in Figure 9.4. That should be music to the ears of many app developers because it is often so hard – and so crucial – to get early feedback from real users.

Developers these days like to follow agile methods across the board, which includes putting apps as quickly as possible into the hands of users and then driving the app in directions suggested by real user feedback. I will return to this topic fully in the final chapter of the book when I look at start-up techniques. Gaining access to these "early adopter" users is a

[2] See my book *Next Generation Wireless Applications*, pages 426–429, for a discussion of on-device portal technologies.

potentially useful asset to developers. I used it myself to good effect with one of O2's own projects that I led as a consultant, called #Blue ("Hash Blue").

The #Blue service enabled users to access online their text messaging streams to and from their device. It was a major refresh of the previous "Phone backup" service called Bluebook, which aimed to provide users with a means to back up all the content on their mobiles, starting with contacts and messages. Bluebook had been defined, designed and delivered using the classic operator approach to all IT projects, namely the requirements-heavy waterfall approach, using fairly traditional client-vendor model. The net result was a product that took a long time to release and was unable to respond to any unexpected user behaviour. For example, it was found that a substantial number of users returned regularly to the site to access their text streams, even though the "requirements insight" had suggested that users would visit very infrequently, in the same way that an IT administrator would be seldom expected (hopefully) to access an offline archive tape to retrieve backed-up data.

What I had wanted to explore was using a much more agile "Lean start-up" methodology that could react to user feedback and measured real-time insights (i.e. via analytics) in order to evolve the site rapidly towards something other than what it was originally designed for. We had thought, for example, that adding a feature to allow message replies (oddly missing from Bluebook[3]) might provide an even more compelling reason to stay on the site. Indeed, we did find suggestions in the analytics that users who had the feature enabled (we used A/B testing to give the feature to only half the users) spent longer on the site.

The point is that we were able quickly to explore different ideas because we "closed beta" launched the service via the O2 Litmus early-adopter community. We got enough users in just one day to get some useful early feedback. That same community is available to any of the developers who work with O2 Litmus. It's one of its greatest features and a testament to the ambition of James Parton, its founder, who decided that the best job that the operator could do was to "get out of the way" of letting developers get on with the business of software.

9.1.6.4 Community Power

Developers very often want to be part of a community. Ever since the Internet, developers have found ways to collaborate with each other, whether out of passion for their craft, for the sake of advancing knowledge and software generally, or for commercial gain. Developer communities have been around for a long time in the form of bulletin boards and IRC chat channels and so forth – developers are often extremely helpful to each other, offering free advice on coding issues and so forth, as seen on popular sites like Stack Overflow, which is a collaboratively edited questions and answers site for developers. In recent times, we have seen an even greater trend for open sharing, often in the form of open source code via repositories like GitHub, SourceForge, LaunchPad and Google Code Projects.

Sites like Stack Overflow are interesting because they reveal a lot of insights as to how developers think. In fact, the insights apply equally to any public space where volunteers might donate their time, ideas and energy. There are typically three reasons why someone offers to assist in a community: love, glory and money. Let's briefly explore all three, because they have important lessons for operators trying to deliver "Community Power" to developers.

[3] Though not that odd if the logic of back-up was slavishly followed.

Love – This is simply doing something for the love of it. There are plenty of these people in all community projects, not just software. They just love what they do and want to help others to feel the love, so to speak. Coders who love to code have no problem with sharing the love. They'll gladly help out with solving problems, offering advice to newbies and so on. In fact, the level of contribution can be quite astounding. It is not uncommon to find voluntary project spaces where some of the volunteers provide vast amounts of work for free, sometimes prepared to tackle very complex problems that require huge sacrifice. This has been demonstrated in some of the recent crowd-sourcing projects, like Innocentive, where they have found scientists who will practically produce a PhD thesis worth of work to solving a challenging problem – they simply love the challenge.[4]

Operators can embrace this approach by trying to bring problem setters together with problem solvers. Doing this between developers is fairly standard. Where I think it gets interesting, and as yet largely untapped, is where problems jump from customer to developer. In any given territory, with its usual low single digit number of operators, any given operator is going to have a large and representative slice of the market. Within that slice are going to be people working in businesses, charities, science labs, and all manner of projects where they are likely to encounter complex problems that need a solution in the shape of an app. Imagine how many coders out there might be willing to provide a solution, just for the love of it, if only they knew about the problem. I think that the opportunity here for operators is obvious and very compelling.

Glory – This is the chance to be admired by others. Let's face it, we all like a bit of admiration for our work and for our smarts. This can be utilized in community sites by recognizing the contribution of others in proportion to their contribution, such as the awarding of stars, points and badges, which is the method used by Stack Overflow. Users can vote on the contributions of others, causing points to be awarded. The whole system works via the "opinion" of the community, following heavily the so-called Theory of Moderation. They have bronze, silver and gold badges, the latter of which require a serious amount of graft and contribution to accomplish. Interestingly, the badges come in all types of flavour, such as:

- *Altruist (Bronze)* – First bounty you manually awarded on another person's question.
- *Disciplined (Bronze)* – Deleted own post with a score of 3 or higher.
- *Good Question (Silver)* – Question scored 25 or more.
- *Guru (Silver)* – Accepted answer and score of 40 or more.
- *Great Question (Gold)* – Question score of 100 or more.
- *Legendary (Gold)* – Hit the daily reputation cap on 150 days.

With badges like that last one, it's easy to see how a developer can really shine amongst his or her peers in a site like Stack OverFlow. This "glory" factor has proven a strong motivator for attracting persistent talent in many types of online communities. Developers are no exception and operators should think about how to exploit the glory factor in a way that is unique and authentic to operator scenarios.

Indeed, an example of exploiting the glory factor in online communities comes in the form of the pioneering "self-help" online community run by the SIM-only MVNO GiffGaff, which

[4] Although I should be clear that Innocentive usually offers a reward to the problem solver.

has the tagline "The mobile network run by you." There is no help line to call GiffGaff for help. If a user wants help, there's only one place to get it, which is the online community of affectionally named GiffGaffers. Here, the members – GiffGaffers – help each other and award each other Kudos points. It's an idea that seems to be working, albeit on a relatively small scale and with a rather limited subset of issues compared with the wide range of problems that developers could pose across the myriad software technologies, platforms, languages and APIs. Nonetheless, the glory factor clearly works and it's a wonder that operators haven't cottoned onto it before. It's a massive opportunity, waiting to be exploited.

Money – Hardly surprising, but some people participate in communities just because they want some money or equivalent reward. Sure, glory points and karma stars are good, but money is often better. We're not talking here about the allure of "Cash Power" where developers want money for their apps. Here we're talking about making money through participation in an online community. This can take place in a number of ways, from paid writing of long and useful blog articles to giving away cash rewards for reaching certain numbers of points etc. The options are endless and well established in a number of successful online communities.

9.2 Examples of NaaS Connected Services

9.2.1 NaaS Case Study – O2 Litmus

Let's take a look at how one operator – O2 – has approached the NaaS opportunity thus far, although the story is far from over. I will then speculate as to how it might have been done differently, or how it might be extended in the future. These are purely my own thoughts, not those of O2.

The essential idea of O2 Litmus is contained in its name – "Litmus." The concept was to open the doors to innovation and allow other communities, especially developers, to create new ideas, applications and services and then test them out – the litmus test – on willing early-adopter customers. This combining of two communities was very much the genius of O2 Litmus.

When developers create products, they are eager to try it out with real users and get feedback quickly. The O2 Litmus community provides such a vehicle. The advantage to O2 is that the innovation happens on their network in every sense:

1. Developers use the O2 Litmus APIs (e.g. Location API) – "NaaS network."
2. Developers use the O2 customer base – "customer network."
3. The apps run on the actual physical network – "data network."
4. Apps are also exposed to fellow developers – "developer network."

By letting the innovation take place on its network, or platform, O2 has intimate access to the innovation in order to observe which ideas resonate with customers, keeping in mind that a company like O2 is heavily influenced by customer insights. The two-sided community approach allows new insights to be revealed as they unfold, which is often a necessary ingredient for innovation with an operator environment where the notion of offering customers something that has not been thoroughly proven is quite alien. But then this is the dichotomy of innovation, as told well by the familiar Henry Ford quote: "Had I asked them [customers] what they wanted, they would have said faster horse-drawn carriages."

It is remarkable that this point continues to be lost on operators even though one of their most successful services ever – text messaging – had **zero** demand from customers initially, and **zero** customer insights.[5] The folk behind O2 Litmus, like James Parton, certainly get this point. It is the main idea behind O2 Litmus – open innovation. As suggested by James on many occasions, "We [O2] don't want to dictate the service ideas to the developers, we just want to give them the raw ingredients [APIs]."

In this way, O2 Litmus is pushing hard at the "Community Power" element for NaaS success. What have been missing are the other ingredients, which I will explore now in terms of what's missing and what could be done about it. Again, these are my own observations and suggestions, unrelated to O2.

Connected Power and the Market Disconnect – This continues to be a problem with O2 Litmus because there is no easy way to move an app from the O2 Litmus community to the main O2 customer base. Applications that pass the litmus test should progress automatically to the market. In essence, this has to mean some kind of connection with an app store or operator portal. The question is how such stores align with other stores already winning in the "off deck" world, such as iTunes and Android Market. The stated aim of the industry is to create a "combined" app store using the WAC standards, but let's see.

Meanwhile, as I have pointed out on many occasions to various operators, and as hopefully this book clearly reveals, the world is much bigger than just mobile apps, which has been the maniacal focus of mobile operators for years, naturally so. Web applications are important. The Web is rapidly evolving to become an "operating system" of sorts and, with platforms like NaaS and the Tropo-type "cloud telephony" platforms, we need to start thinking of "connected services," no longer Web-only services, no longer mobile-only services.

This is a difficult transition for operators, as they tend only to understand and think about mobile apps or mobile access to apps. However, with various NaaS APIs, it is perfectly reasonable – and in fact desirable – to expect network services to be consumed by applications that run on servers in the Web. For example, if an operator can offer some kind of "Call conferencing API" say, then no reason why this shouldn't get embedded into Web-based CRM apps, project management apps, Wikis and so on. The point I am driving towards is that the route to market needs to include the Web itself, not just "on deck" app stores and folders.

Cash Power and its Absence – O2 Litmus does have a revenue share scheme for any apps that get sold via the O2 Litmus "directory" (apparently, it is not to be called an app store[6]). However, there are no other mechanisms for making money, which is due to an absence of business models wrapping the APIs.

There is no doubt that some APIs could make money for O2. I think that this is likely to be the case in the SME and corporate sectors. For example, the ability to detect call-status presence is considered to be a very valuable feature of any modern PBX or CRM solution, but is usually difficult to achieve with mobile calls without the use of a mobile client application.

---

[5] Of course, operators aren't stupid, so this approach deserves an explanation. However, I defer that discussion to the last chapter.

[6] This further underlines the point of the disconnect between O2 Litmus and the route to market. What is the market here? Is it the O2 Litmus website or the O2 deck? The issue is discoverability –O2 customers aren't ordinarily exposed to the O2 Litmus site via their mobiles.

The challenge with using a mobile app is penetration. Typically, such apps will only work on a limited subset of smartphones, which misses a lot of the wider customer base, and will continue to do so for some years to come. Getting reliable and verifiable call-status data from the network has value.

Community Power and Social Coding – When I was invited as a independent "voice of a mobile developer" to present at the inaugural O2 Litmus meeting I gave a presentation that talked about the idea of "Social Coding." I didn't mean the Github model for sharing open-source code, although that works too. What I meant was the wider enterprise of collaborating in a symbiotic or synergistic fashion, which can include a number of ideas:

1. Code sharing – that is, like Github.
2. People/skill sharing – I have Ruby skills and you love JQuery – can we help each other out with snippets, expertise, advice etc.
3. Job sharing – one of our developers works for you, and vice versa (or spread across a pool of people/companies).
4. API collaboration – we'll support your API and you use ours in order to get greater network effect.
5. Community collaboration – help each other in the forums.
6. Service bundling – let's sell our services for one (lower) price.
7. Shared testing and feedback – there's often no one better than another developer to give you feedback. It can be brutally honest, but softened by a suggestion or two for improvements.
8. There is no 8, but there probably ought to be.

Collaboration between developers is not a new idea. It's very much part of the developer culture that has emerged around the Web ecosystem. However, providing a place for some of it to happen more formally, and around a common theme (mobile) or platform (O2), seems a smart move to me.

9.2.2 Update to O2 Litmus Story – BlueVia

Just as I was about to submit the final draft of this book, the global Telefonica developer program, called BlueVia, replaced O2 Litmus. It is a much more ambitious and comprehensive project that addresses many of the concerns that I've just outlined. For instance, BlueVia includes business models with its APIs so that developers can earn revenues from the use of those APIs. The launch APIs are:

1. **SMS send and receive** – This has the usual features, but a unique and useful twist. When sending texts via the API, it is possible to send them on behalf of a user, with the user paying. This is a welcome step and one that I advocated during the GSMA OneAPI seminars at Mobile World Congress in 2009. Integrating SMS into Web services has always been hampered by the need for a developer to pay for the texts. With the BlueVia model, this disappears. Moreover, the developer can get a share of the text messaging revenues.
2. **User Context** – This is one of the first examples of an operator doing what it says it can do – giving access to customer information. The API gives access to customer profile information, like prepay or postpay status and added insights like parental controls, useful

to filter the delivery of services like gambling and adult content. The API also gives access to terminal information, which is useful for tailoring service design and delivery.

3. **Mobile Advertising** – The mobile advertising API allows a developer to insert adverts into a mobile website, gaining revenue share in return. The adverts can be text or image based and are first requested from the API before being inserted into an anchor element in the mobile website.

As you might have realized, sending texts on behalf of a user and giving out user profile information requires permission from the users. This is achieved via the industry standard OAuth mechanism, which has also now been adopted by the OneAPI initiative.

9.2.3 OneAPI – The Interoperable NaaS Play

OneAPI is a set of APIs defined by the GSMA OneAPI group, which has now moved to become part of the Wholesale Applications Community (WAC). The APIs are specified to expose "network capabilities" over HTTP, following Web standards and principles as much as possible. As such, the APIs follow RESTful principles (see Section 2.2 Beneath the Hood of Web 2.0: CRUD, MVC and REST), supporting both SOAP (XML) and JSON data formats. The idea is to standardize on NaaS APIs across the telco industry so that developers only have to develop one API definition for any telco when accessing common services, like sending texts.

The WAC initiative is now positioning OneAPI as a set of server-side APIs to accompany the client-side APIs they are currently defining. The OneAPI has now moved to become part of the WAC, presumably to ensure harmony between the two sets of APIs. After all, some of the client-side APIs might well require server-side NaaS components.

The first commercial implementation of OneAPI can be found in Canada across the three network operators: Bell, Rogers and Telus. Developers can use the common APIs, provided by a single gateway (from Aepona), to access location, payments and messaging services.

If you're expecting more interesting APIs than the usual candidates, then you'll be disappointed. OneAPI version 1.0 only defines the following:

1. Send and receive SMS.
2. Send and receive MMS.
3. Location look-up of one or more network users.
4. Charging API.

Version 2.0 promises a few more APIs, though still following the standard API features that we have seen for years from telcos who have already exposed some NaaS capabilities:

1. Data connection profile (e.g. bearer type).
2. Device capabilities.
3. Click-to-call (or third party calling).

As the OneAPI website claims, the 2.0 APIs are mostly intended to complement the delivery of video services. This seems odd given that operators are becoming increasingly video-phobic because of concerns with data consumption in the new era of capped data allowances. As you can tell, all of the varied and rich API opportunities that we have been discussing throughout this book, such as Social Graph APIs, identity APIs, and so on, are missing from the list.

Proponents of OneAPI will say that this is just a starting point and we have to start somewhere. That's right, but that's the refrain usually given by such efforts, which seldom catch up with the innovation curve, never mind lead it. That said, the availability of common standards for common APIs could only benefit developers, who can simplify software implementations to access common services across telcos.

9.2.4 Hashblue Case Study? – RT# and SMSOwl

Hashblue (actually spelt #Blue, but available at hashblue.com[7]) is an experimental service developed by O2, which I initiated. The impetus of the project was a desire to re-create an existing service called Bluebook,[8] which was a text, picture and contact back-up service. However, whereas Bluebook was mostly developed using an "enterprise IT" mentality with a heavy set of requirements, implemented in waterfall fashion, #Blue was intended to follow "Web patterns" and "Lean start-up" patterns throughout.

Some of the project background is useful, as it serves as an example of innovation, or how to get new stuff done inside an operator, without using the normal processes. We assembled a small team, mostly of external developers from the excellent GoFreeRange crew, who specialize in Ruby on Rails projects. This in itself was a small challenge, having to persuade the procurement team that we wanted to use a specialized shop, rather than the established IT suppliers, none of whom were really set up to work in a slick, agile fashion like a start-up. For example, these traditional suppliers found it hard to initiate the dialogue without a fairly comprehensive requirements spec. However, we only issued a single page spec, which was more a high-level statement of intent – what we were trying to build and why, plus a little bit of how we wanted to operate (i.e. like a start-up).

Once the project was underway, we followed a Scrum-like approach, beginning with our backlog of user stories, which we captured in the Pivotal Tracker tool that seems quite popular with Web developer shops like GoFreeRange, who favoured its use on the project. The most important decision from the outset was to build the API first and then consume it ourselves to build the Web UI, as shown in Figure 9.5. This approach we dubbed "Eat our own dog food."[9] The idea here is that by producing the API first, we get two things:

1. The API itself, which we can then expose to external developers.
2. A good API because we obviously had to debug it thoroughly and design it well enough to support our own service.

---

[7] Hashes aren't allowed in domain names. Also, I don't know if the service will be around when you read this book, as it was only expected to be experimental, not necessarily ever making it to the mainstream O2 portfolio of services. Nonetheless, it serves as a useful example.

[8] The text messaging archive part of this service was planned to be shut down in late 2010.

[9] I know that this phrase can take some people by surprise, especially the thought of eating dog food. If you prefer, why not use the French expression of "Drinking our own champagne."

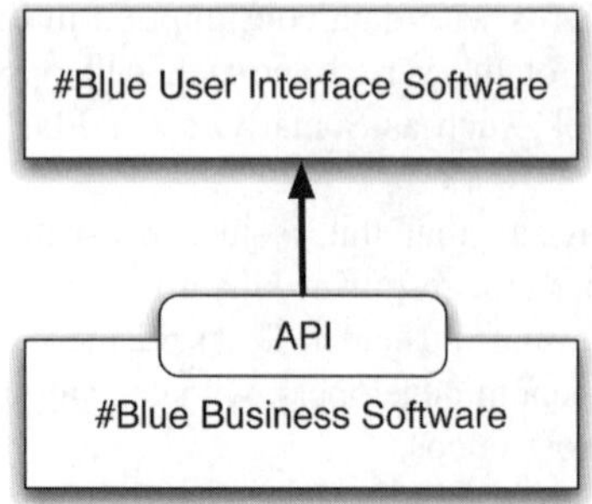

Figure 9.5 #Blue interface built with the API.

From the outset, the API was an important part of the #Blue product strategy, which is why we wanted to expose it early. We wanted developers to use the API to make their own services. We believed that the API would be interesting because it offered a unique capability, which is the chance to access a user's text stream in both directions, no matter what handset they use to send texts, as shown in Figure 9.6.

The possibilities for external apps seemed endless, but I was keen not to suggest what they might be. We wanted to see what developers came up with. We wanted to encourage open innovation on the #Blue platform. Once the API was ready in a very early form, we decided to give access to developers. We exposed the API very early on, even though it was missing some key features that a public API ought to have, such as OAuth authentication.[10] However, this "launch early" strategy proved fruitful in attracting hacks earlier than we might otherwise have hoped for, building credibility for the API and services both externally and internally.

9.2.5 The #Blue Hacks

As soon as the API was available, we promoted it at the WarbleCamp event in London, which was an unofficial Twitter developers hack day (in response to Chirp, the official Twitter developer conference in San Francisco). The event was well attended; such is the popularity of the Twitter APIs. It was no coincidence that we chose a Twitter-developer conference to launch the #Blue API. Firstly, the Twitter APIs are the ultimate APIs in terms of low-friction access and all round success. Secondly, Twitter is a real-time messaging service, very closely associated with texting, so the synergies with #Blue seemed obvious to us. I thought that if we can get some traction with our API here, alongside the Twitter APIs, then we would have succeeded in some small way.

The good news was that a few hackers built apps using the #Blue API. In particular, Adam Burmister (@adamburmister) built a very cool hack called #Owl, which he released at smsowl.com. For me, it was a great example of how to blend the #Blue service with Twitter. Adam's hack monitored a user's text stream for any texts containing the characters "#owl" and then copied those texts out to that user's Twitter stream, as shown in Figure 9.7. It was a blending of person-to-person texting with Twitter, like a "carbon copy to Twitter."

[10] However, it should be noted that Twitter also avoided OAuth for some time, which was clearly a useful step in helping so many developers get started with the API.

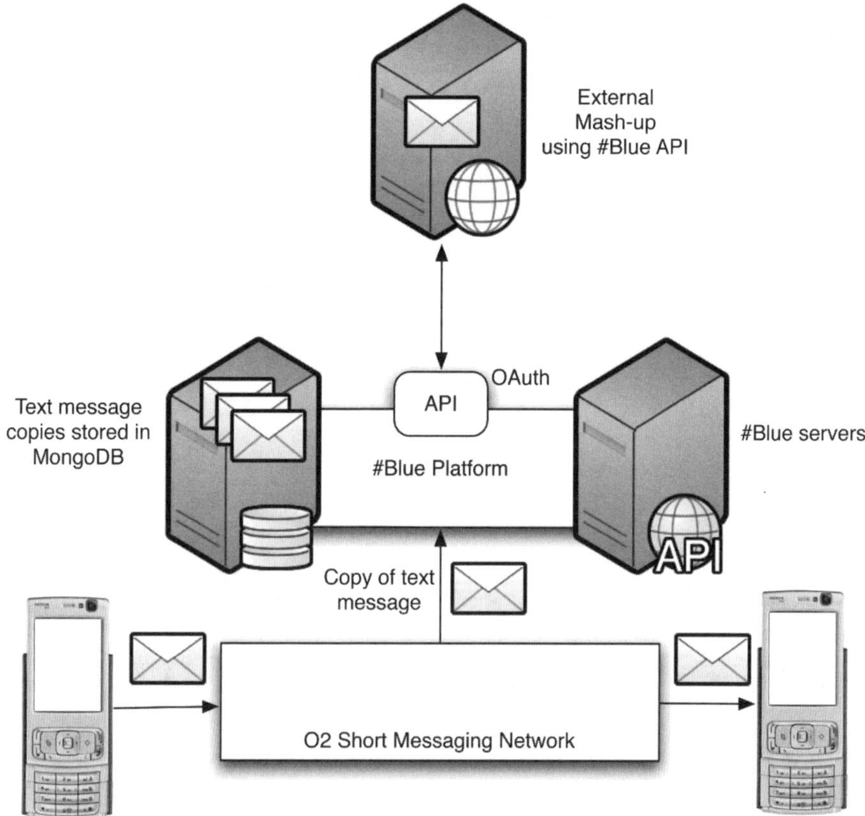

Figure 9.6 #Blue platform and API.

Adam's application proved to be popular with those who saw it. The idea of blending texts with Twitter in this way is uniquely possible because of the way #Blue exposes a capability that only O2 (or an operator) can offer, which is access to the text stream of a user. Adam's idea also started a trend in #Blue hacks. Detecting hash tags and then using these to initiate another action in an external service (e.g. Send a Tweet) proved to be popular.

At the Over-The-Air (OTA) event in London, we saw eight separate hacks built using the #Blue API, even though it was still in its very crude state. This proved that releasing early works. Putting the API in the hands of developers produced results. At the OTA event, we saw the hash tag theme continue. My favourite app was dubbed "Remember the Hash," which sounds slightly alarming. However, the hash in question here is a #todo inserted into the message. The hack pulls messages with this tag and then sends them to "Remember the Milk," which is a popular task-management application in the Getting-Things-Done[11] (GTD) tradition. This sequence is shown in Figure 9.8.

[11] GTD is the title of a popular book by Dave Allen about a task-management methodology – see http://www.davidco.com/.

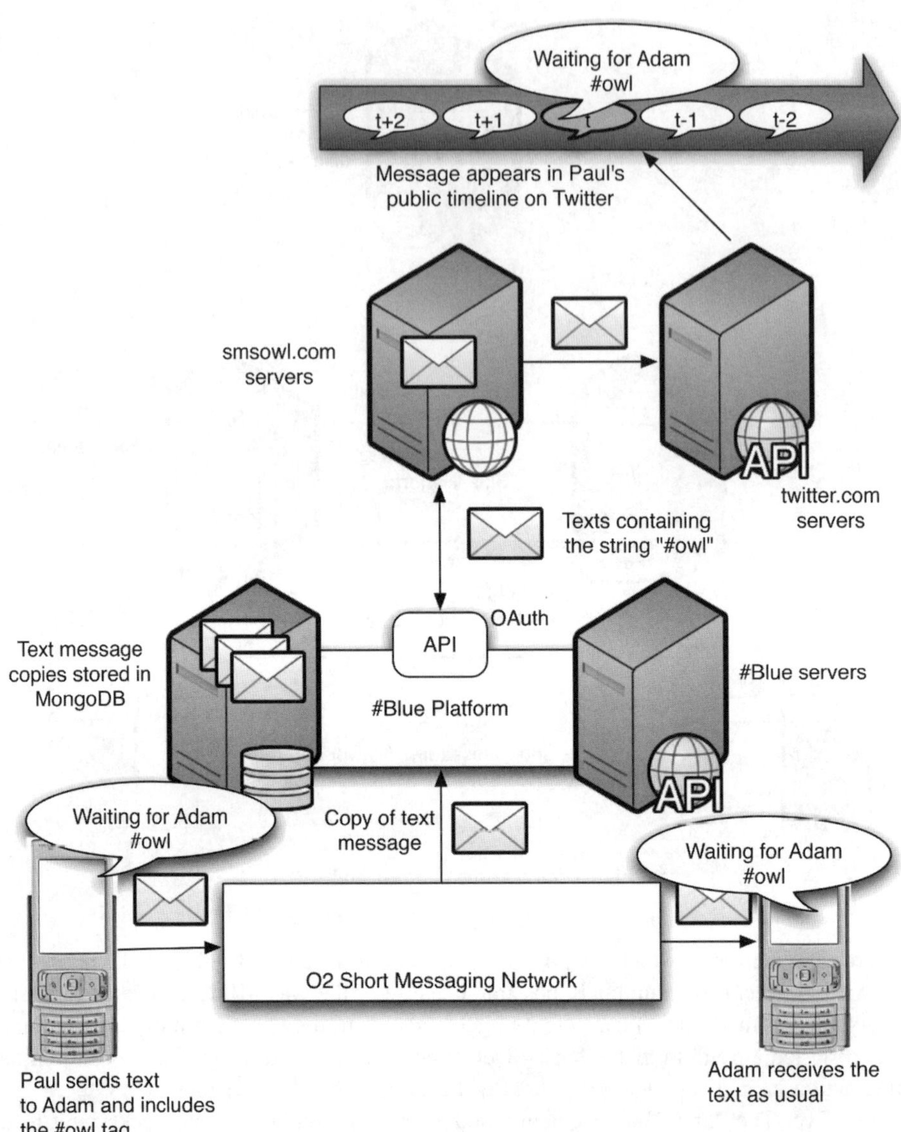

Figure 9.7 Architecture and operation of the #Blue hack called SMS Owl.

9.2.6 The Benefits of #Blue Platform

We promoted the #Blue API at a number of hack-day events where we could support the API and help developers get going. This is a crucial ingredient in API exposure. If you don't turn up to events to promote your APIs, don't expect developers to come flocking. If you don't evangelize your APIs, don't expect success. Operator developer programs would often overlook the importance of promoting technology using real technologists, not marketing

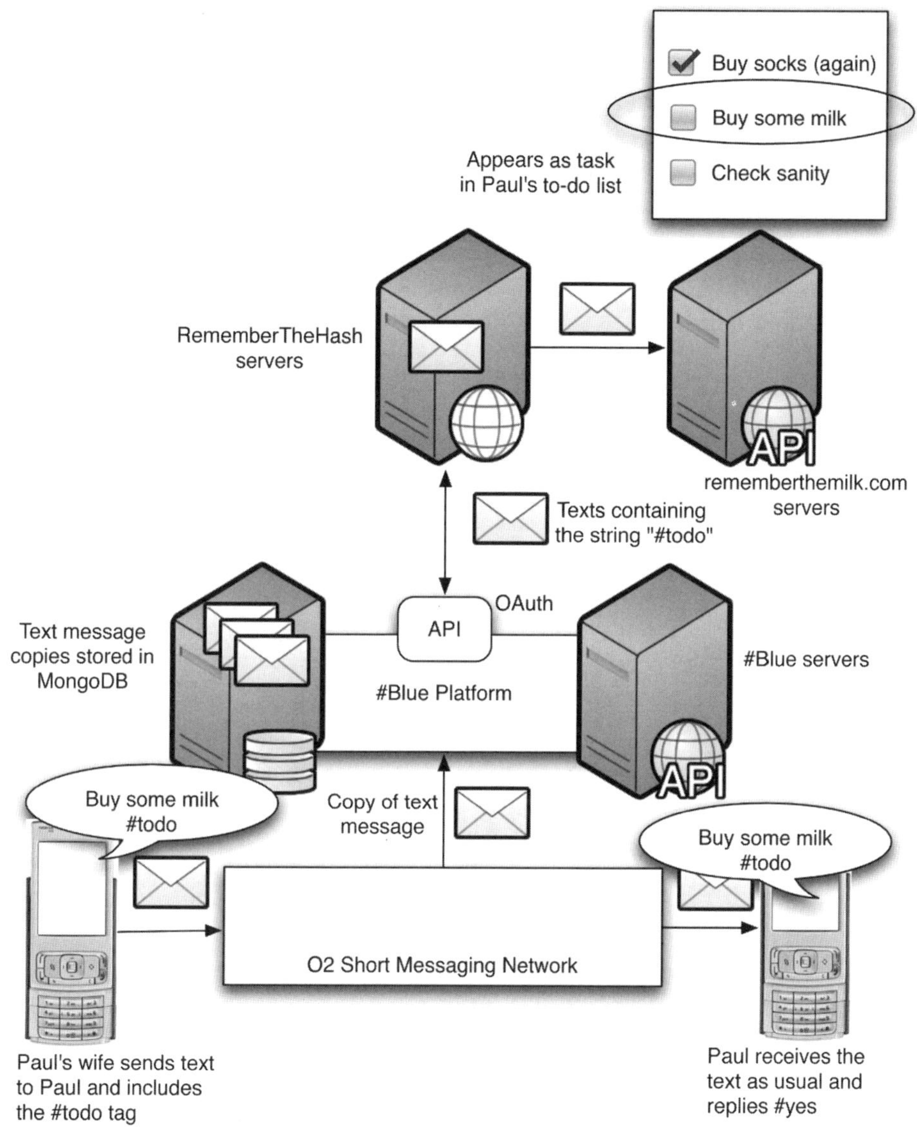

Figure 9.8 Remember the Hash Hack Built on #Blue platform.

managers or "community managers," and other men-in-suit job titles (even when they stopped wearing the suits).

Developers liked the APIs and built apps to their own liking, solving their own problems, or just having fun. We didn't tell them what to develop. We allowed the innovation to flourish. One of the major outcomes and benefits was the emergence of a new category of mobile messaging experience. The simple inclusion of a hash tag in a text can cause all sorts of things to happen on the Web via the #Blue API, whether it's tweeting a text out to the timeline or adding an item to a to-do list. I think the possibilities for this category are quite extensive and a source of new services for O2.

No one in O2 had previously come up with this idea for a service. Even if they did, it would have not survived the "business case" process. Using the platform approach with low-friction APIs that came as part of the development process anyway, we managed to try out new ideas at very low cost. In fact, we didn't try out the ideas at all – developers did! You can search YouTube for a video about the #Blue API.

10

Harnessing Web 2.0 Start-Up Methods for Telcos

```
While cash > 0 and !successful-business-model
   iterate
end
```

- Although highly successful in their own right, telcos can still benefit from understanding how modern Web ventures work, which we explore in detail in this chapter.
- Scalable Web start-ups tend to exhibit a common set of approaches towards exploiting Web 2.0 as a platform for doing business. These approaches span technological, cultural, organizational and commercial concerns, all of which offer important lessons and opportunities for telcos.
- Instead of thinking how to become a "Web venture," a telco should think of how to become a "Web-powered" venture, exploiting the best of Web methods in ways that are unique to telcos.
- Amongst many of the Web-venture attributes described in this chapter, a key one is the use of highly skilled technologists. This is a challenge for telcos whose technical staff, although skilled, are not technologists.
- Many of the uncertainties inherent in Web start-ups are dealt with by an ability to pivot quickly from one idea to another, continually tuning the approach via a set of agile methods and tools.

10.1 Start-Ups and Innovation

This chapter is really about innovation, although I hesitated to use that word anywhere in the title because it has become so lumbered with the baggage of corporate slogan to be made meaningless. I'm not going to dance with definitions of innovation here, nor make the claim that telcos couldn't innovate their way out of a paper bag. Such generalities lead to nowhere useful. So let me be plain about the scope of enquiry, which is to explore the question as to how telcos might benefit from the various techniques and technologies used by Web start-ups

Connected Services: A Guide to the Internet Technologies Shaping the Future of Mobile Services and Operators, First Edition. Paul Golding.
© 2011 John Wiley & Sons, Ltd. Published 2011 by John Wiley & Sons, Ltd.

and successful Web ventures. No one can argue that Web start-ups are invariably innovative, or striving to be so. What I'm interested in here is seeing how that innovation can be applied to telco problems.

In exploring this topic, I'm aiming this chapter at those in the telco industry who have yet to work on a modern Web 2.0 project, product or service. It should also be illuminating for those who have never worked in, or with, a Web-related start-up and who are intrigued by exactly how the "Web start-up" scene works. There is a lot to learn from the Web start-ups machine. And it is a machine, powered by tens of thousands of entrepreneurs, developers and a whole raft of support services, including a good supply of investment money chasing the "next big thing."

To a seasoned stakeholder in a telco, being told to look at the ways scalable Web start-ups tackle business might seem a bit naive. After all, as we all know, telcos are hugely successful and make lots of money. The tier-1 players in particular are exceptionally well qualified in every aspect of running a business. I don't make any claims here that Web start-ups have better techniques and can teach telcos a thing or two about business. However, what they can teach telcos is a thing or two about technology businesses and Web economics, both of which should be of interest to telcos if any of my book thus far has had any impact.

Before I explore how Web start-ups work, let's get one objection out of the way now. Telco folk will often say, like any successful folk, that we've been successful thus far, so why should we change? Why should we listen to what other folks have to say? The answer is simple. We all know that established businesses can fail, despite all their good intentions and smart leaders. There are too many examples of this to argue with. Given that fact, isn't it worth having an open mind to new ideas?

In fact, I'm sure you'll see that most of the ideas are not new at all. Things like "Lean start-up" is really just a play on the older theme of prototyping and skunk-works that many tech companies have deployed before, often with great success. However, some of the ideas will be new to you. I have employed them on real projects inside of telcos with real results.

10.2 What can Telcos Learn from Web 2.0?

In my experience, I have found that there are several ways of looking at Web 2.0 from a telco perspective:

1. New business opportunity
2. Services channel
3. Innovation methodology
4. Web platforms

Let's take a look at these four:

1. **New business opportunity** – Telcos could build or acquire Web 2.0 services as a source of new revenue or engagement with customers. Key stakeholders usually view this with a good deal of trepidation, mostly because they can't make a business case in the traditional sense or they simply don't feel comfortable with Web ventures. There is also a common view that the sorts of revenue opportunities available via the Web aren't attractive when stacked alongside existing revenues, which is just another manifestation of Clayton Christensen's "Innovator's Dillema."

I have also noticed a number of stereotypical reactions and misconceptions of Web ventures, such as the view that they are not real businesses with the same level of customer care and focus as an operator, often leaving customers to care for themselves in the depths of some online support forum labyrinth. Of course, these reactions are typically gross generalizations, the product of lazy thinking. Some Web companies have succeeded precisely because of their maniacal focus on customer service, companies like Zappos, Amazon and Rackspace.

Regardless of operator opinions of Web ventures, most of them have either considered, or sometimes executed, a new business strategy involving a Web venture, although seldom with any success. I struggle to think of any successful Web venture launched by an operator.

2. **Services channel** – Telcos can offer existing services via the Web, which most do already, as in e-commerce and online configuration of services via a customer care portal. However, it has to be said that many operator websites are woefully lacking in modern Web 2.0 features and approach. As I have pointed out to many operators before, the issue here, if nothing else, is relevancy and hygiene. If a website doesn't measure up to a modern Web 2.0 experience, whatever that might be (we will explore later), then it starts to lose relevance and credibility. I will mention some of these hygiene factors in this chapter, plus some new ideas to think about.
3. **Innovation methodology** – As a source of ideas, methods and techniques, the Web 2.0 world has a lot to offer telcos. It is this area that is often overlooked in the various debates that operators have about "utilizing Web 2.0." Blinded by the usual clichés about Web ventures (as mentioned above, such as poor customer support), operators fail to see how it might be possible to improve existing business, or even build new businesses, aided by Web 2.0 methodologies. In my consulting work, I have succeeded in demonstrating how to build new services using the "Web way," which invariably meant using agile processes. However, I will elaborate on Web 2.0 start-up methods in this chapter.
4. **Web Platforms** – I have described the concept of Web platforms in some depth throughout the book. A key question is how can telcos utilize Web platforms? There are two approaches to consider. Either a telco can harness an existing Web platform or it can become a platform, as when we discussed the idea of Section 9.1 Opportunity? Network as a Service. I will explore both approaches in this chapter.

In this chapter, I am mostly intending to explore this last aspect, asking how we can improve telco business generally through the application of Web 2.0 patterns.

10.3 Key Web Start-Up Memes

I'm going to explore some of the key characteristics of scalable Web start-ups, thinking about how they might apply to telcos. What I mean by a *scalable* Web start-up is a start-up with the stated aim of trying to make a significant return on investment, hopefully several orders of magnitude greater. Such a start-up is deliberately pursuing a vector of substantial and dramatic growth, usually exponential growth, if possible. This is the sort of start-up that will pass various stages of funding, perhaps from venture capitalists, in order to achieve growth. The founders are aiming to generate large returns for their stakeholders, way beyond those needed to make a comfortable living for the founders. This is different to so-called "lifestyle" start-ups that are all about the owner/founders seeking a financially independent way of life, such as independent workers, freelancers and sole traders.

Steve Blank describes a scalable start-up:

> A "scalable startup" takes an innovative idea and searches for a scalable and repeatable business model that will turn it into a high growth, profitable company. It does that by entering a large market and taking share away from incumbents or by creating a new market and growing it rapidly.

Clearly, for a start-up to be scalable, this implies that it has some means to scale beyond the direct labour input of its founders and employees. The method for revenue scaling is software that can be accessed by a vast audience – that is, the Web. This is the beauty of the Web. It's always switched on. It can handle lots of users concurrently, whether anyone's in the office, or not.

Below is a list of some of the most important aspects of scalable Web start-ups, followed by an elaboration that includes the applicability or usefulness of these techniques in telco environments.

1. **Use techie people** – This is obvious, but Web start-ups are heavily driven by technologists, who are people who can actually build technology, not just package it. Innovation in Web is invariably technological innovation.
2. **Lean start-up methodologies** – Unlike telcos, Web start-ups are designed from the ground up to deal with the problem that their idea is an unknown quantity. They deal with this challenge by constructing a venture that is as lean as possible, enabling it to move rapidly towards a viable product that resonates with users and, hopefully, supports a viable business model.
3. **Extreme and constant optimization** – Closely related to leanness, Web start-ups use techniques that enable constant optimization of their product, always moving towards better performance against whatever metrics are chosen to measure success.
4. **Co-creation** – Where possible, start-ups will try to utilize the resources of its users in order to build a better product or community.
5. **Exploit big-data** – Increasingly, Web start-ups will exploit various "big data" technologies and approaches to gain a competitive advantage.
6. **Social discovery** – Start-ups will routinely utilize various tools to enable social connection through their venture.
7. **APIs and Developers** – Start-ups think nothing of using data and services from other sites to build their own, and won't hesitate to expose their own data and services via APIs in order to attract developers.
8. **Incubation and Acceleration** – Start-ups can get their ideas to market much quicker than before thanks to the lower technical barriers to entry and now the ecosystems of investment and mentorship that are emerging.
9. **Hack Days, Events and SomethingCool.Conf** – Start-ups can get leverage with their ideas and go viral much quicker than before thanks to the vibrant events scene that is almost a complete re-invention of the traditional business and scientific conference circuit. These are not just gimmicks, but valuable marketing and recruitment tools.

Let's now explore each of these memes in the following sections.

10.4 Tech People

It seems obvious that Web ventures are often run and heavily influenced by techie people who really "get" technology, which can often include a "sixth sense" about where things are headed on the Web. It also includes the ability to build new tech, when required. For example, the guys at Google decided that they needed a new type of database technology to support the efficient operation of their search engine, so they went and built it. A Web start-up almost always includes a strong technical person in the founding team. Even if non-tech guys start the company, or have the fledgling idea, they will almost always immediately seek out a capable tech person who can turn their vision into software.

The issue here is that operators will spend huge amounts of time, resources and energy on various Web exploits, but often using people from within the existing business who aren't technologists. This is part of a pattern that works for an operations business, where the technology is purchased from suppliers and then deployed and operated, but doesn't work for Web projects. From the marketing department to the technology department, the folk trying to get traction with Web ideas simply don't have the technological insights and skills required to be successful. It's common for Web projects to be run by traditional operator folk who wear a standard product or project manager hat, which they mistakingly believe applies to any sort of service. One minute they're running a tariff product, the next they're running a Web product. Seldom have I seen that work.

In my view, operators ought to have someone at the board-level with deep technological skills and experience with the Web. Strategically, the Web is becoming increasingly important in all aspects of doing business in the 21st century. All businesses need a strategy for thriving in the digital economy, which is the successor to the services economy. The strategy should be well informed by digital economy patterns, most of which arise from Web ventures who have invented them as they've gone along. The strategy needs to be understood at the highest levels. Put simply, if an operator really wants to avoid becoming a dumb bit-pipe, then it needs to employ people who might know about something other than running a network, which is, without doubt, the dominant skills base of an operator's technical division, all the way up to the CTO. Again, this works fantastically for a network operations business, but not for innovative ventures related to the Web.

In innovation parlance, operators lack any ability to innovate with technology directly. They usually focus on marketing innovation and process innovation with almost zero inputs from technological innovation, which is merely something they consume after the fact. Technology is an out-sourced commodity. Where operator do possess technological innovation, such as a limited R&D effort, it usually exists on an island that is sufficiently marooned from the motherland that its outputs sink in the vast ocean between the techies and the senior stakeholders in the business. Very often, where telcos do have folk with technological brilliance, they are too far down the food chain to make a difference.

It is interesting to note why this gap exists between operator R&D efforts, which might well be leading edge at times, and the senior business stakeholders. In my experience, it is because of one primary reason: language!

R&D necessarily lives in a world slightly ahead of where we are today. It not only develops and uses technology that is ahead, but it interprets and "thinks" in paradigms that are also ahead of the current business mindset, using a different language – the *language of technology*. I don't just mean different terminology and a different set of acronyms: OAuth versus EBITA.

I mean a whole different way of thinking. After all, thought and language are intimately linked. The problem then arises as how to communicate between two groups of people talking in different languages. Some kind of translation is necessary. As my associates in various telco business strategy departments will say: "you need to put this [technical idea] in a language that *the business* understands."

That's fine, except for two problems. The first is an assumption that "the business" is a group of people whose job doesn't include learning and talking the new language of the future, whether that's technology, digital culture or Chinese. Secondly, it overlooks the possibility that we are dealing with a new language. Translation implies that we intend to stay with what we know, stick with what we've got and stay where we are, which is probably a location in time, not place. That's why we no longer use the word "thou." We wanted to move on, and we did.

This is why I advocate board-level ownership of digital strategy. It is fundamental to business in the 21st century, that it cannot be delegated or relegated to middle-manager folk and business analysts with no ability to translate insights into appropriate actions, which might very well need to be big bets. We can't deny the simple fact that we don't know the future and that we are required to take risks. This requires placing our trust in our instincts or gut, by which I mean that unwritten part of decision making that defies analysis and definition, notwithstanding the current management trend ("fad") to use so-called evidence-based methodologies.

Of course, experienced retailers and marketers of established product lines know their guts well when it comes to doing more of the same, even with a little shift to the left or right. However, nothing informs them of how to take risks into entirely new fields. It's a people problem. If you want to take risks in new areas, you need people who understand and talk the language of those new areas.

Enough of this business philosophy crap.[1] Let's move to the key start-up memes that I want to explore with you. If these are new to you, then think of this as your first lesson in the language of scalable Web start-ups. It's an accessible and exciting language to learn.

10.5 Lean Start-Up Methodologies

Operators deal in a world of services, products, enablers and business models that are well known and understood. On the other hand, a Web start-up is often dealing with the unknown. The objective of the early phase of operation is to search for a minimum viable product (MVP) that resonates with users and has the potential to scale and lead to the realization of a sustainable business model.

This ability to deal with the unknown is, for me, one of the key differences between a scalable start-up and an existing business, like an operator. I have seen over and over again, the frustrating attempts by an operator to contemplate a services future on the Web. When faced with this challenge, I always offer the same advice, which is *to focus on the real problem.* Operators often perceive that the problem is trying to determine what to build. However, I like to shift the emphasis by pointing out the real problem, which is *how to find out what to build.*

[1] It's not so crap of course, but experience tells me that you either get it, or not. No amount of trying to cross the "tech-biz" divide in telcos will yield that much fruit.

This is what an *early-stage* Web venture does. It constructs a machine to figure out how to find out what to really build. Recognizing that the formula for success is unknown, it doesn't try too hard to figure out the answer by an endless series of debates, analyses and presentations, per operator culture. It says: "OK, we're not sure what success looks like, so let's build a venture that can cope with the uncertainty." An operator says: "let's eliminate the uncertainty upfront." This approach is fine for variations to business-as-usual, where a series of analyses might very well predict, with a high degree of certainty, what impact a "My Family" tariff or bundle might have on the bottom line, for example.

To an extent, what a Web start-up is saying is: let's throw spaghetti at the wall and see what sticks, which is completely anathema to operator instincts. In fact, it's so dissonant with operator trading that it is infrequently considered to be unprofessional, ridiculous and lacking in any credibility, as do the folk who suggest such ideas, not living in the real world of hard facts, like revenue predictions, predictable performance, sealable security and so on. However, it is not as foolish as it seems.

A key part of the story that is often missing is a consideration of the underlying business platforms and processes. In an operator environment, most services proceed along a well-oiled roadmap process running atop of a robustly constructed business machine. For operators, Web projects are no exception to the process – they are made to follow the well defined and well understood business pathways as usual. The problem with this battle-grade process is that it is designed and constructed for the deployment of stable services and products, like an extension to network capacity or the introduction of a new voicemail platform, and so on, where variation and flexibility are totally ironed out.

Service and equipment deployments are well known and well understood with exactly defined endpoints. There shouldn't be any need to re-jig the service and start again, except for human blunders. However, a *new* Web idea, with its as yet unknown business model, might well be wide of the mark and require re-jigging not once, but many times. This simply isn't possible with the roadmap process and the entire business infrastructure that surrounds it.

The distinction is that a roadmap process is great at delivery of known products that don't need much in the way of re-work. A little tweaking perhaps, but not a complete re-build. On the other hand, a start-up is all about constant re-building and re-shipping, as many times as possible and necessary until the minimal viable product is found, or until resources run out (usually money). As I say: iterate until cash or crash!

Given this challenge, it is no surprise that the focus in start-up thinking and execution has been on how to maximize the ability to try many ideas as quickly as possible and as cheaply as possible, hence the "Lean" in "Lean start-up methodologies." The idea of finding the minimal viable product has become an area of much interest and debate in the various proposals for running a lean start-up. Most of them boil down to the same buzzword: "agile!"

What really matters is that the start-up team understands that their mission is to hunt for the MVP (which is only the initial target) and so they construct a process that has agility BAKED in. In essence, this involves two things:

1. The ability to change direction, or pivot, as shown in Figure 10.1.
2. The ability to measure the success of each step (e.g. metrics).

The conclusion is unmistakable. Web projects with a high degree of uncertainty in the business model do *not* belong on the traditional roadmap process. They require a much leaner

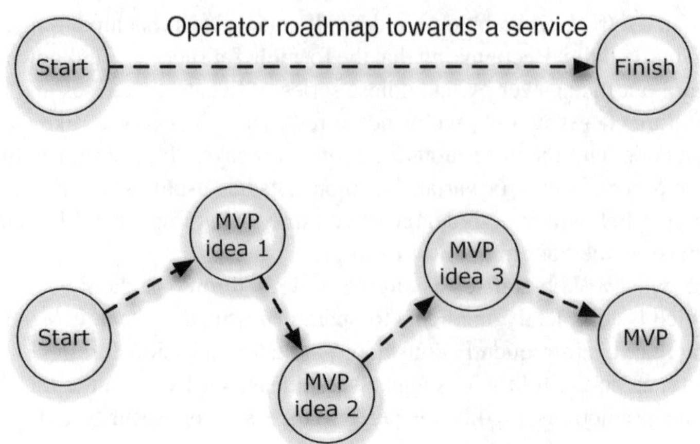

Figure 10.1 Linear roadmap versus Pivot process.

platform and process. How and where this gets solved is down to the operators, but I don't think this is rocket science. It comes down to deploying the right people, methods, tools and platforms, giving them space away from the strictures of the roadmap and its associated bureaucracy. However, in my experience of dealing with lots of operators worldwide, creating an agile environment for Web projects is not the primary challenge.

Invariably, the challenge is getting stakeholder buy-in to pursue an endeavour where the results are unknown, or can't be specified to the same level of detail as traditional products and services destined for the roadmap (e.g. a tariff refresh). Oftentimes, the product kings in marketing are driven in their imaginations by their grasp of the advertising campaign for the service. This requires something very tangible to sell. But this isn't the only mental grasp that will eventually serve to strangle innovation projects that lack clarity. For example, financiers will want a grasp of the business case inside of traditional business frameworks.

Eventually, the stakeholders will ask for a compromise so that they get some of what they want to feel comfortable, allowing the innovators to continue. But this is like a death by a thousand cuts. There is, as well documented in Clayten Christiansen's Innovator's Dilemma, the tendency to subject innovation projects to the same business rationale and measurement as business-as-usual, usually resulting in stillborn innovation, suffocated by compromise and consensus.

If you are interested to know more about the details of the various lean start-up methodologies, then do a search on the likes of Eric Ries, Steve Blank, Dave McClure and David Heinemeier Hansson. Their works (and words) are required reading for any operator stakeholder with a serious interest in launching successful Web ventures using the various techniques that have brought success to Web start-ups.

One thing to mention here is that a Web start-up, despite not having found the business model, doesn't say "to hell with business models." This is another myth that bricks-and-mortar folk often peddle. On the contrary, the Web entrepreneurs know that they are trying to find the business model. Without it, the lean methods are headed nowhere fast, except for the long-shot of an acquisition that can close without a solid business model.

The entrepreneurs will hypothesize about the initial business model and then use the iterations to test the hypothesis. There are some interesting books to help with the business model speculation, including the excellent *Business Model Generation* by Alexander Osterwalder. I have used the techniques in the book to construct initial business models for innovation projects inside of telcos. The "business model canvas" approach is an excellent framework for working up the model from a blank sheet of paper. I love it!

10.6 Extreme and Constant Optimization

10.6.1 Ship Often

As you might guess from the previous discussion about lean methodologies, pivoting is likely to be an ongoing process, which indeed it is. After all, if a machine and process exists to allow for constant change of direction, then it follows that such an approach can be exploited for ongoing service optimization. As Eric Ries pointed out about his Web venture IMVU, his team built a software process that was capable of *deploying* new software releases up to 50 times a day! Make sure that you read that right! I said **deploying**, not thinking, not planning, but deploying! Oh yeah. I also said **50** times a day.

If you haven't worked on a modern software project, then you're perhaps unfamiliar with the build-and-deploy process. The buzzword these days is "Continuous Integration," which does what it says on the tin. Before describing the details, I want you to first look at the problem it's trying to solve, which is sometimes referred to as "Integration Hell."

As Figure 10.2 shows, a software project (Web venture) has a set of software code that makes the product or service run. It's called The Codebase. Developers have to work on the code. To do so, they will take a copy of the current code base and set to work on their piece. Other developers will do the same and eventually all of them, at different times, will submit their changed code back to the repository. However, in order to test the code and deploy it, changes made by each developer have to be tested against, bearing in mind that whilst Developer 1 was working on his code, Developer 2 might have changed hers and then checked it back into the repository, causing possible interactions with Developer 1's code.

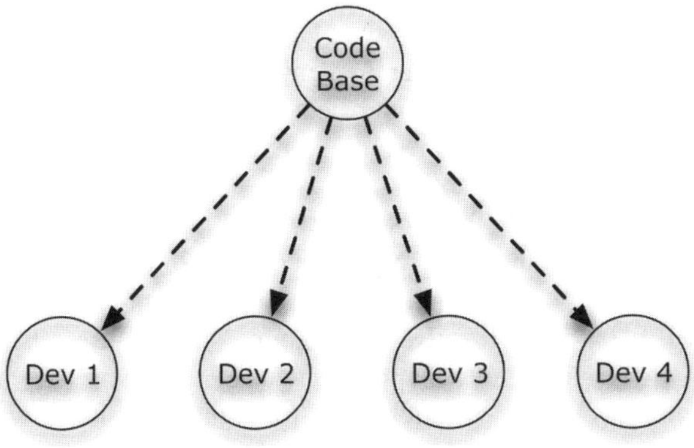

Figure 10.2 Integration hell.

You can probably see where this is headed. Eventually, the repository will change quite a bit from the developers' baselines (at the time they first checked out code) that they will enter what coders sometimes call "integration hell," where the time spent integrating is much longer than the time coding and testing their own code contributions.

In a dire scenario, developers might be forced to toss their revised code and start again; such is the labyrinth of contingencies introduced by the various changes. The solution is to build a process that enables and encourages each developer to integrate as early and often as possible, before integration hell kicks in. The processes and practises to do this are collectively referred to as "Continual Integration." Just like with all buzzwords, there is no single and authoritative definition. Continual integration should be thought of as a set of best practises to keep the code updated and working, seeking to minimize the time between new feature ideas and their deployment. These practises include:

1. Automated unit testing.
2. Automated and frequent deployment to a build server (usually driven by test passes).
3. Automated build on the build server.
4. Automated and frequent integration testing.
5. Agile project management techniques (too many to mention).
6. Agile and transparent project communications across the team.

10.6.2 Always Experiment

It's no use having a means to deploy software ultra fast and efficiently if there's no point or direction to the deployments. Each deployment must edge the project and venture along, pushing it towards a better place than it was at the last deployment. How do we measure improvement? Well, as experienced valley-based software entrepreneur Dave McLure says – "I don't know – that's up to you. It's your f*cking product."

I like Dave's candidness, but he's right. Only the product owners will know their goals. Only they will know what they want to improve. But, what they ought to have, common to all projects, is a *dashboard* of metrics that can be used to measure any improvements that contribute towards validating or improving the business model. Clearly, if you can't measure it, you can't improve it! This seems an obvious point, but is frequently overlooked.

It is up to the entrepreneurs (or "intrapreneurs" inside of a telco) to define measures for success. But, don't get too hung up about this. Don't fret about the metrics. The beauty of running an agile process with continual deployment is that you'll soon get a sense of whether or not your metrics make sense. You should expect the Key Performance Indicators (KPIs) to change, but they must be something that you really can measure and that has a direct causal link with the deployments.

For example, if you want to measure how long visitors dwell on the site before signing-up (or leaving), then that's measurable. Moreover, you can play around with site design (possible cause) and then measure any changes in dwell time (effect). On the other hand, if you fancy a more esoteric KPI, like "brand loyalty," or "fandom," then you must be able to quantify, measure and establish the causal link. As a product manager, you won't be thanked for claiming that your Web venture has increased "stickiness" (or some other fluffy word) without being able to show by how much and why. Establish a measure for "stickiness" and then add it to your dashboard.

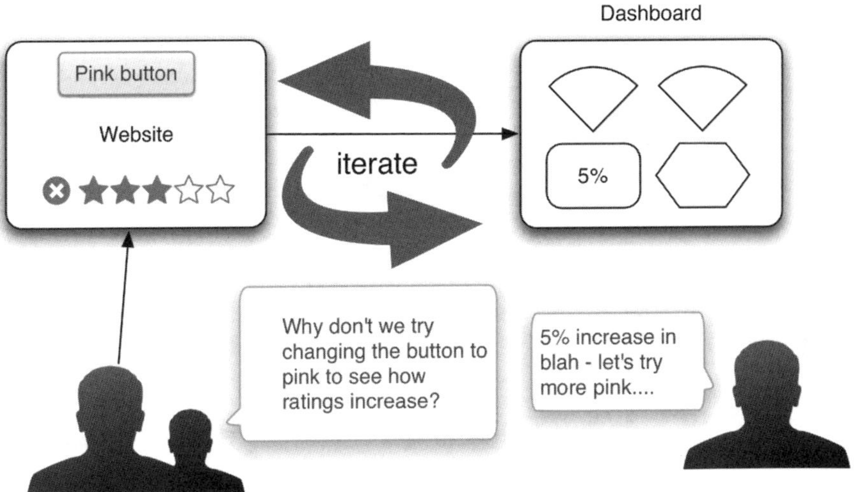

Figure 10.3 Dashboard-driven experimentation.

The process of experimentation through changes and monitoring of KPIs in the dashboard is the means to improve and optimize your site, as shown in Figure 10.3. However, you might be wondering how you gain confidence in the metrics. In particular, how do you know that changing the button to pink drove your KPIs northwards and not some other factor? Also, how do you speed up the process? As a product manager, you will probably have a number of ideas for improvements. These will only increase as you gain more experience with the site and begin to get a feel for how users respond to various features and design propositions.

One answer is to use A/B testing, also called Split Testing, or even multivariate testing (especially if you have more than two variations in play.) It's a simple and powerful idea, made possible by the flexibility of Web software and the ease with which analytics can be gathered in an automated fashion.

As Figure 10.4 shows, upon the behest of the product manager, two variations of the site are created – A and B. These have slight variations from the existing site, which is also left to run as a control for the experiment. When ready to go, the site is updated and now available in three different variations: control, A and B. We now have the basis for our A/B testing.

Our product manager now decides upon a metric that she wants to track during the experiment. This is added to the dashboard, if not already present. The experiment is then left to run for a while. A software mechanism (various can be used) drives visitors to one of the three variants. In our diagram, we're imagining that we want to test the effectiveness of labelling the search box using synonyms for the word search. Another software mechanism is then used to track the metric. In this example, it might be a simple case of measuring how many visitors use search at all during their visit to the site. It might be something a bit more insightful, such as how many times a visitor uses search during a visit, or how long they spend in the search part of the site, and so on.

Of course, the product manager will choose site variations depending on what she is looking for. If driving the use of search upwards is desirable, perhaps because it is monetizable in some way, then experiments to increase search will be aimed at driving visitor behaviour towards

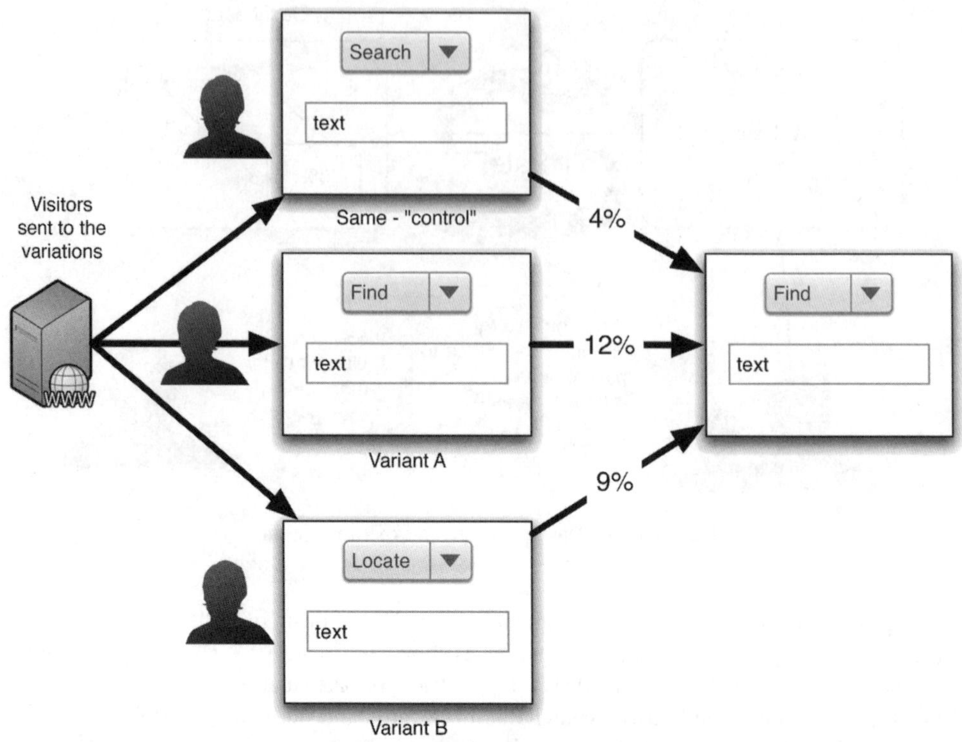

Figure 10.4 Split testing.

using search more often. However, it might be that some direct user feedback (via forums or other means) has shown that users are failing to buy products simply because they can't find them on the site. Or, it might be that the search box is not really a data search, but a physical search for finding where a product is physically located in a store, or across a range of stores. In which case, the label "Search" might be misleading, hence the suggestion to try "Locate," or "Find."

Whatever the reason, the product manager is the one to determine and design the experiments. Generally speaking, the objective is to increase conversion rates, which is the rate at which site visitors are converted into "customers," however that's measured. If it's selling something, then the experiments will be to increase conversion of visitors to paying customers. If the site is a charity site that involves a user survey, then the experiments might be to convert the number of visitors who take the survey.

Analytics will measure the changes in metrics and, hopefully, the experiment will show differences in conversion performance across the site variations, as shown in Figure 10.4. Logic dictates that the variation with the best increase of the desired metric will win and become the new site design that gets deployed across the entire user base. At this point, the experiment is not over. Analytics must be continually monitored to confirm that the site change has the desired effect on the metrics. Once this is confirmed, then the experiment is repeated for a new set of design variations, perhaps to look at how to increase shopping cart loads, rather than enhance the search performance. Or, it could be an ongoing set of experiments to improve search.

There are no rules for how many changes to make and how often to experiment. The beauty of continual deployment and browser-based solutions is that it can be done probably more frequently than product managers can track and users can tolerate. The duration of a split-test experiment depends on the amount of traffic on the site, the complexity of the experiment, and the difference in conversion rates for the variations. Less traffic, more complexity, and very similar conversion rates will suggest a longer duration in order to identify significant trends. Clearly, there's a whole art and science to A/B testing and I recommend reading the book *Always Be Testing: The Complete Guide to Google Website Optimizer*, by Bryan Eisenberg et al. Google Website Optimizer is a tool hosted by Google to support A/B (and multivariate) testing. I suggest checking it out.

10.6.3 Experiment Driven Development (EDD)

You wouldn't forgive me for not letting you in on the latest and greatest buzzwords. Well, it's all about EDD these days,[2] or Experiment Driven Development. It started with an inspiring quote from Eric Ries: "You ought to be able to try out an experiment with just one line of code." Whoa! That would be really powerful, but it's easier said than done, unless you happen to have built your software environment with this goal in mind from day one. So, what is EDD and how is this idea of "writing experiments in code," possible.

Let's begin our exploration of this exciting new idea with a brief tour of something called TDD, or Test-Driven Development. As you can probably appreciate, if you were to write a piece of code to implement a feature, then you probably want to know that it's going to work before you release it. This implies testing, which is a fairly obvious thing to do. No one would dream of releasing code that hadn't been tested in one way or another. Of course, the temptation is to write the code and then test it. Duh! I hear you say. That's obvious. When would you test code, if not after it's ready to test? Well, it turns out that a better idea is to write the test first. Imagine a piece of code like this:

```
test_tax_calculator
  Result = tax.collect(100, VAT)
  assert_equal result, 117.5 # VAT is currently 17.5%
end [of test]
```

We write this test and then run a test harness that calls the above test and checks to see if "result" is indeed equal to "117.5," which the coder/tester is expecting it to be. Of course, as we haven't yet implemented the code, we expect the test to fail. This is indeed what we want. If it didn't fail, we wouldn't have to do anything! If it didn't fail, we wouldn't have anything to check against when we implement the code and then, hopefully, expect the test to pass. This is the so-called Red-to-Green approach to testing code.

In the above example, we might well already have a chunk of code to create objects called Tax, but we might not have the chunk of code to collect the Value Added Tax (VAT). So, we

[2] OK, I'm exaggerating. It's not ALL about EDD because EDD is so new at the time of writing, that this really is the latest and greatest in buzzwords.

might be postulating a test here for an as-yet unwritten piece of code to implement the "collect" method on our Tax class of objects. Or, we might already have a "collect" method that doesn't understand VAT, so we are testing the new parameter VAT. It might already understand other parameters, like "INCOME_TAX." The point is that we are writing tests first to drive our code development (from red/fail to green/pass).

In the past, writing tests ahead of time might have seemed an ugly affair for coders, keen as they are to just start hacking some code and get things going. It's a strong temptation! However, these days there are many frameworks for developing software, such as the fashionable and popular Rails framework for building websites with the Ruby language, which contain the testing apparatus as an integral part of the framework. In other words, testing is so easy, it would be crazy not to test. That's why TDD has become so popular. Moreover, as the name of the game is agility, testing makes a lot of sense. TDD and its later cousin BDD, have encouraged an economical approach to coding, driven by writing small tests followed by small chunks of code to get from red to green in a hurry.

Inspired by the success of TDD, Nathaniel Talbott talked about EDD at the RubyConf conference 2010, suggesting that if we built EDD into our software frameworks, similar to TDD, then we might inspire a culture of experimentation in the same way that we have succeeded, by and large, in promoting a culture of testing in our agile software teams. This idea was picked up by Assaf Arkin who went on to propose and develop Vanity, a framework for EDD compatible with the popular web framework called Ruby on Rails.

Here's an example of what EDD code might look like:

```
ab_test "Price options" do
  description "Which is the better price of all?"
  alternatives 19, 25, 29
  metrics :signup
end
```

This code snippet defines an experiment to test out different pricing options. You can see the pricing variations listed next to the "alternatives" keyword. This sets up the framework to expect to deliver somewhere (i.e. on a Web page), these three different options and then to track the results (metrics) against a variable called ":signup."[3]

It is now up to the programmer to decide how and where he presents the pricing variable within the website, but he will end up doing so via a label ":price_options" somewhere in the code, like so:

```
Buy this widget for $<% = ab_test :price_options %>
```

Ignore those funny <% %> symbols, but what will appear on the site here, in between the symbols, are the pricing options taken from the above code snippet: 19, 25, 29. Another piece of code, that I won't show here (as it's probably getting confusing) will track the responses (stored against the symbol :signup) and collect the statistics that measure how often each price

[3] Note that the full colon in the word :signup is just a peculiarity of Ruby programming language to indicate the name of a variable that has some kind of context and significance elsewhere, at least within the Rails framework. Technically, it is referred to as a Symbol and is closely related to a String in the language inners.

option was acted upon.[4] Vanity then prepares the statistics in a chart that enables the site owner (product manager, most likely) to view the metrics.

So now that we know that metrics are so important, perhaps we should drill a bit deeper into the world of web metrics and which ones are important to track. That's the topic of the next section.

10.6.4 The Metrics Mantra – Startup Metrics for Pirates: AARRR!

No discussion of start-up metrics would be complete without Dave McLure's wonderful set of metrics – AARRR! That's right – if you say it out loud, you will sound like a pirate. These are start-up metrics for product marketing and product management, with due emphasis upon building a measurable feedback loop that the start-up team can use to iterate.

In his various talks and presentations about AARRR, Dave tells us that the most important thing about Web businesses and Web 2.0 businesses is gathering and reacting to the metrics in real time, giving yet more credence to the real-time theme that we have visited often throughout this book. The challenge though, is in finding a way to make useful decisions with the collected data. It is important to understand how to collect and use these metrics, and then couch them with a viable framework for making product development and product marketing decisions. The framework must be based on goals that are measurable!

Dave suggests that one of the biggest errors that many start-ups make is imagining progress is linked with adding more features, which is an approach equally criticized by Eric Ries. Engaging and activating more users is less about features and more related to the user experience at a basic visual (design) level with text, images, call-to-action buttons, and the overall size and placement of these elements. Often start-ups underestimate the value of user experience at the early customer acquisition and activation level, when it is most crucial. After all, without that initial flock to the site, it might well languish in the vast swamp of irrelevance, which every start-up fears. Taking care of user experience early is vital. The impact can be significant. The actual quality of features is more important later on, during the customer retention stage.

Events should be measured in a simple discrete way. This is where Dave's "pirate acronym" AARRR comes in, helping to define five main categories of events that a Web venture should be measuring. The goal is to create a feedback loop that makes the metrics you gather drive a decision process for product features/marketing efforts.

Here are the five main steps to AARRR:

> **Acquisition:** The various ways you acquire users/customers, via the distribution channels that bring them to your product or service (paid or organic search, social media & social networks, apps & widgets, traditional PR, affiliates, etc).
>
> **Activation:** The "happy" first experience where users take some kind of activation step (click on a link, provide an email address, buy something).

[4] I've glossed over the "call to action" aspect of the code, but that's obviously what we want to track in our metrics, such as how many times the item was bought, viewed, clicked, or whatever makes sense to measure for the site being tested.

Retention: Creating repeat and regular visits from your users (weekly emails, blog posts, periodic notifications).

Referral: Users tell other people about your service (viral loop, forward email to a friend, send a link).

Revenue: Users conduct some monetization behaviour, either direct (e-commerce, subscriptions) or indirect (advertising).

10.7 Co-Creation and Crowdsourcing

Yet more buzzwords for you to learn, although I hope you will find them interesting. Keep in mind that we are reviewing the methods used by scalable Web start-ups, using their terminology where possible. It's entirely likely that much of what I'm exploring in this chapter is simply variation on old ideas, perhaps more traditional time-tested ones. Certainly, I'm sure that when I tell you that co-creation means involving the customer in the creation of the product or service, you will sniff at the idea, bringing customer workshops, forums, and surveys to mind. "That's just plain common sense Mr Golding," you might retort. Yep! It is.

However, as with all things on the Web, these ideas are manifested in interesting and useful ways that many traditional businesses might not have thought to deploy. Also, as with all of the ideas in this chapter, don't forget that we are talking about cash-limited *start-ups.* Very small companies are deploying these methods. A one-man business could deploy most of them. Well, that's the point isn't it. The one-man doesn't have to do much. It's his site that does it for him. Welcome to the age of software!

Authors C. K. Prahalad and Venkat Ramaswamy might well have become the unwitting fathers of the co-creation "meme" when they wrote their *Harvard Business Review* article, "Co-Opting Customer Competence."[5] They expanded on their ideas in the book published by Harvard Business School Press, *The Future of Competition*. In the book, the authors discussed popular Web start-ups like Napster and Netflix, where customers were no longer treated as mere visitors to site, mere consumers of the product and services on offer. The visitors were invited to take part in decisions about what the companies should offer. This was a trend that the authors noticed, causing them to claim that value will be increasingly co-created by the firm and the customer, rather than being created entirely by the firm's product management team.

However, the authors went further than merely claiming that co-creation is only about jointly creating products. They observed that it also captures a movement away from customers buying products and services as transactions, to those purchases being made as part of an ongoing experience. The authors held that consumers increasingly expect to encounter freedom of choice as the ability to select how to interact with the venture through a range of experiences, not just product selection. Customers want to define choices in a manner that reflects their view of value, and they want to interact and transact in their preferred language and style.

This idea has been entirely embraced by the start-up community. Indeed, if we return to the earlier discussion of the so-called Lean Start-up methodologies and Experiment-Driven Design, then these are essentially all about co-creation, letting the users drive the product design

[5] "Co-Opting Customer Competence" *Harvard Business Review* January 2000.

choices, not the product managers. At its extreme, it is far more aggressive than traditional customer workshop/focus approaches. These tend to lead to a set of insights, such as one that says how customers will check the weather more frequently in the lead-up to a journey. These insights are typically collected by the product manager teams and then interpreted into a set of service requirements, which in turn are converted into a set of product features.

The features themselves are often not designed or interpreted by the customers. Of course, there is a feedback loop, but one that is often too coarse or slow. No one is knocking this approach, as companies like Apple who position themselves as interpreters par excellence of the consumer mindset execute it with great results. Interpretation is always required, as is design. One cannot assume that users will interpret their own requirements well. They often can't. Nonetheless, Web start-ups make full use of the potential to use their software to hold a much more interactive and ongoing dialogue with the users, taking place across a range of touch points.

It is increasingly common for Web ventures to get users involved with product requirements and interpretation. Users are invited to suggest features via a series of design hooks. These might include the use of third party tools like Get Satisfaction and User Voice, which court users for their feedback in a range of ways, including feature suggestion. The inclusion of these features into the main site is often done by placing a widget on the site, such as the one for hashblue.com, shown in Figure 10.5. The widget is initially accessed via a persistent floating "Feedback" button that hovers on the margins of each page as a constant reminder to users that their voice is welcomed.

An important point is that these user-feedback conversations are usually in public, in full view of the entire user base. This allows at least the chance for ideas to take shape and gather

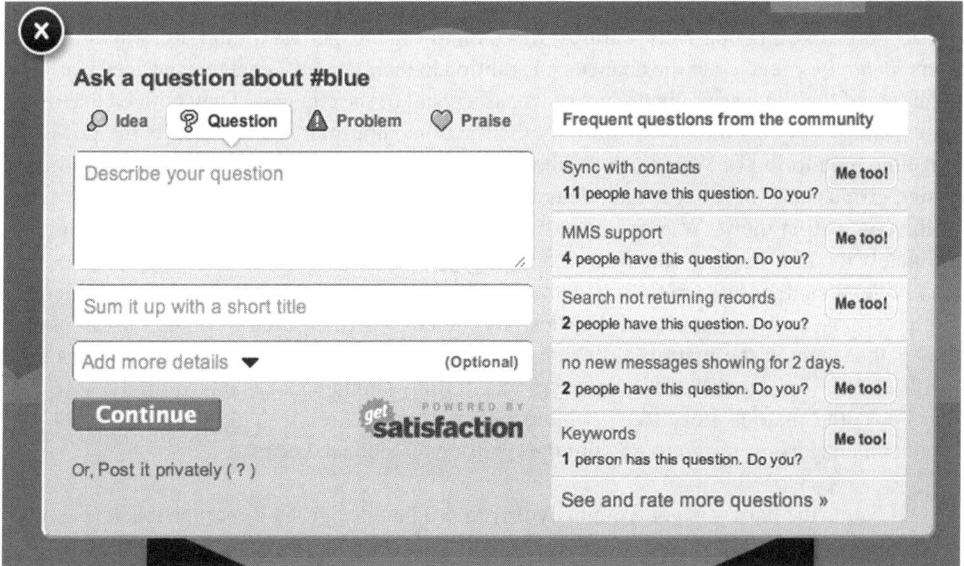

Figure 10.5 Get satisfaction feedback widget for Hashblue.com. Reproduced by permission of © getsatisfaction.com.

weight. What product managers really want is clear direction from the users as to what will make their experience of the website more engaging. The more transparent and open the conversation, the more likely this is to happen, although it will often require strong and skilful intervention by a community manager who monitors the conversations and provides useful feedback that represents the start-up's view. A company that isn't listening won't get much feedback for long.

Co-creation doesn't really work without the other ideas we have been reviewing in this chapter, such as metrics gathering. It might sound strange to the uninitiated, but the user voice can't be relied upon as the sole source of design guidance. That's where the metrics gathering and ideas like Experiment Driven Design work alongside co-creation, seeking to measure and identify trends that supplement and often qualify the user feedback, whether in support or denial of user suggestions.

An interesting example of explicit versus implicit feedback was when Facebook, early in its evolution, decided to cross-post user status updates from one linked-friend to another. Suddenly, users could see what each other was doing on their Facebook pages, hitherto thought of as private spaces by some of their owners, even though they weren't – Facebook walls were always visible to linked-friends, but required a deliberate visit to go check out the updates. With the new design feature, visits were no longer necessary. Instead of me visiting a friend's page to see his updates, those updates were posted to my page so that I got to see them anyway.

This relatively small design change to Facebook caused a massive outcry. The users made their opinions loud and clear, with nearly 1 million users forming a Facebook group to protest about the new feature, claiming that it degraded privacy and control, which is an ongoing debate and area of concern in most social networking sites. Now, did Facebook act like a well-behaved Web venture, following the newly emerging textbook play for co-creation? Did they let the users' interpretation of status updates win out? The answer is yes! And no!

Facebook, like all good Web ventures, was monitoring the metrics for the site, watching how users' *behaviour* reacted to the changes, in addition to their voices. What they noticed was how the usage of the site was going up, not down, as a result of the changes. They noticed that users were voting "yes" for the new feature by their actions, not their words, such is the power of real-time monitoring of Web metrics. The lesson for us is that metrics and experiment-driven design give us new tools to explore ideas along with the users. In my view, these tools are part of the co-creation motif. We are empowering users to vote through their behaviour, which is ultimately the best way to "voice" feedback. If placing too many adverts on a page causes hits to decline, then the users are telling us that the ads are getting in the way of their experience.

I have come across examples of Web design by telcos where the metrics of the site (assuming they gather them in the first place, which they often don't) were telling one story whilst user feedback, via traditional online surveys, was telling another story. Interestingly, just like Facebook, the metrics story told us more about what was really happening. However, the product managers preferred to go with the traditional feedback because:

1. Old habits die hard – if you've mostly driven product design by direct feedback (e.g. user focus groups), then it's hard to ignore, especially if it's been successful in the past.
2. The traditional feedback reinforced the original "customer insight" (why wouldn't it?).
3. The metrics feedback was telling a story that was unexpected. The product managers didn't really know how to react because they didn't have a flexible (lean) execution environment that could pivot towards the new insight.

Metrics and EDD are the tools that enable us to place more confidence in the explicit user voice because, having listened to user suggestions, we have a very reliable and accurate method to try out the suggestions to see if they actually work. After all, we shouldn't forget that we aren't really interested in the user voice at all. What we are interested in is driving up our metrics and how the user voice can help with that process. These are two different things entirely.

A naive start-up might place too much emphasis on user voice, rolling out features from the users' list as though this is being a good Web start-up citizen. It isn't! A start-up could easily waste its precious resources implementing features that it thinks will make a difference, but don't, because the product managers slavishly followed a co-creation mantra in a vacuum of metrics and design responsibility, especially if the features happen to coincide with their own views of product evolution.

The method of seeking ideas and feedback from users has been taken up by a variety of brands, not just Web start-ups. For example, Starbucks have used the Force.com platform to create an ideas market site (http://mystarbucksidea.force.com/) where users are invited to give ideas in three categories:

1. **Product ideas** – New drinks, foods and things to serve in the retail outlets, such as the adopted idea to sell a Christmas Blend version of the instant coffee product called Via.
2. **Experience ideas** – New ways to bring the service to the users, such as shorter queues, pre-ordering via mobile and so on, including the adopted idea of sending email notifications about reward card offers.
3. **Involvement ideas** – New ways to extend the Starbucks community-involvement programs, such as the adopted "12 Days of Sharing" promotion for AIDS awareness (via the various Red-branded products in the Starbucks range).

It's fascinating to see a combination of Web ideas in motion with the Starbucks example:

1. **Platform as a Service** – The ideas site was built entirely on top of the Force.com platform
2. **Community involvement** – Visit the site and you will see that it's a hive of activity and conversation that involves both users and Starbucks staff, all of them visible by name and photograph and writing their contributions with a very human voice, not necessarily the "corporate voice."
3. **Crowdsourcing** – The ideas are submitted and then voted on. The site features a leader board of current "hot" ideas.

Is crowdsourcing and co-creation the answer to everything? Of course not. Are they are replacement for good old-fashioned product managers, interpreters and visionaries? Definitely not! They are just tools, made possible by scalable Web technologies that enable the venture to broaden and enrich the dialogue with customers in ways not so easily possible and accessible to start-ups in the offline world.

10.8 Exploiting Big-Data

We have explored Big Data in some depth throughout this book, but here I want to review its application and relevance to Web start-ups. The trend is simple. Large amounts of on-demand computing power, combined with new lower-friction data analysis technologies and techniques, are giving rise to new entrepreneurial enquiry. Founders of start-ups are now beginning their venture with a question: "What can I do with all this computing power that I couldn't do yesterday?" In other words, what new ventures are entirely made possible by Big Data?

Of course, there are lots of start-ups who seek to offer Big Data technologies and products, such as hosted MongoHQ (see http://mongohq.com) and Cloudera, who will offer supported and hosted versions of Big Data storage solutions. They will also offer professional services to get you going. However, this book isn't really about "Big Data" solutions start-ups, even though they are an increasingly significant and important part of the start-up ecosystem.

A small start-up can spin-up a large cluster of servers in a cloud-computing environment in order to process a large data set for a while before releasing the servers. They will only have to pay for the CPU time, notwithstanding that the realities of doing this are far from easy. Nonetheless, this is the way things are headed.

Big Data is the talk of the town, especially in the Silicon Valley start-ups. As we have discussed in this book, Big Data is not new. Companies like Google, eBay and Amazon have massive amounts of user's data that they have been exploiting via innovative software solutions, often invented for the job (see Amazon's Dynamo discussed in Section 4.2 Some Key Examples of Big Data). Credit card companies have been mining customer data for their competitive advantage for a long time, as have telcos, especially to predict churn.[6] What kinds of attention is Big Data getting from start-up founders and why. Let's take a look:

1. **Cloud computing makes Big Data accessible** – Start ups do not need to have expensive hardware and software to crunch large amount of data. They can use cloud providers and get a large number of clusters running in a matter of days with low cost. Open source (and therefore free) software like Hadoop and HBase paved the way for creating applications cheaply. Flightcaster is an excellent example of a startup that processes massive amounts of data, in this case to predict flight delays days in advance. This is something that you think the big airlines and flight-handling companies should be doing, but along comes this tiny start-up to do it using the latest in Big Data software.
2. **Big Data can add value!** Let's be clear – and many entrepreneurs who get Big-Data are – Big Data is less about the technology (and scale) and more about the value. Mature companies like eBay, Facebook and Google have shown us how to *unlock the potential of massive data* to their competitive advantage – the technology is merely a sideshow, which is why these companies seem happy to publish their software with an open source license. Once infrastructure becomes easy to setup in order to crunch large amounts of data, entrepreneurs are no longer interested in the tech. They want to get the real value out of that Big Data, which is how I defined it earlier in the book (see Section 4.1 What is Big Data and where did it come from?), borrowed from *The Economist*:

> Big Data is about getting economic value out of unthinkably large amounts of data.

Trending Topics (trendingtopics.org) is a good example of a Big Data project undertaken by an individual, showing how the technology, along with some thought, can really unlock

[6] Churn is when a customer leaves the network. It is possible to spot various breadcrumbs in the data trail of a user about to leave.

information hidden in large amounts of data available via the Web. What this example shows clearly though is that Big Data is not magic sauce that can be applied to any data set to create value. The value still has to be mined and then applied to a business model that can monetize the value. What this actually means is the rise of new skills, like machine learning, data mining, and predictive modelling. We still need algorithms and people who can dream them up! However, it's worth reading a book like *Super Crunchers* to understand why Big Data makes the quest for data meaning easier, opening up new opportunities for entrepreneurs.

To put it simply, with only a limited amount of data available, mathematicians have to struggle harder to find ingenious ways to unlock any meaning in the data. I have spent many years working with various pattern detection techniques in my early career as a signal-processing engineer designing chips to speed pattern processing up. Back then (early 1990s) we were concerned with optimal use of limited resources – slower chips, smaller memories, thinner pipes. Now, it is almost as if the processing power and data sizes aren't the limitation.

The limitation today is rapidly approaching that of our imaginations, thinking what to do with all the data. The reason that Big Data makes data processing easier is that we can use relatively brute force techniques because resources are not the limiting factor. If processing a large data set were going to take three months, then we would think very carefully about how we did it and what we were doing it for. When it takes three minutes in the cloud, say, then we can iterate our enquiries and ideas until we strike gold in the data. In other words, Big Data is about the commoditization of pattern detection, which used to be the preserve of top mathematicians and scientists who were endowed with big brains and big budgets.

AdKnowledge.com is another example of a start-up using predictive analytics on large data sets to connect advertisers to the consumers. Cooliris.com uses hadoop and coverts your Web browser into a place that is cinematic and fast. You can browse photo and videos from the Web as well as from your local computer. At Cooliris, it was Kevin Weil, now heading up user analytics at Twitter, who innovated with user growth and advertising-focused analytics on a server cluster running Hadoop, Pig, and Hive, which are open-source implementations of the Google technology stack, central to companies like Facebook and Yahoo.

Datameer.com is a start-up that provides Big Data analytic solutions for big businesses, packing Hadoop-cluster processing in the form of more business-friendly spreadsheet metaphors as a front end to explore large data sets. The solution integrates rapidly with existing and new data sources to deliver sophisticated analytics for Gaming, IT Information Management and Customer Behaviour. Deepdyve.com is the Big Data mining solution for the large scientific community for journal and article search. Enormo.com generates and filters real state listings based on huge data. Gumgum.com does image analytics to provide in-image advertising. There are plenty more examples of start-ups whose foundational principle is all about getting unique value out of Big Data, and this is still very early days in the development of Big Data. The opportunities are exciting and huge!

The strategic threat and opportunity for telcos is when we think of these start-ups as an entirely new industry and platform play – a kind of "Big Data as a Service." If this trend continues, then it means that we might be looking at the creation of a new value chain (or value-net) and ecosystem built entirely around data as the principle commodity that goes in one end and comes out as something entirely different, be that a rich new way to communicate, driven entirely by context, or a set of predictions that are extremely valuable to the customer.

These powerful new capabilities could be exploited by telcos to process the incredibly large amounts of data already flowing through a typical telco network in the form of call records, messages, bill items, calls and so on. In other words, telcos could offer new services based on the value of their data, which is how they are already beginning to think and act with various new business initiatives, such as targeted advertising. At the same time, it might be that these start-ups can find new and much cheaper ways of getting the same value, but from alternative data sets easily scraped from the Web or harvested somewhere else. We don't know where yet, perhaps, but sure enough the availability of Big Data tools and services is going to find out, driven by entrepreneurial endeavours.

10.9 Social Discovery

Web start-ups seek one thing initially...

> Users!

Of course, if you're an existing telco with a big brand, a big advertising budget and millions of existing users, it's not that hard. But, for a new venture, getting users is harder than you might think. Discoverability is the issue – it might be easy to put a site on the Web, but getting visitors to come and sign up is the real challenge. What start-ups try to do – at least the good ones – is to build in a number of mechanisms to make discoverability as easy as possible. Before the rise of social networks, discoverability was all about search engine optimization (SEO), which still has its place. However, in recent times, social discoverability has become more important than search, as Brian Solis indicates in his blog article (http://t.co/MciDEN5):

> USA Today receives upwards of 35 percent of its referral traffic from social networks and just over 6 percent from Google. People Magazine receives 23 percent from social networks and 11 percent from Google. And, CNN earns 11 percent from social versus 9 percent from Google.

If folk aren't talking about your site in their online social networks, then you'll struggle to get noticed and struggle to get users. But that's only one half of the equation. The other is allowing users to tell their friends about the site. There are a number of ways to do this. The first is to include a mechanism for inviting others to the show. This is simple to do and often very effective. Surprisingly, operators are slow to adopt these methods, even though the cost of acquiring a customer in this way is far cheaper than acquiring via the traditional means of advertising.

Another method of allowing friends to connect via a service is social sharing, as shown in Figure 10.6. This is where a user shares content from the site using his or her social network, such as the iLike buttons on Facebook, or the similar "Tweet this" buttons, of varying kinds, that push out content via Twitter. It is remarkable that telcos have failed to exploit similar ideas with their own social network.

It ought to be easy for an existing telco customer to share stuff with his or her friends on the same network. In fact, this method of sharing would help to bring the notion of "friends on my network" into the minds of the network's users. It seems rather obvious that if carrying out a particular task is lower friction with people on the same network, then the outcome should be a propensity for certain groups of users to congregate on the same network.

Social-sharing or social-discovery mechanisms require a social graph to work. Again, the problem with telcos is the absence of an explicit social graph that is accessible via any kind

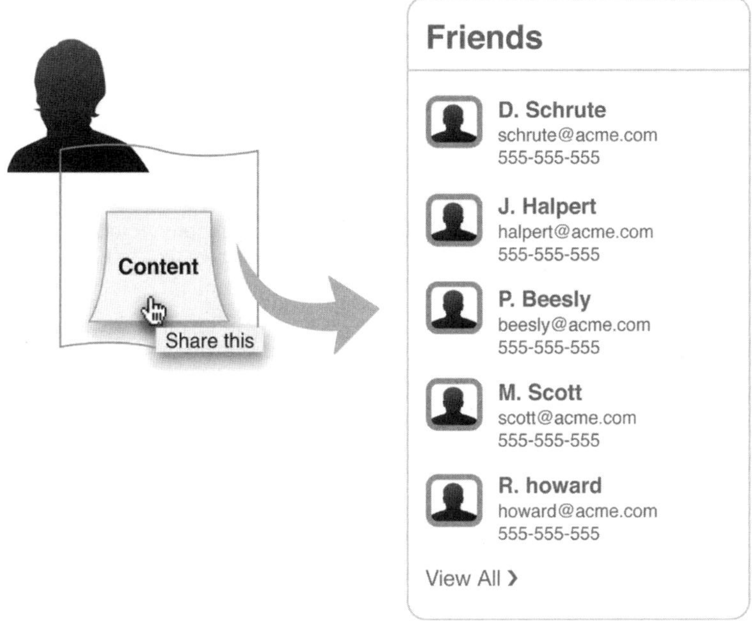

Figure 10.6 Social sharing of content/features.

of API. This is a pity given that social graphs have always been an implicit feature of mobile networks – dig down into the message or call records and you will soon find who calls whom, how frequently and for how long. Indeed, the social graph buried in a mobile network is quite rich. Telcos must find ways to exploit this.

10.10 APIs and Developers

This far into the book, I don't need to describe what an API is, nor why one might be useful. The purpose of this section is to tell you how widespread APIs have become. Operators still tend to think of APIs as "nice to have" add-ons that are mostly a kind of technical afterthought. This is not the case for Web ventures. Web start-ups tend to think of APIs as an integral part of the product. In other words, APIs are part of the product manager toolkit, not a tech-guy's accessory or plaything.

This point is greatly overlooked by operators, who often think about the role of APIs as just an ingredient in a developer community program. This is a mistaken view! APIs ought to be part of a product strategy, which is certainly not to build a "developer community," as though this were the end goal. The strategy is to use external resources (e.g. developers and adjacent user communities) to amplify and accelerate the success of the core product offering. This is how Web start-ups and ventures think about APIs and how they use and exploit APIs. Telcos need to adopt the same attitude and work aggressively to try out all manner of business models and innovation strategies around the API product offering.

That said, APIs are clearly related to developers and develop communities. Many Web start-ups will recognize from the outset that they have two sets of customers: users and developers.

They will treat both with respect and use many of the above principles. For example, start-ups will co-create with both the user community and the developer community. Often, it is the developer community who push for new features that add tremendous value to the users. Even better, users and developers will interact, such as via forums or events (see below) and create a mini-ecosystem around the product or platform, bringing unexpected innovations to the mix.

Of course, it is much easier for start-ups to think about developers because some of the guys in the start-up are developers. As I pointed out in the previous chapter, telcos will often attempt to manage developer communities using people who aren't developers. This is something that gets picked up easily by developers who will often interpret this as an inauthentic attempt to court their interests. Bottom line – if you want to work with developers, then you'll need to get some developers!

10.11 Incubation and Acceleration

Start-ups can build their ideas much more quickly and cheaply than ever thanks to the lower barriers to entry brought about by powerful software techniques and tools plus low-friction online services and ecosystems of various kinds for example, Web frameworks, SaaS, open source, social coding, PaaS, etc. While the Web has lowered many of the technical barriers for start-ups to get up and running with a product, many of the commercial barriers still remain. However, this is changing too with the emergence and success of various incubation and acceleration ecosystems.

Incubation and acceleration is the process of taking a start-up with an idea, or with an initial product, and getting it to a viable commercial position quicker than it might otherwise manage by itself. Incubators have popped up all over the world, offering to take early stage start-ups on a journey from initial idea to viable business. The process of incubation differs from one incubator to the next, but is essentially the same:

1. **Recruitment** – Cast a wide net initially and then select a number of promising candidates to undergo the incubation program. Incubator leaders deploy various selection criteria, but almost always focus on the quality of the people rather than the idea. The old investor's adage seems to apply – good people with a bad idea will do better than bad people with a good idea.
2. **Acceleration** – There are various ways to do this, but it is essentially a program of intense effort combined with various methods of intervention, guidance and mentoring by others outside of the business. The intense effort is essential and can only come about when the entrepreneurs are fully engaged with their business and trying to make it successful. Incubation schemes almost always require the candidates to be dedicated full-time to their ideas. In some cases, this means being prepared to live and work in and around a specific location where all the incubated teams co-work.
3. **Investment** – Start-ups need money, but the money on offer in an incubation program is typically only enough to pay for food and living quarters, often not that far above the "bread line." The reason that start-ups endure such limited compensation is because of the lure of greater funding at the end of the incubation period, which is also, by necessity, relatively short (e.g. 8–12 weeks). Often the start-up capital in the incubation scheme comes from a

number of angel-investors backing the scheme. These investors know how to access greater amounts of capital, typically from institutional Venture Capital sources.

4. **Mentorship** – This is, without doubt, a key component of an incubator program. All start-ups get access to mentors of varying kinds: business folk, tech folk, marketing folk, and so on. The richness of the mentoring network is important. A start-up might expect to "recruit" two to three mentors to their cause throughout the incubation period. Some of those mentors, though often not all of them, will go on to take a non-executive position on the board of the company. When I say: "recruit," I mean that the start-up will usually have a chance to interact with a large pool of potential mentors before settling on a few that gel with the team. Gelling is important and, naturally, must be mutual. Mentors will usually find affection for one or two start-ups in the program. If the affection is reciprocated, then all is well and good. Mentors are usually experienced and entrepreneurial folk recruited from a wide range of backgrounds.
5. **The Pitch** – At the end of the program, all start-ups get a chance to pitch their product to a comprehensive audience of mentors and potential investors. The idea isn't to get investment there and then, but to win a business card or two from an investor who would like to open a dialogue.

Incubator programs are increasingly popular, if somewhat fashionable. Some schemes are well known and attract a lot of interest from start-ups:

1. Y-Combinator
2. Tech Stars
3. Seed Camp
4. Springboard

I am a mentor with the UK incubator Springboard, which is loosely based on the Tech Stars format. My role is to mentor start-ups wherever I might be able to provide advice that relates to the mobile industry and tech.

Working in and around incubators, somewhat inspired by Y-Combinator, I set up an incubator scheme at O2, where I was hired as a consultant to explore the application of alternative "Web" methodologies and disruptive practises to conventional telco projects. As part of that remit, I looked at several Web-related project proposals where O2 was struggling to get traction. To cut a long story short, I eventually launched the O2 Incubator. It is still early days, but initial results have proven fruitful in identifying how to utilize incubation principles in the telco industry, which I hope to share in future publications.

10.12 Hack Days, Events and Barcamps

Developer days, start-up gigs and entrepreneurial events have become increasingly popular in the Web 2.0 era, presenting an opportunity for telcos to engage with the start-up scene. Start-ups get to promote their ideas – and themselves – much quicker and louder than before thanks to a vibrant events scene that is almost a complete re-invention of the traditional business and technical conference circuit. These are not just gimmicks, but valuable marketing and recruitment tools.

10.12.1 Hack Days

Hack days are where developers are invited not just to talk about their ideas, but to actually code them. A common format is to arrive early on day one and perhaps sit through a few opening talks to set the theme and scene, providing inspiration for the event. Hack days usually focus on a single theme, which might be industry related, such as mobile or music, or technology related, such as Android or Ruby. Big software companies will hold hack days dedicated solely to their wares, such as the Google IO event.

For an open event, participating companies with an API might be invited to pitch their APIs to the developers, explaining briefly how they work and what they do. Documentation for the APIs will be provided via accompanying Wikis or developer websites, along with API keys (to gain access to the APIs). When I participated in the London WarbleCamp event, which was the unofficial UK Twitter developer's conference, I had the chance to pitch O2's #Blue API, which allows access to text-messaging streams on a per-user basis. I published the API documentation on the event's Wiki page and, as is fashionable at such events, tempted the developers with the promise of prizes for the best hack with the API.

The event was successful. We had a number of hacks with the #Blue API. As an aside here, a "hack" means an informally written piece of code that explores a particular idea, crudely constructed as an experiment rather than a product. It doesn't mean an attempt to break into a secure computer system, per the popular urban myth surrounding the term. The benefits of promoting APIs at a hack-day event are:

1. **Concept check** – Testing of the API and the underlying service or product concept. In the case of #Blue, developers gave feedback that the streaming texts API was an interesting idea with lots of potential. We got a lot of ideas verbally in addition to the actual hacks. It was relatively easy to get a quick sanity check for the idea and the API approach.
2. **API test** – There's no better way to test something than to give it to real users, hoping they will flex it and push it to the limits. Testing of the API mechanics is one thing, but a hack event, if successful, will also tell you something about the design of your API. Does it include the right functions? Does it return data in the best way? Does it enable developers to build useful apps? Is it understandable? It is usable? Is it useful? Real developers trying to build something atop of your API, not by the product developers sitting in a lab, best answer all these questions. By this point in the book, you should be totally on the page of opening up your products early in order to get real feedback that will guide your attempts to pivot towards a great product. Maybe a hack day will get you started.
3. **Ideation** – Developers who take the time to use your API are going to have some degree of enthusiasm for the underpinning product idea. You should harness this energy by inviting developers to give you ideas about the product and the APIs. Have an open mind and keep your ears open too. It's amazing how much developers will try to help you build a great product. Ignore them at your peril! I can't tell you how often I've seen bombastic and I-know-best product managers ignore the advice and feedback of real developers and real users. Dumb!
4. **Publicity** – If you manage to attract interesting developers to your API, then the journey doesn't end there. You should shout out your successes. Don't feel shy. Developers love the attention, even the ones who pretend that they don't. As I've explained already, one of the reasons that developers build stuff for free (which is what they're doing at a hack day) is

the prospect of being recognized for their efforts and expertise. Shout it out! Make a noise. Blog it and make a big song and dance.

In fact, it's a good thing to say that the developers using your APIs are great coders (assuming they are). Not only does it reward the developers, but it adds even more credibility to your API and product. Hopefully, the developers hacking with your API will talk to their friends and peers about it, spreading the good news on their blogs and Twitter streams. It's all great news for your API. It's worth noting that some developers are attending hack days in order to show off what they can do as part of a marketing campaign for their venture.

10.12.2 Barcamps

Barcamps and "Unconferences" are a variation on the hack day theme. The Barcamp idea began in the Silicon Valley area and rapidly grew into an international trend. At a Barcamp, individuals (rather than companies) take it upon themselves to organize "user-generated" conferences (also known as unconferences). Unlike organized conferences with preset agendas and invited speakers, Barcamps are completely open for anyone to propose a session and then hold it. The first Barcamps focused on early-stage Web applications, and were often related to associated themes, such as open source technologies, APIs and open data formats. They rapidly grew into events aimed at start-ups generally, including entrepreneurial and commercial elements.

Its originators documented the Barcamp theme and format. Having said that the participants rather than the "organizers" run these events, you might be wondering how they work without falling into chaos. Well, there are some guidelines for the format and proceedings, beginning with priming participants with what to expect, as follows:

1. **Pre-register on the event wiki** – A Barcamp must have a wiki, which is the central place for advertising and documenting the proceedings, plus a place where participants can sign-up. You will need to know how many folk are planning to attend. This will help with getting the right location and in printing of name badges and possibly even T-shirts to promote the event.
2. **Bring a laptop** – Attend any Barcamp and you'll notice a sea of glowing laptops. If you're only used to more conventional conferences, then the laptops will distract you at first. Participants are expected to document their sessions on the site Wiki, in real-time as much as possible. This is one of the ways that users engage in the process, which leads me to the next "rule" of Barcamps...
3. **Be ready to get involved** – It's one of the "rules" that all participants must be exactly that – participants! Attendees must be willing to stand up and talk or to engage in the sessions. They can do this verbally, of course, or via one of the many "back channels," such as the Wiki or Twitter. It is common for the organizer to suggest a hash tag for the conference so that all participants can group their comments on Twitter by inclusion of the hash tag, like #hack99 or #musichackday, or whatever.[7]

[7] There aren't any rules as to how you set up and use a hash tag. Of course, you probably want to avoid collisions with other hash tags that are already widely in use, but this is seldom the case for a Barcamp, as it's always possible to find a meaningful and unique hash tag that works for the event.

You're probably wondering where the Barcamp name comes from and what it means. Some of the early events were held in bars and lasted overnight, requiring participants to camp-out at the venue. Of course, this isn't a requirement, which is why the more general name "unconference" has been adopted. It isn't easy to find a place where users can camp out, plus many prospective attendees, who might be interested in the topic, aren't so keen on discussing it all night long. Camping and drinking aren't so important. What really seems to matter is the chance to participate and get those creative juices flowing, feeding off other participants.

If you've ever attended a formal conference with a structured agenda, you might have noticed how the real value of the event emerges during the informal chats around coffee in the lobby, away from the auditorium. In essence, the unconference idea is trying to amplify the value of the "coffee chat" part of a structured conference. However, this overlooks the appeal of an organized conference, which is often the quality of the speakers and some of the topics on the agenda. Therefore, as you might expect, hybrid events are often the answer, kicking off the day with a structured agenda and invited speakers, moving into unconference mode as the day progresses. I have attended (and spoken at) many such hybrid sessions, such as the excellent OpenMIC Barcamp sessions organized by Chris Book in the UK (open-mic.org.uk).

How can telcos benefit from Barcamp events? I think in two ways, both of which I have tried and tested:

1. Participating in various industry-related Barcamps.
2. Holding Barcamps for employees as a refreshing alternative to other meeting styles already (un)popular in corporate life.

Let me add a word or two of caution. Firstly, it's important to grasp that Barcamps are very democratic. That's why they work so well. Everyone has a chance to set the agenda, including the folk who don't usually get a voice in bigger events, but who might well have great insights, talent and contributions. My advice is to go with an open mind and open ears, seeking these gems in the crowd, listening to what they have to say and learning from them. Don't go with a I-have-something-to-sell agenda, expecting only to broadcast your wares rather than take part in a genuine dialogue.

If you attempt to hijack a session or hold your own session with a strong my-agenda-matters attitude, then expect it to backfire. Barcamps are usually very focused on the topic and not the commercial trappings around it, unless a session (or entire barcamp) has a commercial theme, like how to make money from apps. Also, managers might do well to stay away in case they stifle open participation from the do-ers.

Secondly, if you plan on holding a Barcamp inside your organization, then be prepared for all kinds of negative reactions. People in corporate life aren't used to the format. They expect, and usually want, an agenda upfront. The absence of an agenda might cause too much disruption from the outset. My advice is to use the hybrid approach. Have a bit of structure at the start and then work your way into the unconference part of the day. Take that part easy too. Don't go all out for a camping session in the office, or wherever you stage the event. That might be too much to handle and the whole event will derail, poisoning the Barcamp approach forever.

Try to set the tone, which is the open participatory format. Make sure everyone has a voice, which might not happen if all the managers turn up, stifling the enthusiasm of the wider staff. A Barcamp session is only worth holding inside an organization if you think that by letting everyone participate, you will get a lot more value from the session versus a more conventional conference or meeting. Therefore, you need to work hard to make sure that participation really takes place. My advice is to start small. Hold your first Barcamp with a few people, just enough to enliven the debate. Don't forget – avoid the presence of too many managers.

Index

Connected Services: A Guide to the Internet Technologies Shaping the Future of Mobile Services and Operators, First Edition. Paul Golding.
© 2011 John Wiley & Sons, Ltd. Published 2011 by John Wiley & Sons, Ltd.